JONES AND BARTLETT SERIES IN BIOMEDICAL INFORMATICS
SERIES EDITOR JULES J. BERMAN

R *for Medicine and Biology*

D0901372

Jones and Bartlett Series in Biomedical Informatics
Series Editor Jules J. Berman

R for Medicine and Biology
(© 2010, ISBN 978-0-7637-5808-0)
Paul D. Lewis

Python for Bioinformatics
(© 2009, ISBN 978-0-7637-5186-9)
Jason Kinser

Ruby Programming for Medicine and Biology
(© 2008, ISBN 978-0-7637-5090-9)
Jules J. Berman

Perl Programming for Medicine and Biology
(© 2007, ISBN 978-0-7637-4333-8)
Jules J. Berman

Biomedical Informatics
(© 2007, ISBN 978-0-7637-4135-8)
Jules J. Berman

Other Related Texts

Exploring Bioinformatics: A Project-Based Approach
(© 2010, ISBN 978-0-7637-5829-5)
Caroline St. Clair and Jonathan Visick

Medical Informatics 20/20: Quality and Electronic Health Records through Collaboration, Open Solutions, and Innovation
(© 2007, ISBN 978-0-7637-3925-6)
Douglas Goldstein, Peter J. Groen, Suniti Ponkshe, and Marc Wine

JONES AND BARTLETT SERIES IN BIOMEDICAL INFORMATICS
SERIES EDITOR JULES J. BERMAN

R *for Medicine and Biology*

Paul D. Lewis
Swansea University

JONES AND BARTLETT PUBLISHERS

Sudbury, Massachusetts

BOSTON   TORONTO   LONDON   SINGAPORE

World Headquarters

Jones and Bartlett Publishers
40 Tall Pine Drive
Sudbury, MA 01776
978-443-5000
info@jbpub.com
www.jbpub.com

Jones and Bartlett Publishers
Canada
6339 Ormindale Way
Mississauga, Ontario L5V 1J2
Canada

Jones and Bartlett Publishers
International
Barb House, Barb Mews
London W6 7PA
United Kingdom

Jones and Bartlett's books and products are available through most bookstores and online booksellers. To contact Jones and Bartlett Publishers directly, call 800-832-0034, fax 978-443-8000, or visit our website, www.jbpub.com.

Substantial discounts on bulk quantities of Jones and Bartlett's publications are available to corporations, professional associations, and other qualified organizations. For details and specific discount information, contact the special sales department at Jones and Bartlett via the above contact information or send an email to specialsales@jbpub.com.

Production Credits
Publisher, Higher Education: Cathleen Sether
Acquisitions Editor: Molly Steinbach
Associate Editor: Megan R. Turner
Editorial Assistant: Caroline Perry
Production Director: Amy Rose
Production Assistant: Ashlee Hazeltine
Senior Marketing Manager: Andrea DeFronzo
V.P., Manufacturing and Inventory Control: Therese Connell
Composition: Atlis Graphics
Cover and Title Page Design: Scott Moden
Cover Image: © Paul D. Lewis
Printing and Binding: Malloy, Inc.
Cover Printing: Malloy, Inc.

Library of Congress Cataloging-in-Publication Data
Lewis, Paul D., 1969-
    R for medicine and biology / Paul D. Lewis.
      p. ; cm.
    Includes bibliographical references and index.
    ISBN-13: 978-0-7637-5808-0 (pbk. : alk. paper)
    ISBN-10: 0-7637-5808-6 (pbk. : alk. paper)   1. Medical informatics.   2. R (Computer program language)   3. Bioinformatics.   4. Medical statistics--Software.   I. Title.
    [DNLM: 1. Medical Informatics Applications.   2. Software.   3. Programming Languages.
W 26.55.S6 L675r 2010]
    R858.L495 2010
    610.285--dc22

                                                                      2009009810

6048

Printed in the United States of America
13 12 11 10 09   10 9 8 7 6 5 4 3 2 1

To Tracy, Sam, and Georgia

Contents

Introduction

Welcome to *R for Medicine and Biology*. In a nutshell, this book is:

- An introduction to the R software package
- A guide to managing, analyzing, and visualizing your biomedical data
- An instruction manual for building a free, open-source, expandable data management and analysis platform

What Is R?

Because you are already reading this book, you are no doubt aware that R is some kind of statistical analysis software package. That is correct. Search for this question on the Web, though, and you will find different answers. You will find that R is also described as a programming language. However, many experts talk of R in terms of being a complete environment—an environment that allows:

- Simple and complex data analysis
- Data visualization
- Data management
- Simple or complex programming
- Connection to other software, as well as being a programming language in its own right

You should think of R as an interactive computing environment. It has become popular for many different reasons:

- It is multiplatform, running on Windows® and Linux® varieties, as well as other operating systems.
- R is totally free, available for download under the GNU General Public License.
- R is open-source and constantly being developed by a core team.
- There are a number of updates each year.
- It contains more data analysis methods than any other software package.
- It can be expanded by you, the user, using packages or your own programming.
- R is also a community and is continuously being expanded for you by a dedicated group of users worldwide.

Expanding the functionality of R by users is the key to its success. Expanding R is easy, as the system is modular. To start using R, you just need the base system, but then you can easily add freely available packages to suit your needs. These packages already total over a thousand in number and cover a huge array of analytical, graphical, and data-handling functions.

Because R is based on a computer language, it allows you to write your own functions. You can write tiny functions just to automate chunks of analysis or complex functions, depending on your competence. If a data analysis method is not currently implemented in R, you could write the code for it yourself. You could even make it available to the community via a new package. This ability to program added functionality means R can also be viewed as a data analysis software development tool. The language has syntax that is similar to other programming languages and supports object-oriented programming.

R in Medicine and Biology

R has a large number of users in the fields of biology, medicine, and bioinformatics. Many packages are devoted to analysis and visualization methods required in these fields. This is because R—as this book will attest—is ideally suited for working with biomedical data and has functions for methods that you won't find in other software packages.

Many biomedical publications in the literature have listed R as the data analysis tool in the Methods section, so you can be confident of its pedigree. R will also provide you with publication-quality graphics, avoiding the need for third-party imaging software. If you do use R for publishable results, be sure that you cite the project.

Who Should Read This Book?

This book is for anybody who uses biomedical data and wants to learn why R is the most powerful free data analysis platform available. It is intended for those from many different biomedical domains who need a guide to managing, analyzing, and visualizing their data using a single resource. It is designed for those who wish to learn how to analyze their data themselves. It is also for those on a tight budget who cannot afford commercial software but want to learn how to put together the most comprehensive and free data-analysis environment very quickly.

It targets people from a broad spectrum of biomedical backgrounds. You might be a physician, biologist, pathologist, nurse, biomedical scientist, statistician, bioinformatician, medical informaticist, mathematical modeler, psychologist, professor, PhD student, MD student, undergraduate student, or perhaps just somebody eager to learn R using some easy-to-follow examples. Whether you view data analysis as a dark art or a walk in the park, this book will be useful because it targets rookies and experts alike.

The only real assumption is that readers will be users of (and have some familiarity with) Microsoft® Windows.

Why This Book?

There are already some excellent books on R. Some of these books are relatively basic, whereas others fill specific niches in the field. So, why publish another book?

As someone who analyzes biomedical data for a career, I have lost count of the number of times doctors and biologists have knocked on my door asking for help. I have also lost count of the times that their requirements have gone beyond the means of mainstream commercial data analysis packages. These specialists often arrive, data in hand, unable to take the next step. I always wished I could point them in the direction of R with a step-by-step guide to handling and analyzing their data. Usually, I end up analyzing their data for them, and they learn very little about how you get from data to a result. Hence, the great need for writing this book.

R for Medicine and Biology will provide you with a single reference for R. You will see that R is not just a tool for analyzing biomedical data. We will explore how R can be combined with other tools such as database management systems and data mining software to become a "control center" for storing, processing, sharing, analyzing, and visualizing your biomedical data.

As you read the chapters, with a bit of dedication, you will learn:

- How to obtain, install, and run R
- How to expand R using packages
- How data is structured in R
- Common functions for creating and processing data
- How to generate publication-quality graphics for your data and results
- How to import and export biomedical data from many different file types
- How to connect R to different databases to store and retrieve biomedical data
- How to combine MySQL and R to create comprehensive biomedical databases from disparate sources
- How to create anonymous, heterogeneous biomedical datasets for research studies
- How to analyze your data with descriptive and inferential statistical methods
- How to create your own functions by programming R
- How to perform multivariate analysis to explore your data
- How to analyze survival and surveillance data
- How to combine the data mining software Weka with R to generate predictive models from your biomedical data
- How to visualize and statistically analyze medical image data
- How to retrieve, process, analyze, and annotate publicly available microarray data
- How to retrieve and analyze array comparative genomic hybridization (CGH) data
- How to store and share data using Extensible Markup Language (XML)

The wide range of topics in this book partly reflects the spread of biomedical informatics as a discipline to cover diverse data-generating domains of medicine and biology. Hopefully, this book will demonstrate that it is possible to create a true, free biomedical informatics environment based around R, regardless of your background. By reading this book, you will not only become a competent R user, but also gain an understanding of the processes of data analysis and how different methods can be applied to generate the desired result. If you are a doctor or biologist in particular, learning the skills detailed in the book for data management and analysis using R will give you an advantage in your research. The ability to manage and share data in the way described can greatly speed up research and sharing of data and results.

How This Book Is Organized

The book falls into three sections. Chapters 1–4 walk you through installing R and packages, as well as provide the foundations for working within the environment. Chapters 5–9, as well as Chapter 20, focus on how you may use R as a biomedical data handling platform. Chapters 10–19 detail the use of R for basic statistical and graphical application to complex data analysis of different types of biomedical data. These are the "applied" chapters, which focus on how R can be used with specific biomedical datasets from defined biomedical domains.

The R code used throughout the book is kept as simple as possible. Indeed, veterans of R may be mortified by the use of long-winded coding throughout the book. I felt it was better to use code that was understandable by novices rather than be economical and potentially confusing.

The datasets used in the book are meant to represent typical, real-life, simple as well as complex, heterogeneous data from patients representing real-world scenarios that range in size. Each dataset is described in detail in Chapter 5 and is available for download at http://www.jbpub.com/biology/bioinformatics. Simply click on this book—*R for Medicine and Biology*—and then follow the link to "Student Resources." You will then be able to download the Zip file.

If you have no experience using R, the first few chapters should get you up and running quickly. If you are familiar with R, you can jump straight into the more applied chapters. Each applied chapter is a step-by-step guide to performing the tasks, laid out in a way that it should be easy to replace the example data with your own data. More than anything, the book has been structured so that it can be used as a quick reference as you work through your analyses in R.

I would encourage readers to experiment as they work through each chapter. Get familiar with the concepts and methods by playing with the code at each step. Use the questions at the end of each chapter to test your knowledge and gain more experience in the R environment. Practice makes perfect! Plug your own data into R as you play. Remember that if you use R to produce publishable results in your research, you should cite the R development team accordingly.

Acknowledgments

I'd better start my acknowledgments by admitting that the idea for this book wasn't mine at all but the brainchild of Jules Berman. I can't thank Jules enough for his encouragement and input, but more than anything, for being a leader and visionary in the field of Pathology Informatics. Many thanks to the really friendly and supportive production team at Jones and Bartlett including my editors, Shoshanna Goldberg and Amy Rose, as well as Caroline Perry and Molly Steinbach. Two of the chapters in this book would not have been written were it not for the kindness of Dr. Stefanie Brassen and Dr. Kristian Unger in providing high quality data and images. I'm also grateful to Nathan Hill and Oliver Lyttleton for proofreading as well as their suggestions. Last but not least, I'd like to thank my wife and best friend Tracy for her love and support throughout this book-writing adventure.

1 R Installation and Getting Help

Installing R on Windows is simple, but as with most software, you need to ensure that you have the right version for your operating system. New releases on R are only tested on versions of Windows that are still supported by Microsoft. Currently, there is only a 32-bit version of R available for Windows. If you are running Windows 95, 98 Me, or NT4, you will need to use an R version that is 2.6.2 or older. All versions since 2.7.0 will only run on Windows 2000 or newer.

The instructions that follow for installing R are for Windows XP. Make sure that you have enough disk space available, as a full installation can take up to 60Mb. Also remember that datasets you create can also take up quite a bit of drive space.

Before we get our hands on R, let's first have a brief look at the history of the software's development.

1.1 A Brief History of R

We can trace the roots of R back to the mid 1970s, when John Chambers and others at AT&T Bell Laboratories developed a statistical analysis platform called S (for a brief history of S, see http://www.stat.bell-labs.com/S/history.html). The original S libraries were written in Fortran, but a decade later these were translated to C code, making the functions much faster. S developed into a commercial platform called S-PLUS that is now marketed by Insightful.

Despite S becoming industry standard among statisticians, the costs prohibited its use by a wider academic audience. Hence, Ross Ihaka and Robert Gentleman at the University of Auckland developed R as a free system with a language based on S (for a brief history of R, see http://cran.r-project.org/doc/html/interface98-paper/paper.html). The semantics of R are based on another language called Scheme, itself derived from Lisp (commonly used in the world of artificial intelligence). The name R derives from the initial letter of the two original developers, as well as a play on the name S (R comes before S).

R was developed from the mid 90s prior to the first stable release in 2000. At the time of this writing, we are already at version 2.7.1. R is now developed by the R Development Core Team and, along with other contributors, is often referred to as the R Project.

The R Foundation for Statistical Computing is the not-for-profit organization that governs administration and continuing support for R (http://www.r-project.org).

New and old releases of R are made available from the Comprehensive R Archive Network (CRAN) website (http://cran.r-project.org/mirrors.html/). CRAN also provides access to the user-submitted packages. Not all R packages are available through CRAN, as some will be available only through the developer's own website. With the head node based in Vienna, Austria, CRAN has more than 60 mirrors worldwide, allowing you to obtain a local download.

In 2001, a new R-based project called Bioconductor (http://www.bioconductor .org/) was begun, with several objectives aimed at improving the development of analytical tools for biological data. Bioconductor has grown enormously and greatly benefits analysis of biomedical data.

1.2   Downloading R for Windows

The first thing to do is navigate to the main CRAN web page that contains the link to download the different versions of R: http://cran.r-project.org/.

In the "Download and Install R" box, you need to click on the link for "Windows." Alternatively, you could scroll down the page to find the link to a number of CRAN mirrors: http://cran.r-project.org/mirrors.html.

You will then be directed to the "R for Windows" web page. Under the heading "Subdirectories" will be a hyperlink called "base," which you need to click. This will take you to the "R-2.x.x for Windows" page, where you will need to look for the hyperlink for the "R-2.x.x-win32.exe" file. By clicking this link, you will be able to save this executable file to your computer. The executable file is the installer program for the precompiled R binary distribution. The current version at the time of this writing (R-2.7.1) is about 30Mb in size for a standard installation.

Although your operating system dictates which version of R you can run, you should ensure that you keep up-to-date with new releases. New releases contain new features as well as bug fixes. Packages developed on newer releases of R will not always work on older releases, and it is strongly advised that you download the latest version of R before you work through this book.

1.3   Installing R

To install R, double-click on the saved installer executable file. The following steps detail the dialog boxes that appear during the installation process for R-2.7.1. Older and newer versions may vary slightly.

You may at first be confronted with a "Security Warning" dialog box stating that the publisher could not be verified. This can be ignored, and you can proceed to click the Run button. Select your language preference in the "Select Setup Language" dialog box.

The next dialog box is the "Welcome to the R for Windows Setup Wizard." You can click on the Next button to take you to the "GNU General Public License," which you must agree to before installing R. The next dialog box allows you to select the location where R is installed. The default will likely be a folder with the name of the R version you are installing in another folder called R in Program Files (e.g., c:\Program Files\R\R-2.7.1). You can, of course, change this. If you are installing on Windows Vista, there have been some suggestions in the R forums that you should avoid installation into the Program Files folder and instead install directly to the main drive (e.g., c:\R\R-2.7.1). A minimum of just over 20Mb will be required for just the basic installation.

The next screen allows you to customize your R installation by selecting different components. These components include help files and online PDF manuals. The full install would take more than 50Mb of hard drive space, so you could deselect some options, although this is not advisable. Proceed to the next screen and ensure the option for startup customization is "No (accept defaults)." The last two dialog boxes will ask if you would like to create a Start Menu folder, desktop icon, Quick Launch icon, and registry entries (leave at default).

One more click on the Next button, and R will install to the destination folder.

1.4 The R Folders

The main R-2.x.x folder should contain, among others, a series of subfolders called "bin," "doc," and "library." This main folder is often referred to as the home folder and will likely be the default working directory (where R will look for files) when you run R. The bin folder contains the executable file called "Rgui.exe," which is what we use to open the R graphical user interface. If you specified not to create a shortcut during the installation process, you will have to navigate to this folder to launch R. The R graphical user interface (GUI) is dealt with in the next chapter. R can also be run without the GUI using a command line interface (Rterm.exe).

The doc folder contains a folder called "html," which houses the web browser interface–based search engine for searching the documentation in R folders (more on help tools in the next section).

The library folder is where all the R packages are held. By default, R will install a collection of packages for you, which includes standard packages, such as "base," "cluster," "graphics," "MASS," and "utils." The library folder is also the place where the extra packages that you download will be placed. If you ever need to locate a package (such as its help facilities), you need to look here.

1.5 R Resources and Help

The R Project and CRAN websites will be your ultimate resource for R. The R Project website has hyperlinks to the R frequently asked questions (FAQ) web pages. There are two FAQ sites useful for Windows: the main R FAQ site maintained by Kurt Hornik (http://CRAN.R-project.org/doc/FAQ/R-FAQ.html) and the R for Windows FAQ page maintained by Brian Ripley and Duncan Murdoch (http://CRAN.R-project .org/bin/windows/base/rw-FAQ.html). Both FAQ sites are packed with information on how to use R.

The main website also has a link for the R manuals, which can be downloaded in PDF format or browsed online. One superb manual, entitled "An Introduction to R," was originally a book written by Bill Venables and David Smith. Other manuals give detailed insight into the workings of R, programming R, as well as installation and administration. Another link points to a web page housing a list of more than 50 links to books and book chapters on R.

The R Project frequently provides an online newsletter (http://CRAN.R-project .org/doc/Rnews/). Each edition contains a number of short articles on R usage and development. The newsletter began in 2001, and the online archive contains some invaluable articles for analysis of biomedical data. It is well worth sifting through these articles as you learn R.

There are some excellent resources elsewhere on the Internet, particularly on the personal websites of folk who provide some great introductory tutorials. If you wish to search for information on R in a search engine, be sure to use other keywords, such as "cran." The R Seek website is a Google-based search engine that allows you to search for keywords across many R-related sites. As with most open-source developments, R also has its own wiki (http://wiki.r-project.org/).

One of the major contributors to the success of R has been the availability of mailing lists (https://stat.ethz.ch/mailman/listinfo). There are a number of mailing lists, but the two most useful lists are R-help and R-devel. You can subscribe online by following the link for "Mailing Lists" on the main R Project web page. You can email a question to a list with the view of receiving a reply from one of the hordes of users already out there. Hypertext Markup Language (HTML) versions of the mail archives are also available. Chances are that any question you have will already have been answered and can be retrieved by searching these archives. R also has its own Nabble forum (http://www.nabble.com/R-f13819.html), allowing you to submit questions on any problem. Another informative resource is RSeek (http://www.rseek.org).

If you need to seek an answer to a problem from a mailing list, you must remember a few golden rules (List 1.1).

List 1.1 Tips for asking questions on the R mailing lists.

- Read the manuals and FAQ pages of the R Project.
- Read any help files that come with a function you are stuck on.
- Ensure that the question has not already been answered elsewhere.
- Give as much information about the problem as possible.
- State your operating system and version of R.
- Restrict questions to the R software, and avoid using the list for basic statistical problems.

Follow these rules and you will be pleasantly surprised how quickly you will get a response. The most important things about asking a question on R are politeness and courtesy. You will notice that many questions are answered by people from the core development team, such as Professor Brian Ripley of Oxford University. These people dedicate many hours of their spare time helping users, and you should always show your appreciation and gratitude.

1.6 Summary

Installing R on Microsoft Windows is simple, where binary installation files may be downloaded from a number of CRAN mirrors. A number of updated releases are provided by CRAN each year, and users should try to keep their version up-to-date. In the next chapter, we will begin to use the R GUI and learn how we can extend the functionality of our R environment using freely available packages.

2 The R Environment and Packages

At a high level, we can think of the R "environment" as being the graphical user interface (GUI) used to interact with R's functions, as well as the folders that include the functions, help, and data sources. The RGui program provides us with an easy-to-use Windows interface. Through the RGui, we can talk directly to functions held within packages in library folders that sit on your hard drive. Through the RGui, we can load and save data, perform statistics, mine data, generate high-quality plots, and talk to other software, as well as communicate with the CRAN website and third-party developers to install new packages.

2.1 The RGui

The RGui itself is quite a basic "Windows style" multidocument interface. It is multi-document in the sense that it can have a number of windows open within the main R window. When you start R, you will see the R Console window, which is where you type commands to communicate with R functions (Figure 2.1). The main window will also contain graphical output as it is created, and you can have many plot windows open simultaneously.

2.1.1 The R Console Prompt

The R Console is your interface to R. You communicate with R by typing or pasting commands at the R Console command line represented by an angled bracket ">" called the prompt.

Performing data analysis and processing tasks using a command line is probably a new concept to most users. Those experienced with Linux or Microsoft DOS will be familiar with the concept. You type the command at the prompt, press the Return key, wait for R to process the command, and then view the result. As simple as this process sounds, it is surprising the number of people who get put off by the thought of having to use a command line for anything. Yet with a bit of practice you will quickly realize how much more powerful a command-line environment is compared to a button-based window GUI.

As a first (and simple) example, type a command to print "hello" at the prompt to the console screen (Code 2.1).

```
> print("hello")
[1] "hello"
```

Code 2.1.

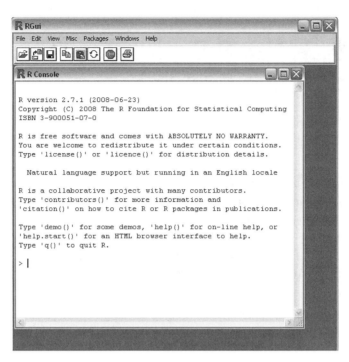

FIGURE 2.1   The RGui and R Console windows.

All you have done is send the `print()` function, parsing the string "hello" as an argument, to R which has then returned this word to the screen. The R functions you will use can be much more complex than this or just as simple.

After R carries out the function call, a new prompt appears after the output and awaits your next command.

2.1.2 An R Session

Working with R involves continual typing of commands at the prompt. If you need to repeat or modify lines of code that you have previously typed, you can navigate through these using the up and down arrow keys. You don't have to repeatedly type the same code. You can move along a line of code using the left and right keys and delete code using the backspace or delete keys. To insert new code, just start typing the text at the desired position.

You can scroll up and down the Console window, copy lines of code (Ctrl+c) and paste them at the prompt (Ctrl+v). When you copy from the console, you may often copy the prompt character, but you can use the "Paste commands only" option to avoid pasting this back to a new prompt, thus avoiding an error. To clear the Console window, you can press the Ctrl and l keys together.

These functions are also made available by right-clicking the Console window and choosing the menu item or selecting the "Edit" item in the main window menu.

If for any reason you need to stop R when it is in the middle of carrying out a process, you can press the Esc key or choose "Stop current computation" from the Misc menu.

In a session, you will create different R objects (virtually everything in R is some kind of object, and these will be introduced in the next chapter). Collectively, objects make up a workspace. To list the names of objects in the current workspace, you can select the "List objects" item in the Misc menu or type:

```
> ls()
```

Using the `objects()` function has the same effect. You can delete these objects from the workspace by selecting "Remove all objects" from the Misc menu or typing:

```
> remove(object name)
```

To close R at any time, you can type:

```
> q()
```

2.1.3 Saving and Retrieving R Sessions

A typical session in R can see you typing perhaps hundreds of commands at the prompt. You can save the collection of lines by selecting "File" from the menu and clicking the "Save History" item. The history of commands can be retrieved by selecting the "Load History" item. A history is saved using the ".Rhistory" file extension. The workspace can also be saved as an ".Rdata" file. Again, saving and loading a workspace is done using the options in the File menu.

You can save histories, workspaces, and objects (using functions) to any folder you wish.

2.1.4 The Working Directory

It is a good idea to create a new folder for each project. However, to allow R to see the location of files in a folder, you need to set the Working Directory. This can be achieved by choosing the "Change dir" item from the File menu or using the `setwd()` function:

```
> setwd("c:/R_Project_Folder")
```

Note the use of a forward slash as opposed to a backslash when writing the path. To determine the current working directory, type:

```
> getwd()
```

To list the contents of the working directory, type:

```
> list.files()
```

2.1.5 The R Editor and Scripts

At some point you will realize that a lot of lines of code you write are often repetitive, and it may be more convenient to create a file to contain the code. You can write lines of code as a script using the R Editor, which is opened by selecting "New script" from the File menu.

The R Editor is like Notepad and allows you to type in code as text and then save it. Saved scripts can be loaded and run in R by using "Source R code" from the File menu. The output (if any) will appear in the R Console just as it would if you had typed in the commands.

To open a script for editing, you select the "Open script" item from the File menu. Scripts are usually created using the ".r" file extension.

2.1.6 Changing the Appearance of the R Console

The R Console window can be customized to suit your taste. Apart from altering settings for windows, such as size, appearance, and number of rows and columns visible, you can change the color scheme.

Often, an R session will involve many hours being glued to your screen, and a comfortable color scheme is desirable. To change the appearance of R, you select the RGui Configuration Editor by clicking the "Gui preferences" item in the Edit menu.

2.2 R Packages

What gives R a huge advantage over other data analysis software is the wealth of add-on packages available. A basic R installation comes with a range of standard packages, and then the functionality of R can be extended using additional packages available from CRAN or third parties. To view the names of all packages installed, you can type:

```
> library()
```

2.2.1 The Anatomy of an R Package

A package is a collection of files and folders. The standard R packages can be found by looking in the library folder of the main R-2.x.x folder. To get a feel of what is included in a package, open the cluster folder. Inside this folder you will see eight sub-folders. The five additional files are all created when a package is made and hold information about the package contents.

The R folder is the one that contains interpreted code for different functions. The libs folder contains a file called cluster.dll, which is a dynamic link library file containing code from another language (probably C or C++). The data folder contains data objects that can be loaded into R. The chtml and html folders contain help files in either CHM format or as HTML web pages. When you call a help file from the console, the corresponding chm file will open. The R-ex folder contains example scripts from the help files.

2.2.2 Loading an R Package

To access the functions or data in a package, you first need to load it. You can either load a package into R using the "Load package" item in the Packages menu or by using the `library()` function with the package name as an argument:

```
> library(cluster)
```

To retrieve the names of all packages currently loaded, type:

```
> search()
```

2.2.3 Adding More Packages

Adding new packages is straightforward if you have an Internet connection.

If you select the "Install package(s)" item from the Packages menu, a window will open that allows you to choose a CRAN mirror website. Once you select a mirror, a new window will open listing all available packages at the CRAN repository. You can select a package, and it should install. Once installed, a subfolder will be created in the library folder. Alternatively, you can type the following command to achieve the same result:

```
> install.packages()
```

If you know the name of the package, you can parse this as an argument:

```
> install.packages("package name")
```

To install more than one package at a time, you can parse the package names as a list:

```
> install.packages(c("package name 1", "package name 2"))
```

Sometimes, a package will require other packages to be installed, and you can ensure they are by setting the appropriate argument when you perform the download:

```
> install.packages("package name", dependencies = True )
```

Some packages may not be held in the CRAN repository but reside on the developer's personal website. In this case, the package should be available as a zip file and can be downloaded to your own PC. Then the package can be installed by selecting the "Install package(s) from local zip files" item from the packages menu. There is also an argument called "repos" that allows you to specify a Uniform Resource Locator (URL) where the package is held.

Finally, each package will have a namespace. A namespace is like an ID that tells R which package a function belongs to. This helps avoid loading a wrong function that has the same name as another function. To access a function using the namespace, you use the double colon operator:

```
> package::function
```

2.3 Summary

R provides a simple-to-use GUI on the Windows platform. Communicating with R is facilitated via a command-line interface within the GUI. Functions in R are held in packages, and many other packages are available that greatly enhance the power of the whole environment.

3 Basic Fundamentals of R

A little bit of time is needed to familiarize oneself with R's syntax, the concept of objects and functions, as well as how data is handled. The best way to learn R is to jump straight in and try things.

3.1 What Is an R Object?

Whether you type in data or read it from a file, the data must be stored in memory by R. Think of an object as being the container for a piece of data or lines of code. You give an object a name that can be used every time you need to access that piece of data.

Objects are not just named data structures, however. An object can also be a function. To store data in an object, we make an assignment. To put data into a named object, we use the assignment operator "<-":

```
> x <- 1
```

Here, we have created a data structure object called "x" and assigned the number one to it. You could also use the `assign()` function to achieve the same:

```
> assign("x", 1)
```

Assignments may also be made using the "=" operator.

As we shall see, objects that are data structures can be a lot more complex than just a single value.

3.2 What Is an R Function?

Functions contain lines of prewritten code and perform some type of task. Functions can be used to gather information about the R environment, change properties of the environment, or most commonly, perform some task related to one or more data structure objects.

A function also has a name and often has one or more arguments passed to it. As an example, Code 3.1 demonstrates how we can create three objects called "a," "b," and "c" by assigning values to them and then calculate the sum of the values by parsing the objects as arguments to the function `sum()`.

```
> a<-1
> b<-4
> c<-10
```

```
> d<-sum(a,b,c)
> d
[1] 15
```

Code 3.1.

We also put the result returned by the sum() function into an object called "d." To retrieve the contents of d, we just typed its name.

Thus, if we want R to do something, we either ask it to assign something to an object or perform a function that is stored in one of the packages.

3.3 Some More Basic Syntax

It is critical that you use the correct syntax when typing commands. If you don't use the correct syntax, you will get an error message printed in the console. Look what happens when we attempt to output the word "hello" without using the print() function (Code 3.2).

```
> hello
Error: object "hello" not found
```

Code 3.2.

R didn't know that you wished to "print" the word, as the command for the print() function wasn't used. Also, with no function specified, R has assumed the word "hello" is a reference to an object that does not exist. For an object to exist, of course, it must first be created.

R is also case-sensitive. Data structures and functions can only be referred to by name if the correct case is used. Code 3.3 shows what happens when you try to call the print() function by using a capital P instead of a lowercase p.

```
> a<-5
> Print(a)
Error: could not find function "Print"
```

Code 3.3.

If you begin to type a line of code at the prompt and then press the Enter key, R will just move you to the next line and expect you to finish what you were typing. The prompt will change to a "+" symbol, after which you can continue to type. This is useful for typing in long lines of code where you want to keep all the text visible. An example is shown in Code 3.4 where the print() statement is incomplete after the Enter key was pressed. The code is simply completed on the next line.

```
> x<-1
> print(
+ x)
[1] 1
```

Code 3.4.

When writing collections of lines of R code, such as in scripts, it is useful to annotate your code. You can add comments either before code or in lines of code themselves using a hash ("*#*"):

```
> x <- 1        #This is a comment and ignored
```

3.4 Getting Help

One of R's best attributes is the provision of help for just about everything. Every package should come equipped with complete help files relating to every function contained within it. This is a stipulation when a developer builds a package. Most help files are well written and comprehensive, but you will find the odd package created by a lazy developer.

Calling for help in R can be done in a few ways. At the command line, you just need to type "?" and a function name:

```
> ?sum
```

Here we are asking R to open the chm file for the `sum()` function that is held in the standard base package (Figure 3.1).

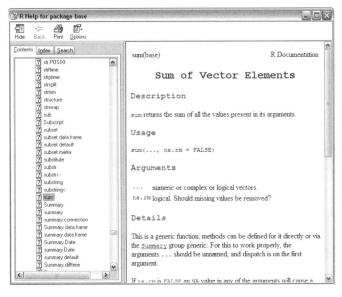

FIGURE 3.1 chm help file.

Let's say you want to view the help file of a function but were not quite sure of the function name. You can search for a term using the `apropos()` function that will return all functions that include that term (Code 3.5).

```
> apropos("mean")
 [1] "colMeans"       "kmeans"         "mean"           "mean.data.frame"
 [5] "mean.Date"      "mean.default"   "mean.difftime"  "mean.POSIXct"
 [9] "mean.POSIX1t"   "rowMeans"       "weighted.mean"
```

Code 3.5.

Examples of functions may also be provided in a package. Code 3.6 shows how you can use the `example()` function to show an example of using the `mean()` function (note that the function name appears before the cursor in an example). You do not have to worry about the meaning of the output in this example, but just relax with the knowledge that R examples usually provide plenty of information about the function syntax.

```
> example(mean)
mean> x <- c(0:10, 50)
mean> xm <- mean(x)
mean> c(xm, mean(x, trim = 0.10))
[1] 8.75 5.50
mean> mean(USArrests, trim = 0.2)
  Murder  Assault UrbanPop     Rape
    7.42   167.60    66.20    20.16
```

Code 3.6.

You can also obtain help by looking in the set of R HTML files included in the html folder of doc. This can be done manually, or you can bring up an index web page by typing:

```
> help.start()
```

From here you can enter a search page, which allows you to search for any term held within R's help documents.

You can look at the Help menu in the RGui to select these functions, as well as access the R FAQ and manuals in PDF format.

3.5 A Quick Primer on Types of Data We Encounter

Before we look at how data is created, processed, and stored in R, let's take a quick look at the nature of data we might encounter in biology or medicine. When we are analyzing biomedical data, we often have data that is a measurement of some variable. The measurement level of a variable can be any one of a number of different types.

Data is *discrete* if the values in the variable are a distinct set of observations, such as the number of patients in a registry who have breast cancer.

Data is *continuous* if the values are taken from a finite (or infinite) interval; a good example is height. Continuous variables are always numerical in nature. They could be integers or real numbers.

Interval scale data, where one unit has the same magnitude anywhere on the scale, is also numeric but differs from continuous data in that zero does not imply zero measurement. An example is a temperature scale.

The other data types we consider are *categorical* in nature, where the values in the variable fall into non-overlapping categories.

Nominal data values are those that can be either a number or a string that merely acts as codes for an observation, such as alive = 0 and deceased = 1.

Values on an *ordinal* scale possess an ordered relationship where the data is ranked or forms part of a rating scale. As an example, the scores for antibody staining that range between 0 and 7 to represent intensity are ordinal, where 0 represents no staining and 7 the most intense. Values on an ordinal scale do not have to be of the same magnitude, unlike interval scale data.

3.6 Objects Explained

We know that R stores data structures as named objects and can now explore these in greater detail. The first thing to understand is that data can be of a certain *mode*. The term mode describes how the data is stored in memory. In R, data modes can be numeric, character, logical, complex, or raw. Numeric objects store numbers; characters contain strings of alphanumeric values; complex objects contain complex numbers; raw objects contain sequences of bytes. In this book we will only consider objects of mode numeric, character, and logical.

An object can be described by the type of *class* to which it belongs. An object is merely an instance of a particular class.

Numeric objects can also be of different types. Numbers can be integer or real. In R, an integer number is of *type* "integer" whereas a real number is of type "double." It is important to ensure that your data variables are of the correct type when performing mathematical functions.

The mode only describes the type of data held within an object.

There are also different types of object structures (List 3.1). These structures can range from simple to quite complex.

List 3.1 R's data structure objects.

- Vectors
- Factors
- Array
- Matrix
- Data frame
- List

3.6.1 Vectors

A vector is the most basic of data structures. It is a sequence of data that can be numbers, characters, or even logical. At its simplest, a vector is just a single item of datum called a scalar. We say that a scalar is a vector with a length of 1. Examples of creating numeric, character, and logical scalar objects are shown in Code 3.7.

```
> x<-1
> x
[1] 1
> x<-"hello"
> x
[1] "hello"
> x<-T
> x
[1] TRUE
```

Code 3.7.

Vectors with more than a single element can be created using the c() function (Code 3.8).

```
> v<-c(4,2,3,8,2,2,5)
> v
[1] 4 2 3 8 2 2 5
> v<-c("a", "b", "c", "d", "e")
> v
[1] "a" "b" "c" "d" "e"
```

Code 3.8.

Elements of a vector must be of the same type and mode. When you create a numeric vector, by default, it will be of type double. The two examples in Code 3.8 show that the vector "v" only contains numeric values in the first instance or character in the second instance. When we specify a character value, it must be enclosed in either single quotation marks (') or double quotation marks (").

To determine the mode structure of an object, we can use the is() function (Code 3.9).

```
> x<-c(1,2,3,4)
> is(x)
[1] "numeric" "vector"
> x<-c("a", "b", "c", "d", "e")
> is(x)
[1] "character" "vector"
```

Code 3.9.

Sometimes we may have a variable in a dataset that contains missing data. If elements in a sequence of data are missing, we can let R know by defining the element as "NA" (Code 3.10).

```
> v<-c(4,2,3,8,2,2,NA,5)
> v
[1]  4  2  3  8  2  2 NA  5
```

Code 3.10.

To retrieve the value at any position in a vector, we have to refer to the *index* of the value. We can index any element by specifying the subscript in square brackets ([]) (Code 3.11).

```
> v[4]
[1] 8
```

Code 3.11.

To change the value in an element, you again use the index to specify the position and just assign the value:

```
> v[4]<-10
```

Logical vectors contain values of TRUE and FALSE. Code 3.12 shows how we can create a logical vector by testing a condition in a numeric vector.

```
> x<-c(2, 4, 7, 4, 3, 8)
> y<-x<7
> y
[1]  TRUE  TRUE FALSE  TRUE  TRUE FALSE
> x<7
[1]  TRUE  TRUE FALSE  TRUE  TRUE FALSE
> is(y)
[1] "logical" "vector"
```

Code 3.12.

3.6.2 Factors

A factor is a type of character vector that has its elements defined by groups. We can create a factor of character data using the factor() function. The following example shows the creation and display of factor information for eight patients categorized into three different groups (Code 3.13).

```
> x<-c("groupA", "groupB", "groupA", "groupC", "groupC", "groupB",
+ "groupA", "groupA")
> x
[1] "groupA" "groupB" "groupA" "groupC" "groupC" "groupB" "groupA" "groupA"
> x<-factor(x)
> x
[1] groupA groupB groupA groupC groupC groupB groupA groupA
Levels: groupA groupB groupC

> levels(x)
[1] "groupA" "groupB" "groupC"

> table(x)
x
groupA groupB groupC
     4      2      2
```

Code 3.13.

Note how the factor() function creates "levels," where a level represents the individual types of element in the vector. The code also shows that the levels for a factor can be retrieved using the level() function. Furthermore, by applying a function called table() to the factor, it will output a table showing the counts of elements in each level.

3.6.3 Matrices

A vector is a one-dimensional sequence of data. Often, data needs to be entered as two-dimensional structures. For example, we may have twelve numerical values that need to be recorded as three groups of four values each. To create a matrix, we use the matrix() function, parsing the vector as an argument, as well as specifying the number of rows in the matrix (nrow) and number of columns (ncol) (Code 3.14).

```
> mat<-matrix(c(4,12,1,5,21,7,10,7,2,19,24,3), nrow=4, ncol=3)
> mat
     [,1] [,2] [,3]
[1,]    4   21    2
[2,]   12    7   19
[3,]    1   10   24
[4,]    5    7    3

> class(mat)
[1] "matrix"
```

Code 3.14.

The example in Code 3.13 outputs the matrix as a two-dimensional table with both the columns and rows numbered. We have also demonstrated the use of the class() function to tell us that the mat object is of the class matrix.

To retrieve the value of an element in a matrix, we need to specify the row and column numbers (subscripts) in square brackets and separate the row number and column number by a comma (Code 3.15).

```
> mat[2,3]
[1] 19
```

Code 3.15.

Here, we have retrieved the value (19) of the element in the second row and third column. Let's look at another way to create the structure of a matrix using the dim() function (Code 3.16).

```
> mat<-array(c(1:15), dim=c(5,3))
> mat
     [,1] [,2] [,3]
[1,]    1    6   11
[2,]    2    7   12
```

```
[3,]    3    8    13
[4,]    4    9    14
[5,]    5   10    15
```

Code 3.16.

We created a two-dimensional structure by parsing a vector of a sequence of numbers. A sequence of numbers is created by separating the start number and end number with a colon. The function call ends with an argument for the dim() function, where we parse a vector of two numbers referring to the number of rows and columns of the matrix we wish to create. The empty 5 × 3 matrix is then filled with the sequence of values from the vector.

The dim() function can be used separately to convert a vector to a matrix (Code 3.17).

```
> x<-c(1:15)
> dim(x)<-c(5,3)
```

Code 3.17.

We might have a number of existing vectors that we wish to combine into a matrix. Let's say we have a series of systolic blood pressure measurements for five patients. Two other vectors could be patient height in centimeters and weight measurements in kilograms. To combine the vectors into a matrix to create a single dataset, we can use the cbind() function (Code 3.18).

```
> bp<-c(132, 144, 151, 120, 136)
> ht<-c(183, 162, 181, 168, 165)
> wt<-c(192, 210, 240, 187, 212)
> mat<-cbind(bp, ht, wt)
> mat
      bp  ht   wt
[1,] 132 183 192
[2,] 144 162 210
[3,] 151 181 240
[4,] 120 168 187
[5,] 136 165 212
```

Code 3.18.

3.6.4 Arrays

So far, we have considered vectors and matrices. Both these data structures are actually one-dimensional and two-dimensional arrays. If our data is multidimensional, however, we need to use the array() function to create an appropriate data structure.

Like vector and matrix structures, array structures also must contain data of the same mode. The example in Code 3.19 shows how we can create a 3 × 4 × 2 three-dimensional array for a sequence of 24 numbers.

```
> arr<-array(c(1:24), dim=c(3,4,2))
> arr
, , 1

     [,1] [,2] [,3] [,4]
[1,]    1    4    7   10
[2,]    2    5    8   11
[3,]    3    6    9   12


, , 2

     [,1] [,2] [,3] [,4]
[1,]   13   16   19   22
[2,]   14   17   20   23
[3,]   15   18   21   24

> class(arr)
[1] "array"
```

Code 3.19.

Notice the structure of the array in the output and how numbers are ordered by dimension. Although the syntax in the code is the same as that for creating a matrix, when we output the class using class(), we see that we have indeed created an array.

Again, we can change any value in the array by indexing it and assigning a new number:

```
> arr[2,2,1]<-12
```

3.6.5 Data Frames

One drawback about matrices is that all values must be of the same mode. As we shall see in this book, biomedical datasets have variables that are often of different types. Take the matrix we created previously for blood pressure. The matrix is numeric, but we would like to add another vector of a character mode. The vector is a series of characters indicating the blood groups for each of the five patients:

```
> bg<-c("O", "O", "A", "B", "AB")
```

If we tried to create a matrix using cbind(), all the existing values would change to mode character. (Try it.) Instead, we can use one of the most powerful data structures in R, called the data frame, which can be composed of vectors that have different modes from one another. In fact, the data frame is a collection of vectors that are all of the same length. The R data frame is a perfect structure for holding mixed-type biomedical data where each vector represents a variable of interest.

Let's create a data frame to hold the blood pressure data that also includes the blood group data (Code 3.20).

```
> bp<-c(132, 144, 151, 120, 136)
> ht<-c(183, 162, 181, 168, 165)
> wt<-c(192, 210, 240, 187, 212)
> bg<-c("O", "O", "A", "B", "AB")
> df<-data.frame(bg,bp,ht,wt)
> df
```

```
    bg  bp  ht  wt
1    0 132 183 192
2    0 144 162 210
3    A 151 181 240
4    B 120 168 187
5   AB 136 165 212
```

Code 3.20.

The individual vectors have their names displayed as column headers. To see what is held in a data frame and retrieve the vector names, you use the names() function (Code 3.21).

```
> names(df)
[1] "bg" "bp" "ht" "wt"
> df$bg
[1] 0  0  A  B  AB
Levels: A AB B 0
> df$bp
[1] 132 144 151 120 136
```

Code 3.21.

The code also shows that each vector is held as a component within the data frame called "df." Component is the technical name for the vector in the frame, and a component is accessed using the "$" character. By typing "df$bg," we return the values held in the "bg" component. Similarly, typing "df$bp" returns the values in "bp."

Since components are vectors, you can access values in these by index (Code 3.22). Alternatively, you can just index the data frame without referring to a component.

```
> df$bp[4]
[1] 120
> df[1,4]
[1] 192
```

Code 3.22.

The summary() function can supply statistics about the data held within each data frame component. The information will depend on each vector mode, but includes the minimum, maximum, and mean values for numeric data as well as counts for factor vectors (Code 3.23).

```
> summary(df)
  bg           bp              ht              wt
A :1    Min.   :120.0   Min.   :162.0   Min.   :187.0
AB:1    1st Qu.:132.0   1st Qu.:165.0   1st Qu.:192.0
B :1    Median :136.0   Median :168.0   Median :210.0
0 :2    Mean   :136.6   Mean   :171.8   Mean   :208.2
        3rd Qu.:144.0   3rd Qu.:181.0   3rd Qu.:212.0
        Max.   :151.0   Max.   :183.0   Max.   :240.0
```

Code 3.23.

You can add names to each row in a data frame. This is useful for data analysis and plotting results. For example, you can add labels as row names for each of the patients in the blood pressure data set (Code 3.24).

```
> rownames(df)<-c("p1","p2","p3","p4","p5")
> df
   bg  bp  ht  wt
p1  0 132 183 192
p2  0 144 162 210
p3  A 151 181 240
p4  B 120 168 187
p5 AB 136 165 212]
```

Code 3.24.

You can even change the names of the vectors in the same way using the colnames() function.

3.6.6 Lists

Another powerful data structure in R is the list. Actually, a list is literally a collection of objects. Vectors, matrices, or data frames held in a list don't have to be of the same length. In fact, a list is a great way of collating lots of data and information about a subject, and is well suited to biomedical data analysis.

The example in Code 3.25 shows how we can create a list that contains the vector for the blood pressure measurements. We have also added a new character vector that describes the patient's hospital.

```
> a_list<-list(bp, "Swansea Hospital")
> a_list
[[1]]
[1] 132 144 151 120 136
[[2]]
[1] "Swansea Hospital"
> a_list [[2]]
[1] "Swansea Hospital"
```

Code 3.25.

It is important to remember when using lists that objects held within are indexed using double square brackets ([[]]). Also note how these brackets must be used to retrieve objects that are not actually named.

Alternatively, you can create names for the objects held in the list, as shown in Code 3.26.

```
> a_list <-list(bloodgroup=bg, bloodpress=bp, height=ht, weight=wt,
hospital="Swansea", doctors=c("Dr Lewis", "Dr Hill"))
> a_list
$bloodgroup
[1] "0"  "0"  "A"  "B"  "AB"
```

```
$bloodpress
[1] 132 144 151 120 136

$height
[1] 183 162 181 168 165

$weight
[1] 192 210 240 187 212

$hospital
[1] "Swansea"

$doctors
[1] "Dr Lewis" "Dr Hill"
```

Code 3.26.

3.6.7 Some Useful Functions for Objects

The following functions are useful for extracting information from an object or manipulating it in some way.

`typeof(), mode()`
We have already encountered the class function for determining the class of an object. Two other useful functions are `mode()` and `typeof()` for telling us the object storage mode and type (Code 3.27).

```
> a_vector<-c(1.1, 2.1, 3.0)
> typeof(a_vector)
[1] "double"
> mode(a_vector)
[1] "numeric"
```

Code 3.27.

`attributes()`
The `attributes()` function provides information such as the dimensions and names of vectors held within an object (Code 3.28).

```
> x <- cbind(bg=bg,ht=ht)
> rownames(x)<-c("p1","p2","p3","p4","p5")
> attributes(x)
$dim
[1] 5 2

$dimnames
$dimnames[[1]]
[1] "p1" "p2" "p3" "p4" "p5"

$dimnames[[2]]
[1] "bg" "ht"
```

Code 3.28.

as()

The as() function is useful for converting (or coercing) one object type to a different class (Code 3.29). You can use as() with much larger objects such as data frames (as.data.frame).

```
> as.integer(c(1.2, 5.2, 7.3, 4.5))
[1] 1 5 7 4
> size.data[,c(2)]<-as.factor(size.data[,2])
> size.data$er<-as.factor(size.data$er)
```

Code 3.29.

apply(), tapply(), lappyly()

The apply() family of functions allows you to perform specified functions across array objects. To apply a function, we need to parse the array as the first argument of apply() (Code 3.30). We then need to specify whether we wish to apply the function over rows (1), columns (2), or both rows and columns (c(1,2)). Finally, we specify the function name. In this example, we parse a matrix called mat and apply the mean() function across the columns.

```
> result<-apply(mat, 2, mean)
> result
[1]  3  8 13
```

Code 3.30.

Let's say we had systolic blood pressure measurements for each of eight patients held in a factor called x and wanted to know the mean value in each group. To do this, we can use a function called tapply(), which allows us to parse a vector of measurements, the factor vector, and the function that we wish to apply (Code 3.31).

```
> labels<-c("groupA", "groupB", "groupA", "groupC", "groupC", "groupB",
+ "groupA", "groupA")
> bp<-c(132, 144, 151, 120, 136, 147, 161, 145)
> result<-tapply(bp, labels, mean)
> result
groupA groupB groupC
147.25 145.50 128.00
```

Code 3.31.

lapply() may be used to apply a function to each vector held in a list, returning the result in a list (Code 3.32). The sapply() function is the same as lapply(), but returns the result as a vector.

```
> bp<-c(132, 144, 151, 120, 136)
> ht<-c(183, 162, 181, 168, 165)
> a_list<-list(bp,ht)
> result<-lapply(a_list, mean)
> result
[[1]]
[1] 136.6

[[2]]
[1] 171.8

> result<-sapply(a_list, mean)
> result
[1] 136.6 171.8
```

Code 3.32.

3.7 Vector Math

As you would expect in a data analysis platform, R has many functions for performing math on vectors. We will only demonstrate some more common functions on vectors, although such functions can also be applied to matrices, arrays, and data frames by using specified indices.

Actually, before we look at using math with vectors, it should be shown that R is an excellent calculator in its own right (Code 3.33).

```
> (247+18)/41*12
[1] 77.56098
```

Code 3.33.

Code 3.34 shows us how R can perform arithmetic using standard operators (+, −, *, /, ^) on values held within scalars.

```
> x<-12
> y<-3
> z<-x/y
> z
[1] 4
> z+1*10
[1] 14
> z^2
[1] 16
```

Code 3.34.

A mathematical operator can also be applied to all values in an object simultaneously (Code 3.35).

```
> x<-c(2, 4, 7, 4, 3, 8)
> y<-c(0.1, 0.4, 1.7, 2.2, 0.4, 1.0)
> x*y
[1]  0.2  1.6 11.9  8.8  1.2  8.0
> mat<-matrix(c(4,12,1,5,21,7,10,7,2,19,24,3), nrow=4, ncol=3)
> mat_new<-mat/2
> mat_new
     [,1] [,2] [,3]
[1,]  2.0 10.5  1.0
[2,]  6.0  3.5  9.5
[3,]  0.5  5.0 12.0
[4,]  2.5  3.5  1.5
```

Code 3.35.

There are a number of built-in mathematical functions in R. Examples of more common functions are shown in Code 3.36.

```
> x<-5
> y<-c(4, 7, 19, 21, 10, 18)
> z<-c(1.2, 3.1, -4.7, 8,2, -7.7)
> q<-exp(x)                    # exponent
> q
[1] 148.4132
> q<-sqrt(x)                   # square root
> q
[1] 2.236068
> q<-min(y)                    # minimum value
> q
[1] 4
> q<-max(y)                    # maximum value
> q
[1] 21
> q<-sum(y)                    # sum of values
> q
[1] 79
> q<-mean(y)                   # mean of vector scores
> q
 [1] 13.16667
> q<-median(y)                 # median of vector scores
> q
 [1] 14
> q<-var(y)                    # variance of vector
> q
 [1] 50.16667
> q<-sd(y)                     # standard deviation of vector
> q
 [1] 7.082843
> q<-prod(y,z)                 # product of two vectors
> q
 [1] 4331665756
> q<-abs(z)                    # absolute values
> q
 [1] 1.2 3.1 4.7 8.0 2.0 7.7
> q<-sin(y)                    # sine
> q
 [1] -0.7568025  0.6569866  0.1498772  0.8366556 -0.5440211 -0.7509872
> q<-cos(y)                    # cosine
> q
 [1] -0.6536436  0.7539023  0.9887046 -0.5477293 -0.8390715  0.6603167
> q<-tan(y)                    # tangent
```

```
> q
 [1]  1.1578213  0.8714480  0.1515895 -1.5274985  0.6483608 -1.1373137
> q<-log(y, 10)                # log base 10
> q
 [1] 0.602060 0.845098 1.278754 1.322219 1.000000 1.255273
> q<-log(y)                    # natural log
> q
 [1] 1.386294 1.945910 2.944439 3.044522 2.302585 2.890372
4.8. Some more useful functions
> pi                           # pi
> q
 [1] 3.141593
```

Code 3.36.

3.8 Some More Useful Functions

This last section is a miscellaneous list of functions routinely used in R. Many of these functions are used later in the book, so they are listed here in alphabetical order so as to provide a useful reference.

attach()

The attach() function allows you to "attach" a data frame with named vectors to the R environment (i.e., the session in which you are working) so that all variables can be accessed without having to refer to the data frame object:

```
> attach(df)
```

detach()

Similar to attach(), detach() can remove a data frame from the environment.

length()

This function returns the number of values in a vector (Code 3.37).

```
> q<-length(y)
> q
[1] 6
```

Code 3.37.

order()

This function returns a vector of numbers for the position of elements in the parsed vector, given an ascending order (Code 3.38). If the "descending" argument is specified as TRUE, the order is decreasing.

```
> x<-c(2, 4, 7, 4, 3, 8)
> q<-order(x)
> q
[1] 1 5 2 4 3 6
> q<-order(x, decreasing = TRUE)
> q
[1] 6 3 2 4 5 1
```

Code 3.38.

paste()

Create a character vector by concatenating two other vectors. An example is given in Code 3.39 where the term "patient" is concatenated to the numbers 1 to 10, with no character between, as stated by the "sep" parameter.

```
> x<-paste("Patient", 1:10, sep = "")
> x
 [1] "Patient1"  "Patient2"  "Patient3"  "Patient4"  "Patient5"  "Patient6"
 [7] "Patient7"  "Patient8"  "Patient9"  "Patient10"
```

Code 3.39.

print()

This function prints to the screen the contents of the object that is parsed to it as an argument.

range()

This function returns the minimum and maximum values of a vector (Code 3.40).

```
> y<-c(4, 7, 19, 21, 10, 18)
> range(y)
[1]  4 21
```

Code 3.40.

rbind()

This function is similar to cbind(), but binds together vectors to form rows in a new matrix (Code 3.41).

```
> y<-c(4, 7, 19, 21, 10, 18)
> z<-c(1.2, 3.1, -4.7, 8,2, -7.7)
> mat<-rbind(y,z)
  [,1] [,2] [,3] [,4] [,5] [,6]
y  4.0  7.0 19.0   21   10 18.0
z  1.2  3.1 -4.7    8    2 -7.7
```

Code 3.41.

rep()

You can create a vector of repeated values of a specified number of instances using rep() (Code 3.42).

```
> x<-rep("A",10)
> x
 [1] "A" "A" "A" "A" "A" "A" "A" "A" "A" "A"
```

Code 3.42.

sample()

This function allows you to select a random sample of values from a specified vector. You can return a random order of values from the vector of the same length, or select a fixed number of random values (Code 3.43).

```
> y<-c(4, 7, 19, 21, 10, 18)
> sample(y)
[1]   4 21  7 10 19 18
> y
[1]   4  7 19 21 10 18
> z<-sample(y, 2)
> z
[1] 21 18
```

Code 3.43.

seq()

We saw earlier how a sequence can be created in a vector. You can also use the seq() function to specify a range, given set increments (Code 3.44).

```
> x<-seq(0,0.5,0.1)
> x
[1] 0.0 0.1 0.2 0.3 0.4 0.5
```

Code 3.44.

sort()

This function allows you to sort a vector in ascending or descending order (Code 3.45).

```
> y<-c(4, 7, 19, 21, 10, 18)
> y
[1]   4  7 19 21 10 18
> z<-sort(y)
> z
[1]   4  7 10 18 19 21
```

Code 3.45.

t()

The t() function is used to transpose a matrix or data frame, as shown in Code 3.46.

```
> mat<-matrix(c(4,12,1,5,21,7,10,7,2,19,24,3), nrow=4, ncol=3)
> mat
     [,1] [,2] [,3]
[1,]    4   21    2
[2,]   12    7   19
[3,]    1   10   24
[4,]    5    7    3
```

```
> mat_t<-t(mat)
> mat_t
     [,1] [,2] [,3] [,4]
[1,]    4   12    1    5
[2,]   21    7   10    7
[3,]    2   19   24    3
```

Code 3.46.

3.9 Summary

Working with R requires knowledge of how data can be structured. The fundamental concept of a structure in R is the object. There are different ways of structuring data, depending on the object type. Data structures range from simple vectors to complex data frames and lists. Many functions are available in R for manipulation of data and mathematical procedures.

3.10 Questions

1. What is an R object?

2. How many ways can data be assigned to an R object?

3. Why do you think it is important to ensure that data objects are of the correct type?

4. What is the relationship between vectors, matrices, and data frames?

5. Why might a data frame be more suitable than a matrix for holding heterogeneous biomedical data?

4 Plotting Data

As to be expected in any comprehensive statistical analysis platform, R has powerful graphic facilities for displaying data in simple and complex ways. The resolution quality of the graphic images is also of a high standard, and plots from analyses may be produced at publication quality.

Biomedical users will be reassured that many different types of plots are available for presenting data and analysis results. Furthermore, most plots can normally be obtained by a simple function call, and the user doesn't have to worry too much about the nuts and bolts of how a graph is produced. A little knowledge, though, is useful to produce the ideal plot, and it is worth the user being aware of the different types of graphic systems that exist. With a basic understanding of how a plot is output, the user can create multiple, combined, and even customized graphs to suit their needs.

This chapter provides a simple overview of the main plotting procedures in R so the reader can learn the basics for generating plots, customizing elements of the plot, such as axes, and ways to output the plot. Later chapters will use plotting functions, either provided by the base packages or packages that we will install specific for a chapter. In many plotting functions, the parameters passed as arguments to enhance the display are the same, and we will look here at many of these.

4.1 Graphical Systems in R

When you download R and the base packages, you will be provided with two basic graphic systems for generating plots. Although described as basic, these are powerful graphics systems in their own right. The first package, called graphics, contains the functions that many people could use for basic data analysis without the need for any other package. The graphics package produces traditional static plots using functions that output plots directly.

The second and newer graphics system is called grid. It is described as low-level because, rather than produce actual plots as output, the functions work by allowing plot objects to be built and then printed to the output medium. The grid system is more flexible and powerful, but also requires the user to invest some learning time. There is a base package called lattice that allows the generation of trellis graphics using the grid system. It is important to remember that the settings and parameter options within the graphics package functions are completely different from those of the grid package functions. Most of the graphic plotting functions used in this book, regardless of the package in which they are contained, are based on the base graphics system.

The most widely used function for producing static graphics in R is the `plot()` command. The `plot()` function is classed as high-level since it produces a complete graphic in your output destination of choice. In its basic form, you only have to pass the data itself to be plotted without any further arguments. A high-level graphics function such as `plot()` can be supplied with function-specific arguments or standard graphic arguments (these can also be set for the current graphics device using the `par()` function). There are also low-level graphics functions such as `text()` and `lines()` that allow users to customize their plot. Low-level functions do not produce a plot but simply allow you to add graphic enhancements to one.

4.2 Choosing a Graphical Output Device

Both graphics and grid systems rely on another base package called grDevices. Functions in this package allow you to output plots to different graphics devices, including the monitor screen or files of varying formats, such as PDF, PostScript, bitmap, jpeg, png, and LaTeX. Each device type has its own function call, which starts a corresponding device driver.

In Windows, by default, R will output a plot to the screen within the main RGui window. This is because when R starts up, it makes Windows the current graphic device (called "windows"). When the current graphic device is windows, plots are displayed within a plot window that is separate from the R Console window in RGui. You can actually have a number of devices open at the same time, including multiple plot windows as well as other device types. Imagine a scenario where you want to generate multiple versions of a graph for a variable in a dataset, view them on-screen side by side, and save the most recent graph each time to the same PDF and bitmap image files for storage.

The current graphic device can be determined using the `dev.cur()` function (Code 4.1).

```
> dev.cur()
windows
      1
```

Code 4.1.

`dev.cur()` lists the name of the device type and also the number of the device that is current, which in this case is one. To change the output device type, you just type the function corresponding to the device. As an example, to open a new PDF device, you would type:

```
> pdf()
```

Another plot window can be opened on-screen using the `windows()` function:

```
> windows()
```

This creates an empty plot window in which a new plot can be displayed. Since you can have many devices open at the same time, it can become easy to lose track of them, but by using the dev.list() function, you can display each one along with its allocated number in the list (Code 4.2).

```
> dev.list()
windows   windows    jpeg      pdf
      1         2       3        4
```

Code 4.2.

To switch between devices, you can use the dev.set() function by parsing the device number as an argument (Code 4.3).

```
> dev.set(3)
windows
      3
```

Code 4.3.

When plot windows are closed manually, the corresponding devices are also closed, but the function dev.off() forces any device to close by specifying the device list number (Code 4.4). If nothing is parsed as an argument to dev.off(), the current device is closed.

```
> dev.off(3)
bmp
```

Code 4.4.

You can close all open graphics devices by using the graphics.off() function:

```
> graphics.off()
```

To save and copy a plot from a windows device, you can right-click the plot window and select an option. You can also send a plot to a printer in this way.

One thing to remember when generating a plot in PDF or jpeg formats is that the plot will not be displayed as output in the RGui but will be saved to the current working directory. Therefore, it is probably wise to create your plots in a plot window, as we shall see, before changing the graphics device to an alternative format.

4.3 Generating Basic Graphics with plot()

The plot() function is generic in the sense that the type of plot produced depends on the object that is passed to it by argument and what class the object represents. We will see many different types of plots that can be produced in R in later chapters, but first we should take a look at the nuts and bolts of the plot() function.

As an example, we can apply the `plot()` function to a vector representing the number of deaths from lung disease per month in the United Kingdom between 1974 and 1979. The dataset is called "ldeaths" and is provided in the datasets package that ships with R, which is also loaded when R starts (Code 4.5).

```
> plot(ldeaths)
> ldeaths
     Jan  Feb  Mar  Apr  May  Jun  Jul  Aug  Sep  Oct  Nov  Dec
1974 3035 2552 2704 2554 2014 1655 1721 1524 1596 2074 2199 2512
1975 2933 2889 2938 2497 1870 1726 1607 1545 1396 1787 2076 2837
1976 2787 3891 3179 2011 1636 1580 1489 1300 1356 1653 2013 2823
1977 3102 2294 2385 2444 1748 1554 1498 1361 1346 1564 1640 2293
1978 2815 3137 2679 1969 1870 1633 1529 1366 1357 1570 1535 2491
1979 3084 2605 2573 2143 1693 1504 1461 1354 1333 1492 1781 1915
```

Code 4.5.

The resulting plot is a line graph showing the change in frequency of lung disease death over the time period (Figure 4.1). Apart from being the default option, one of the main reasons a line graph is drawn without us specifying any graph type is that the ldeaths object we pass to `plot()` is a numeric vector. The style of graph produced by the basic `plot()` function depends on the "type" argument (where "l" for lines is the default option). Options for the `plot()` type argument are shown in List 4.1.

List 4.1. Options for the type argument in the `plot()` function.

- "p" = points
- "l" = lines
- "b" = plot both points and lines with no overplotting

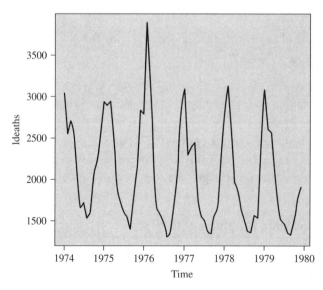

FIGURE 4.1   Line graph showing the change in frequency of lung disease death over time.

- "c" = draw the line but not where points would be
- "o" = plot both points and lines with no overplotting
- "h" = plot histogram-like vertical lines
- "s" = draw line as steps
- "n" = draw only axes

If two numeric vectors are passed to the `plot()` function, a scatterplot is produced by plotting points:

```
> plot(x,y)
```

Other high-level plotting functions in the graphics package that we will use later in the book include those to draw pie charts, bar charts, histograms, and stem-and-leaf plots.

4.4 Displaying Multiple Plots in a Single Graphics Device

Quite often there is a need to draw more than one graph in a single plot window, such as when you are comparing multiple variables or different ways of describing the same variable. You can draw multiple plots to a single graphics device by specifying the number of plots in terms of rows and columns using a function called `mfrow()`. `mfrow()` can be parsed as an argument to the `par()` function prior to calling the plot function. We will look at the `par()` function in more detail in the next section. Assuming we are using a "windows" device to create a plot window with four different plots set out as two rows and two columns, you would type:

```
> par(mfrow = c(2, 2))
```

Each time you subsequently use the `plot()` function, it will fill the row by column slots in the plot window. As an example, we can use the lung disease deaths data and plot four of the different type options (Code 4.6) and view the plots side by side (Figure 4.2).

```
> plot(ldeaths, type="p")
> plot(ldeaths, type="l")
> plot(ldeaths, type="h")
> plot(ldeaths, type="s")
```

Code 4.6.

Once the window is full and the `plot()` function is called again, R will replace that plot window with a new window with the same number of specified rows and columns. To return to a single graph plot window, the user can close the current window (and consequently the device) or type:

```
> par(mfrow = c(1, 1))
```

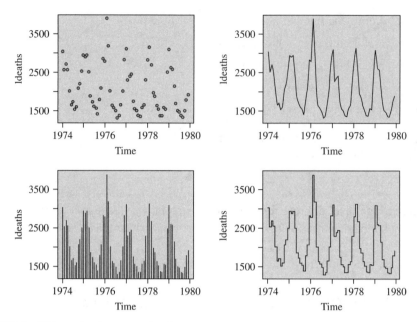

FIGURE 4.2   Different drawing type options in `plot()`.

4.5 Using Standard Graphical Parameters

For presentation purposes you will, no doubt, want to customize a plot rather than accept the default display provided by R. Standard graphic parameters allow you to do just that. Using these parameters you can alter many different features, such as the size

Table 4.1   Some Useful Standard Graphical Arguments for Plotting.

`ann`	Allow annotation of axes, title, etc. (Boolean)
`bty`	Determine how the plot border should be drawn
`cex`	Scaling of point size and text
`cex.axis`	Scaling of axis label text size
`cex.lab`	Scaling of axis title text size
`cex.main`	Scaling of plot title text size (also cex.sub)
`col`	Specify default plotting color (also col.axis, col.lab, col.main, col.sub)
`family`	Select the font family (arial, courier, etc.) for text
`font`	Select font (bold, italic, etc.) for text (also font.axis, font.lab, font.main, font.sub)
`lab`	Specify number of tick marks on each axis
`las`	Specify orientation of axes labels
`lty`	Line type (solid, dashed, dotted, etc.)
`lwd`	Line width
`mfrow`	How many plots should be drawn in one plotting window (also mfcol)
`pch`	Choose style of point to plot (circle, cross, star, etc.)
`tck`	Draw gridlines or tick marks
`xaxt/yaxt`	Plot *x* or *y* tick marks and labels

and shape of points, the font properties for text in labels and axes, as well as adding gridlines. Each parameter has a set of values that may be specified within the plotting function. When parsed in this way, the parameters are only set for the current plot and do not affect future plots. The parameters can also be set permanently for a device by using the `par()` function whereby all future plots will inherit the parameter changes.

There are many standard graphic arguments, and some useful ones are shown in Table 4.1. The reader should refer to the help file for `par()` to get a full description of all the parameters available:

```
?par
```

You will see a number of examples of how these parameters are used throughout the book.

4.6 Adding to a Plot with Low-Level Graphics Functions

In addition to using standard graphic arguments to improve the look and feel of a plot, you can apply one of the many low-level functions to add elements to a graph. Some of these low-level functions are shown in Table 4.2.

To alter a plot using one of these functions, you simply generate the graphic with a high-level function like `plot()` and follow this with a low-level function such as `axis()`.

The `plot()` function does possess its own arguments to add titles and manipulate axes, but some other high-level plotting functions do not.

To add a title and subtitle to a graphic using `plot()`, you could use:

```
> plot(ldeaths, main = "the plot title", sub = "the sub title")
```

The same result would be achieved using the title function after the graphic has already been generated:

```
> title(main = "the plot title", sub = "the sub title")
```

You can also create *x*- and *y*-axes labels by parsing character strings to the xlab and ylab arguments of `title()`. Alternatively, for the `plot()` function, axis labels can be specified directly by parsing strings as the xlab and ylab arguments:

```
> plot(ldeaths, xlab = "the x axis", ylab = "the y axis")
```

Table 4.2   Low-Level Plotting Functions.

`axis()`	Add an axis
`title()`	Add a title and subtitle
`legend()`	Add a legend
`text()`	Add text, such as point labels
`points()`	Add new point(s) to a plot
`lines()`	Draw line(s) between points specified by vectors
`polygon()`	Draw a polygon with *x*, *y* coordinates specified by vectors
`abline()`	Draw a diagonal line by specifying slope/intercept, a horizontal line (*h*), or a vertical line (*v*)

A legend can be added to a plot using the `legend()` function and providing arguments for the position, legend text, and the points:

```
> legend("right", "points", pch=1, title="Legend")
```

Properties related to how an axis is drawn can be changed directly by argument to the `plot()` function, as well as using the low-level `axis()` function. The display of axes can be turned on and off using the `plot()` axes argument:

```
> plot(ldeaths, axes=FALSE)
```

Standard graphic arguments, such as the label font, line width, and color, may be specified for an individual axis in a plot:

```
> axis(2, col=4, lty="dotted", lwd=2)
```

Axis tick mark size can be set using the tck standard graphic parameter whereby the smaller the proportion of "1" that is parsed as `tck`, the smaller the tick mark will be:

```
> plot(ldeaths, tck=0.01)
```

Setting the tck parameter to "1" will provide full gridlines within the plot:

```
> plot(ldeaths, tck=1)
```

To control tick marks on a single axis, you can use the `axis()` function once the plot is drawn and provide a value to the tick parameter:

```
> axis(2, tick=1)
```

It is straightforward to add additional points to an existing plot using the `points()` function and parsing vectors of values for both *x* and *y* coordinates:

```
> points(c(110,120,122),c(80,91,104))
```

You can also draw lines between points by parsing *x* and *y* coordinates to the `lines()` function:

```
> lines(c(110,120,122),c(80,91,104))
```

Similarly, you can parse coordinates to draw polygon shapes on an existing plot, which is useful for highlighting clusters of points:

```
> polygon(c(110,120,122,140),c(80,91,104,87))
```

Another useful line function is `abline()`, which allows you to draw horizontal (*h*), vertical (*v*), or diagonal lines given an intercept (*a*) and slope (*b*):

```
> abline(h=84, v=125, a=65, b=10)
```

One of the most useful and commonly used low-level functions is `text()`, which permits the addition and positioning of text labels to points:

```
> text(c(124,118,130,127,103,141,114), c(75,80,95,77,68,105,84),
c("A","B","C","D","E","F","G"),pos=2)
```

This example adds the letters of the third (character) vector at the *x* and *y* coordinates specified by the first two vectors.

Finally, we shall look at an example of how we can use many of the functions described to significantly enhance the appearance of a plot. We can generate a basic scatterplot for two vectors representing blood pressure readings from seven patients. The first vector is a set of measurements for systolic arterial pressure (mmHG), and the second vector is a set of measurements for diastolic arterial pressure (mmHG):

```
> plot(c(124,118,130,127,103,141,114), c(75,80,95,77,68,105,84))
```

The resulting scatterplot is shown in Figure 4.3. As you can see, the plot is basic, displaying the points at each coordinate as well as axes. Since we parsed vector values directly to `plot()`, it is these values that have been displayed as axis labels rather than a vector name, which is both untidy and uninformative. We will improve the scatterplot with some of the arguments just described (Code 4.7).

```
> plot(c(124,118,130,127,103,141,114), c(75,80,95,77,68,105,84),
xlab="systolic", ylab="diastolic", xlim=c(90,150), ylim=c(60,110))
> text(c(124,118,130,127,103,141,114), c(75,80,95,77,68,105,84),
labels=c('A','B','C','D','E','F','G'),pos=2)
> title(main="Arterial pressure measurements")
> axis(side=1, col="gray", tck=1, lty="dotted")
> axis(side=2, col="gray", tck=1, lty="dotted")
> box()
```

Code 4.7.

With just a few lines of code, we have greatly improved the plot format (Figure 4.4). First, we have increased the size range of both axes so that points are not cramped near them. We have added axes labels as well as a plot title. Gray, dotted gridlines have been drawn using the `axis()` function followed by the `box()` function to redraw the plot border in black. The plot points also have labels positioned to their left. The plot could be improved even further by use of color and even the addition of a legend.

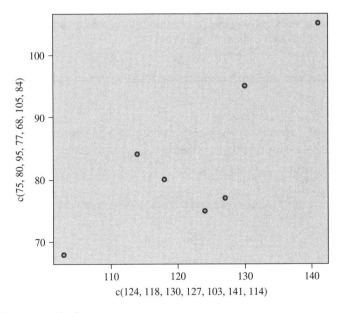

FIGURE 4.3 Basic scatterplot for two vectors.

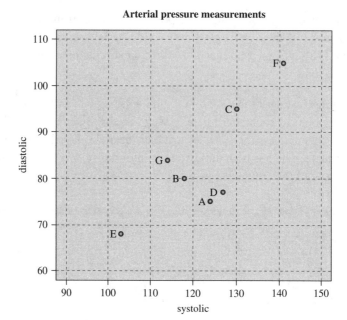

FIGURE 4.4   Improved scatterplot with title, gridlines, and labels.

4.7 Interacting with Plots

Once a plot is generated on screen, you can interact with it by using two functions called `identify()` and `locator()`. The `identify()` function, once called, allows the user to move the cursor over the plot and then label a point by clicking either on it or near to it. The point can either be labeled by increasing number or by parsing a vector of labels (Code 4.8).

```
> x<-c(124,118,130,127,103,141,114)
> y<-c(75,80,95,77,68,105,84)
> identify(x,y,labels=c('A','B','C','D','E','F','G'))
```

Code 4.8.

The `locator()` function can be used to retrieve the *x* and *y* coordinates of a point in a plot (Code 4.9). This function also allows the user to click a point with the mouse and return the coordinates to the console.

```
> locator()
$x
[1] 117.9887 129.9335 123.9611

$y
[1] 80.04370 94.92767 75.11524
```

Code 4.9.

4.8 Summary

R has excellent graphics facilities. With minimal effort, you can soon become familiar with high-level plotting functions and low-level support functions. All aspects of graphics can be controlled by passing a large range of parameters to plotting functions.

4.9 Questions

1. What are the major differences between the two mainstream graphics systems in R?

2. How can we open a new plot window?

3. What is the most commonly used function for creating graphics in R?

4. How can we change the font style of a plot?

5. What function allows us to add lines a to plot?

6. What `plot()` function parameter allows us to change the plotting color?

7. How can we remove a title from a plot?

8. How can we generate, in a single plot window, a series of six plots with a layout of three rows and two columns?

5 Example Datasets

5.1 Background

Throughout the book, we will use example medical and biomedical datasets as we explore the different packages and functionality of R. We will use both artificial and real datasets from a number of sources. Some of the basic datasets that we will use already exist in R packages and will be described at the time of use. In Chapters 18 and 21, we will work with publicly available microarray and comparative genomic hybridization (CGH) datasets, which we will obtain at source. Full data descriptions are given in those chapters along with instructions for downloading the files from their original sources.

The remaining datasets that accompany this book may be downloaded from the Jones and Bartlett website (see Introduction for more details). These datasets include a comprehensive breast cancer patient dataset, a collection of tissue microarray datasets, and a collection of functional MRI (fMRI) medical image sets (List 5.1).

List 5.1 Dataset folders available for download.

- Analysis Data Files
- Patient Dataset
- Tissue Microarray Datasets
- fMRI Datasets
- Example Data Files

The "Analysis Data Files" folder contains a copy of all the data files that you will create over the next few chapters. The patient dataset is large and comprehensive, and we will use the data throughout the book. This chapter serves as a reference for the variable data contained within the patient dataset and also provides an introduction to the tissue microarray and fMRI datasets. The "Example Data Files" folder contains a set of patient data in different file formats, which we will use in the next chapter. These files are also listed at the end of the chapter.

Before you use the datasets described, it must be pointed out that these were designed for ease of use. Each data source is simply data that lacks any descriptors (metadata). In reality, a careful and conscientious person generating data files should use standards (if available) for storing and describing data and generate documentation for each dataset. We will tackle this issue in the last chapter.

5.2 The Patient Dataset

All of the files that make up the patient dataset can be found in the folder called "Patient Dataset." Within this folder are three subfolders called "Patient Records," "Variable Codes," and "Analysis Data Files." If you open the Patient Records folder, you will see that the files come in a variety of different formats. You should notice though that the file names include the terms "demographic," "pathology," "diagnosis," "treatment," and "staging." The file "patient.mdb" is a Microsoft Access database that also contains these individual datasets as tables with a single source.

The file names actually refer to five categories of patient data that exist for a dataset of more than 62,000 breast cancer patients diagnosed in the United States between 1990 and 1997. Thus, all files contain variable data for exactly the same group of patients. The dataset originates from The Surveillance, Epidemiology, and End Results Program (http://seer.cancer.gov/) of the National Cancer Institute (NCI). Most of the data, including pathology, diagnosis, and treatment, is real and provides us with an excellent heterogeneous biomedical dataset.

The demographic data, however, is partially artificial, as the original dataset from SEER is completely anonymous. In the demographic file "demographic.csv" you will see that each patient has a date of birth, which is completely made up. Each file with patient record data also contains a column called "patientid." This field contains a

Table 5.1   Patient Dataset Variables.

Demographic data:	patientid	unique patient identifier (artificial)
	dateofbirth	patient date of birth (artificial)
	maritalstatus	marital status at diagnosis
	race	patient ethnicity
	ageatdiagnosis	age at diagnosis
	alivestatus	patient alive or dead
	survivaltime	survival time from date of diagnosis
Diagnosis data:	patientid	
	yearofdiagnosis	year of diagnosis
	histology	histologic type of tumor
	primarysite	site of primary tumor
	numberofprimaries	number of primary tumors
Pathology data:	patientid	
	grade	tumor grade
	nodesexam	number of lymph nodes examined
	nodespos	number of positive lymph nodes
	extent	extent of disease
	nodalstatus	status of lymph node involvement
	size	size of tumor
	pgr	progesterone receptor status
	er	estrogen receptor status
Staging:	patientid	
	stage	stage of tumor
Treatment:	patientid	
	surgery	surgery regime received
	radiotherapy	radiotherapy received

Table 5.2	Numerical Variables in the Patient Dataset.

survivaltime	months
yearofdiagnosis	1990–1997
numberofprimaries	1–4
nodesexamined	0–77
nodespos	0–50
tumorsize	millimeters

unique identifier for each patient, allowing the data from all sources to be linked. This identifier acts like a hospital record number of a patient but is purely fictitious, as the original data is anonymous. In a later chapter, we will look at ways to de-identify patient data prior to analysis; hence the need for an artificial patient identification number.

Variables for the complete patient dataset are shown in Table 5.1. You may be wondering why the data is provided in separate files of different formats. In later chapters we will look at how we can use R to organize data and integrate biomedical data from different sources stored in different ways. These chapters will highlight the fact that R is not just a data analysis tool, but can also provide us with a comprehensive data management platform.

The variables are a mixture of both numeric and categorical types. Numeric variables and their particular measurements are shown in Table 5.2. Survival time for each patient is recorded in months from the date of diagnosis. Patients who have died of breast cancer have a survival time up to the point of death, whereas survival time recorded for alive patients spans from date of diagnosis to the last known date of assessment. The number of primary tumors observed for the majority of patients is one, but other patients present up to four primaries. The range for the number of nodes examined in a patient is 0 (none examined) to 77, and the range for the number of positive nodes observed is 0 to 50. Tumor size is measured in millimeters, with a range of 1 to 200.

A number of the categorical variables in the patient dataset have had the data recoded. Original data retrieved from SEER can be highly textual and is not practical for creating data analysis input files. Therefore, all patients have been given a categorical value for each variable that required recoding. The codes representing the categorical values and corresponding meanings or annotation can be found in a series of text files in the "Variable Codes" folder.

Categorical variable meanings in the demographic dataset (List 5.2) are quite straightforward.

List 5.2 Categorical values for variables in the demographic dataset.

- Variable: maritalstatus

  1 = divorced/separated

  2 = married

  3 = single

  4 = unknown

  5 = widowed

- Variable: race

    1 = black

    2 = white

    3 = other

- Variable: alivestatus

    0 = alive

    1 = dead

The meanings of each category in variables in the diagnosis dataset require some explanation for those unacquainted with the terminology (List 5.3). The primary variable describes the topography or location of the tumor in relation to the breast. The term "NOS," which occurs in descriptions for other variables, means "not otherwise specified." The histology variable describes the morphology of the tumor.

List 5.3 Categorical values for variables in the diagnosis dataset.

- Variable: primary

    500 = Nipple

    501 = Central portion [subareolar]

    502 = Upper inner quadrant

    503 = Lower inner quadrant

    504 = Upper outer quadrant

    505 = Lower outer quadrant

    506 = Axillary tail

    508 = Overlapping lesion

    509 = NOS

- Variable: histology

    8010 = CARCINOMA, NOS-carcinoma in situ/NOS

    8050 = PAPILLARY CARCINOMA, NOS-In situ/NOS

    8070 = SQUAMOUS CELL CARCINOMA-In situ/NOS

    8140 = ADENOCARCINOMA-In situ/NOS

    8201 = CRIBRIFORM CARCINOMA

    8211 = TUBULAR ADENOCARCINOMA

    8480 = MUCINOUS ADENOCARCINOMA

    8500 = DUCT CARCINOMA-Intraductal carcinoma, noninfiltrating/ infiltrating, NOS

    8501 = DUCT CARCINOMA-Comedocarcinoma, noninfiltrating/NOS

    8503 = DUCT CARCINOMA-Papillary adenocarcinoma, intraductal with invasion/noninfiltrating intraductal

8510 = MEDULLARY CARCINOMA-Medullary carcinoma, NOS

8520 = LOBULAR & OTHER DUCTAL CARCINOMAS-Lobular carcinoma in situ/NOS

8521 = LOBULAR & OTHER DUCTAL CARCINOMAS-Infiltrating ductular carcinoma

8522 = LOBULAR & OTHER DUCTAL CARCINOMAS-Intraductal/infiltrating duct & lobular in situ carcinoma

8530 = INFLAMMATORY CARCINOMA

8541 = PAGET DISEASE, MAMMARY-Paget disease, infiltrating ductal carcinoma

The pathology dataset also requires a brief description of variable meanings (List 5.4). The grade variable shows the respective tumor grade. Grade I refers to tumors with cells that are often well differentiated. These tumors are often referred to as low-grade tumors and are generally considered the least aggressive in their behavior. Grade II tumors are described as moderately well differentiated and are termed intermediate-grade. Grade III and Grade IV tumors are poorly differentiated and undifferentiated, respectively, and are classed as high-grade, generally being the most aggressive in behavior with the poorest prognosis. Thus, the grade of a tumor is used to predict outcome or prognosis and may also be used to plan a patient's treatment.

The tumor extent variable describes the "spread" of disease from the site of origin. The nodal status refers to the involvement of lymph nodes and the extent of involvement. The pathology group also contains data for two biological markers for breast cancer: estrogen receptor (ER) status and progesterone receptor status (PgR). The presence or absence of ER and PgR in combination can dictate how a patient may respond to hormonal therapy. If a patient is positive for both ER and PgR, they would probably have a good response to hormonal therapy. There is still a chance of a good response to therapy if a patient is ER-positive and PgR-negative (or vice versa), but the chances are reduced by about a half. If a patient is negative for both ER and PgR, there is only a small chance of responding to therapy.

List 5.4 Categorical values for variables in the pathology dataset.

- Variable: grade

  1 = G1

  2 = G2

  3 = G3

  4 = G4

- Variable: extent

  10 = Confined to breast tissue and fat, including nipple and/or areola

  20 = Invasion of subcutaneous tissue

  30 = Invasion of (or fixation to) pectoral fascia or muscle

40 = Invasion of (or fixation to) chest wall, ribs, intercostal, or serratus anterior muscles

50 = Extensive skin involvement

70 = Inflammatory carcinoma, including diffuse dermal lymphatic permeation or infiltration

- Variable: nodalstatus

0 = No lymph node involvement

Size of largest metastasis in axillary node(s), ipsilateral:

1 = Micrometastasis (<0.2 cm)

2 = >0.2 to <2.0 cm, no extension beyond capsule

3 = <2.0 cm with extension beyond capsule

4 = ≥2.0 cm

5 = Fixed/matted ipsilateral axillary nodes

6 = Axillary/regional lymph nodes, NOS; lymph nodes, NOS

7 = Internal mammary node(s), ipsilateral distant lymph nodes

8 = Cervical, NOS; contralateral/bilateral axillary and/or internal mammary supraclavicular (transverse cervical); other than above

9 = Unknown; not stated

- Variable: pgr

1 = Positive

2 = Negative

3 = Borderline

- Variable: er

1 = Positive

2 = Negative

3 = Borderline

The staging data contains a single categorical variable called stage. The stage of a tumor takes into account the tumor size (T), the involvement of lymph nodes (N), and whether the tumor has spread (M). Staging is used to assist in the decision of treatment for the patient. The staging system used to create this particular data in SEER is that of the American Joint Committee on Cancer (AJCC) and referred to as a TNM system (For a summary of TNM staging see: http://www.cancer.org/docroot/CRI/content/CRI_2_4_3X_How_is_breast_cancer_staged_5.asp). Once values for T, N, and M have been calculated, they are combined into one of four groups in America (I, II, III, and IV) or three groups in other countries, such as the United Kingdom. These groups also have subgroups (List 5.5), which can denote early- or late-stage cancer according to the individual values of T, N, and M.

List 5.5 Categorical values for variables in the tumor staging dataset.

- Variable: stage (tumor stage)

10 = I: early stage

21 = IIA: early stage

22 = IIB: early stage

31 = IIIA: early stage

32 = IIIB: late stage

40 = IV: late stage

The treatment regimen decided for a breast cancer patient depends on the staging and grading of the tumor, whether the patient has gone through menopause (in the case of females), as well as biological factors, such as ER receptor status. Most breast cancers will be treated with surgery by partial or total mastectomy. Sometimes chemotherapy or hormonal therapy is given prior to surgery to shrink the tumor (neoadjuvant therapy). Radiotherapy may be given post surgery to destroy remaining cancer cells. Most patients will then receive chemotherapy or hormonal therapy to reduce the likelihood of the cancer returning (adjuvant therapy). SEER provides data for surgery and radiotherapy received but not hormonal therapy (List 5.6).

List 5.6 Categorical values for variables in the treatment dataset.

- Variable: surgery

   0 = None

   1 = Incisional, needle, or aspiration of other than primary site

   2 = Incisional, needle, or aspiration of primary site

   10, 18, 20 = Partial mastectomy

   30, 38 = Subcutaneous mastectomy

   40, 48 = Total mastectomy

   50, 58 = Modified radical mastectomy

   60 = Radical mastectomy

   80 = Mastectomy, NOS

   90 = Surgery, NOS

- Variable: radiotherapy

   1 = Beam

   3 = None

In the "Patient Dataset" subfolder you will also see a subfolder called "Code Descriptions." In this folder are a series of files that provide interpretations of the coding used for different variables. As you can imagine, a surgery code of "50" means very little, but by looking up the value of the code we would find that the patient underwent a modified radical mastectomy.

5.3 Tissue Microarray Datasets

Anatomic or anatomical pathology is a field of medicine concerned with the diagnosis of disease based on macroscopic, microscopic, and molecular examination of organs, tissues, and cells. Histopathology, the microscopic examination of tissue, generally involves the examination of a biopsy or surgical specimen that has been placed in a fixative to stabilize and preserve the tissue structure. The fixed tissue may then be embedded in a paraffin wax so that thin sections can be cut for the pathologist to examine.

To visualize cellular components, the sections are stained with histochemical dyes such as hematoxylin and eosin. Proteins and other molecular types may also be stained in a section by immunohistochemistry (IHC) where antibodies, specific for an antigen, are applied. IHC is now widely used for the diagnosis of many cancers where specific diagnostic markers (usually proteins) are known. The quest to discover new diagnostic as well as prognostic IHC markers for many diseases by research groups throughout the world is relentless.

A technology has been developed recently called tissue microarray (TMA) that greatly facilitates this quest. TMA paraffin blocks are created by simply taking tiny cores of tissue from a standard histology block and placing these cores in an array format. TMA permits many tissue samples to be screened by an antibody at once and is thus high-throughput. TMA blocks are sectioned just like standard histology blocks and, once stained, may be assessed by a pathologist. Although tiny, each core should contain enough cells to reveal the pattern of expression for the marker in the donor tissue. Another advantage of TMA is that precious tissue is not wasted or depleted due to the small size of the cores used. By using TMA technology a researcher may carry out retrospective studies for one or more markers in a single experiment at low cost.

The data generated for TMA is a set of scores, generated by the pathologist, for each core that is dependent on the marker used. Types of scoring systems vary for markers. Categorical scores may be given for some markers, such as "0" if no cells show a positive stain (i.e., protein not expressed), "1" if between 0% and 1% of cells stain positive, "2" if between 1% and 10% of cells stain positive, etc. This generates proportion scores. Intensity scores are another system where, for example, "0" indicates negative staining, "1" is weak intensity of stain across the cells, "2" is intermediate intensity, "3" is strong intensity. Another system involves just giving an estimate of the percentage of cells that stain positive. The range between 0% and 100% is not categorical in this case but numeric. Scores can also be given depending on the cellular location of the stain, such as nuclear or membrane. Scoring systems may be combined, such as the commonly used Allred scoring for estrogen receptor, where proportion scores and intensity scores are summed together to give a total score used to determine whether a breast cancer patient is ER-positive (score equal to or greater than 2) or ER-negative (less than 2).

The folder called "Tissue Microarray Datasets" contains files describing the results of two tissue microarray experiments.

The file "esophageal_tma.txt" contains immunohistochemical scores for five protein markers (cyclin D1, E cadherin, p16, p53, and p63), age, and survival data for patients with squamous cell carcinoma (SCC) and adenocarcinoma of the esophagus.

The "breast_tma.txt" file contains immunohistochemical scores for three protein markers called proteinA, proteinB, and her2 (human epidermal growth factor receptor 2).

5.4 fMRI Datasets

The "fMRI Datasets" folder contains three datasets. These datasets are contained within five files of different medical imaging formats (List 5.7). The data files are related to three functional MRI experiments used in Chapter 17 and described in more detail there. The file dicom.dcm contains a three-dimensional image and header information in the Digital Imaging and Communications in Medicine (DICOM) medical image format. The files finger.img, finger.hdr, word.img, and word.hdr are image and header files of the Analyze image format. The finger and word image files are both four-dimensional in that they each contain a time series of three-dimensional images of the brain.

List 5.7 fMRI dataset files.

- dicom.dcm
- finger.hdr
- finger.img
- word.hdr
- word.img

5.5 Example Data Files

In Chapter 6 we will explore how we can import and export files into R using various functions. The files in the "Example Data Files" folder contain six files with identical data but different file formats (List 5.8).

List 5.8 Example files for demonstrating file import and export in R.

- example.txt
- example_commented.txt
- example.xls
- example.csv
- example.sav
- example.mtp

5.6 Summary

The core theme throughout this book is using R with biomedical data. The datasets that accompany the book typify real biomedical data allowing for comprehensive analysis in subsequent chapters.

6 Importing and Exporting Data in R

6.1 Background

As medical data gets collected for a study it must be stored somehow, somewhere in a location where it can be easily retrieved by the researcher. Storage locations can vary from a folder on your personal PC's hard drive to a dedicated computer server accessible perhaps by remote access. More often than not, the file format for the data depends on many factors, including (among others) the computer literacy of the researcher/data collector, the size and complexity of the data, the intended analysis software, and data security requirements. Entering data in R by hand is convenient for small datasets, but for any large dataset, this is impractical. To analyze our data in R and store results, we generally have to read the data from files and export or save objects to files.

Reading and importing data into R is extremely easy, with many functions available to access a range of file formats. A common file format for storing data is an ASCII format, tab-delimited text file. Text files are often created using a simple text editor, such as Microsoft Notepad, and often have the ".txt" file extension. Commonly, this format has the data stored as a table with data separated in rows by a tab character. As we shall see, tabulated data from text files can be read into R using either the `read.table()` or `scan()` functions. Conversely, writing data frames and matrices to text files can be performed using the `write.table()` function.

Over the years, nearly every dataset I have received from medics, eager to analyze their data, has arrived sitting in one or more worksheets in a Microsoft Excel workbook (.xls format). This, of course, reflects the ease of using spreadsheets to enter data and the widespread use of Excel. Excel also has great export facilities for storing data in different formats, such as tab-delimited text and comma-separated values (CSV) formats, which can be imported into many different data analysis packages. Indeed, data stored in table format in text files and spreadsheets have the same rows by columns shape, and it is just as easy to paste data from a worksheet into an empty text file. Data from Excel worksheets can be read directly into R using functions from the xlsReadWrite or RODBC packages. Unfortunately, currently there is no support in R for the open-source OpenOffice Calc spreadsheet software. You could, of course, save data from OpenOffice in tab-delimited, CSV, or other acceptable format before importing into R. You may also have data that has been saved during a previous analysis using a different data analysis software package. R has a package called design that enables you to import data files in formats from commercial and open-source software such as SPSS, Minitab, SAS, Stata, WEKA, and Octave. The flexibility of R in dealing with different file formats should already be clear, and we haven't yet mentioned

Extensible Markup Language (XML). We won't discuss XML as a data storage format here, as it will be covered in detail in Chapter 22. Slightly more adventurous medics I have worked with stored their data in tables in a Microsoft Access database (mdb). R has powerful connection capabilities to relational databases, and this will be discussed in the next chapter.

In this chapter we will explore how we can import and export data to and from R using the packages mentioned. Although we will cover different file formats, we will use the same example medical dataset throughout the chapter. File formats for the dataset will include text, spreadsheet, database, and formats of popular statistical software. The experience gained by working through this chapter will also leave the reader ready for the next few chapters that deal with building heterogeneous biomedical datasets and processing the data prior to analysis. The CRAN site has a very good manual on data import and export in R (http://cran.r-project.org/doc/manuals/R-data.html) as well as an early article in R News on using databases.[2]

The packages used in this chapter can be downloaded from CRAN and installed by typing:

```
>install.packages(c("xlsReadWrite", "RODBC", "foreign",
"xlsReadWrite")
```

6.2 Working with Text Files

Text files, such as tab-delimited and CSV formats, may be read into R using the read.table() function. Actually, the only difference between a tab-delimited file and a CSV file is the character that separates the data—a tab or a comma. Indeed, you could choose from a large number of different characters to use as separators in text files, but it is best to stick to convention, and we will only deal with these two formats here.

By default, read.table() recognizes white space (tabs, spaces, etc.) as separators. If you do have data separated in a text file with a different character, such as a semicolon, read.table() accepts an argument called "sep" allowing you to state the separator. For example, for a semicolon-delimited file, you could use:

```
> read.table("a_data_file.txt", sep = ";")
```

If a separator has been specified, white space used in data values will be recognized as white space. To remove white space from within data values, you can parse an argument to read.table() called strip.white and set it to TRUE.

The read.table() function takes a number of arguments, which gives you flexibility in how the data can be formatted and how much data you may wish to read in. The function imports data into a data frame object. To read in the tab-delimited "example.txt" file as a block of data, you could just parse the file name:

```
> data<-read.table("example.txt")
```

Now look at the structure of the data table (Code 6.1).

```
> data

       V1    V2       V3       V4     V5          V6   V7  V8  V9
1  patientid grade nodesexam nodespos extent nodalstatus size pgr er
2  pid00001    3      32       3     10           6   60   2   2
3  pid00002    2      13       1     10           6   15   1   1
4  pid00003    3       8       0     10           0    8   1   1
5  pid00004    3      20       0     10           0   10   2   2
6  pid00005    2      16       8     10           6   15   2   1
7  pid00006    3      19       0     10           0   48   1   1
8  pid00007    3       3       0     10           0   32   2   2
9  pid00008    2      13       0     10           0   15   1   1
10 pid00009    3      21       0     10           0   22   1   1
11 pid00010    2      15       0     10           0    6   2   1
```

Code 6.1.

If you open the "example.txt" file you will see that it actually contains a table of eight variables, each with a variable name (grade, nodesexam, nodespos, extent nodalstatus, size, pgr, and er) and 10 cases all with a patient ID number. Looking at the data object we have just created, we can see that the `read.table()` function has read all the rows in the text file, including the variable headers, into the data frame, giving 11 rows of data. In this case, variable headers are set by R to V1, V2, V3, etc., and row names are simply numbered. Because we have variable headers and row names in our dataset, we have to let the `read.table()` function know this by setting the "header" argument to true (T) and the "row.names" argument to 1 (the row names are the first column, patientid):

```
> data<-read.table("example.txt", header=T, row.names=1)
```

We can then reload the data from "example.txt" into a data frame after setting these arguments, as shown in Code 6.2, and note that the data object is now quite different, with appropriate headers and row names. Alternatively, if the column names are not included in the file, they may be specified as a vector by the `col.names` argument. Similarly, you can use the `row.names` argument to specify row names as a vector rather than the number of a column in the text file.

```
> data
          grade nodesexam nodespos extent nodalstatus size pgr er
pid00001    3      32        3      10          6      60   2   2
pid00002    2      13        1      10          6      15   1   1
pid00003    3       8        0      10          0       8   1   1
pid00004    3      20        0      10          0      10   2   2
pid00005    2      16        8      10          6      15   2   1
pid00006    3      19        0      10          0      48   1   1
pid00007    3       3        0      10          0      32   2   2
pid00008    2      13        0      10          0      15   1   1
pid00009    3      21        0      10          0      22   1   1
pid00010    2      15        0      10          0       6   2   1
```

Code 6.2.

If your file contains character data enclosed in quotes, including ' '" and '" you don't have to parse this information to read.table(), as this is set by default. Numbers in a text file are read as numeric values by default, whereas non-numeric values are read as factors. The default value for missing data is "NA" and although you could use the na.strings argument to set a character string for missing data, it is once again best to stick to convention. Another potentially useful read.table() argument is dec, which allows you to specify the character that represents a decimal point. This is particularly useful for continental Europeans who use a decimal comma in data rather than a decimal point. One character that you may want to avoid using in your data is the backslash ("\"). This character is used in some programming languages as an escape character that precedes another character, which, in combination, provides a command that needs interpreting. For example, "\n" means "new line." If, for some reason, you did want an escape character to be recognized, you can set the allowEscapes argument to TRUE.

Comments are also permitted in text files, and if your comments are preceded by the hash character ("#"), you do not have to specify this character. If comments are preceded by a different character, this can be set using the comment.char argument. As an example, try loading the "example_commented.txt" file:

```
> data<-read.table("example_commented.txt")
```

When calling the data object, the output displayed should be identical to that seen in Code 6.2. However, if you open up the file you can see that the data is preceded by the comment:

```
#this is a comment
```

Thus, by default, the comment has been ignored.

The read.table() function relies on another function called scan() to read in data. One of the advantages of using scan() instead of read.table() is that you can read in variables as a list of vectors and dictate the mode of each variable as it is read in. To do this, you can pass a list structure as an argument (called what) containing example vectors that tell R the mode for each variable. As an example, let's say that we want to read in all variables in our "example.txt" file as numeric mode, except the first row label variable, which we want to read in as character mode. The R code for this would be:

```
> scan("example.txt", what = list("",0,0,0,0,0,0,0,0), flush =
TRUE,skip=1)
```

The list passed to the what argument contains a dummy character followed by eight dummy numeric characters. The flush argument tells scan() to move to the end of a line after reading the last of the fields, permitting comments to be added (and ignored) at the end of each line. The skip argument tells scan() to ignore the first line in the text file. In our code we declared all but the first variable as numeric, so if we tried to read in the first line of the example.txt file, which is all characters, R would throw an error. By using the skip argument in scan(), as shown, you should see the output in Code 6.3.

```
[[1]]
 [1] "pid00001" "pid00002" "pid00003" "pid00004" "pid00005" "pid00006"
 [7] "pid00007" "pid00008" "pid00009" "pid00010"

[[2]]
 [1] 3 2 3 3 2 3 3 2 3 2

[[3]]
 [1] 32 13  8 20 16 19  3 13 21 15

[[4]]
 [1] 3 1 0 0 8 0 0 0 0 0

[[5]]
 [1] 10 10 10 10 10 10 10 10 10 10

[[6]]
 [1] 6 6 0 0 6 0 0 0 0 0

[[7]]
 [1] 60 15  8 10 15 48 32 15 22  6

[[8]]
 [1] 2 1 1 2 2 1 2 1 1 2

[[9]]
 [1] 2 1 1 2 1 1 2 1 1 1
```

Code 6.3.

To read in CSV files, you could simply pass "," as a value for the `sep` argument in `read.table()`. There is, however, a variation of `read.table()` called `read.csv()` that facilitates this. To read in the "example.csv" file type:

```
> data<-read.csv("example.csv", header=T, row.names=1)
```

For European-style data, where a decimal comma is used, you can use a variation function called `read.csv2()`, which sets the `sep` argument to "," by default.

Writing tabular data to a text file is also straightforward using the `write.table()` function. Load in the "example.txt" data into an object called "data" as described previously. You can then save this data back to a file called "adatafile.txt" by typing:

```
> write.table(data, "adatafile.txt")
```

If you then open "adatafile.txt" you will notice that the data is delimited by a space character, and any non-numeric data values, including variable headers, are enclosed using quotes. The variable headers are also not aligned to their corresponding variables. To prevent the use of quotes and to separate data with a tab character, you can use the following arguments:

```
> write.table(data, "adatafile.txt", quote=FALSE, sep="\t",
row.names=TRUE,col.names=NA)
```

With the `row.names` argument set to TRUE and the `col.names` argument set to NA (to correctly align the variable headers), a tab-delimited text file is created in the correct format. CSV files may be created in a similar way using the `write.csv()` function. Any missing values will be output as NA by these functions.

6.3 Importing from Microsoft Excel Worksheets

Given its widespread use, Microsoft Excel probably contains the data of most nonstatistical/data analysis folk, whether they be a medic or treasurer of the local gardening club. As you will have probably guessed by reading the previous section, a quick way of getting data from Microsoft Excel into R is exporting it as a tab-delimited or CSV text file. There is, however, a package called "xlsReadWrite" that has a nice function called `read.xls()` that will read an Excel worksheet into R:

```
> library(xlsReadWrite)
> read.xls("example.xls")
```

Versions of Excel supported by xlsReadWrite are from 1997 to 2003. Note that we are loading in data from a worksheet here as opposed to a workbook. When you create a new workbook in Excel, it will, by default, contain three worksheets to which you can add or delete at will. By default, `read.xls()` reads in the first worksheet in a workbook, but you can change this by specifying the number of the worksheet you wish to read in using the `sheet` argument:

```
> read.xls("example.xls", sheet=2)
```

Like `read.txt`, `read.xls` has a number of arguments that can be changed to suit the dataset. By default, you load data into a data frame object, but if preferred, a matrix may be created by setting the `type` argument to the mode of choice (e.g., double, integer, character, etc.). Also by default, the first row is read as variable headers, but this can be changed by setting the `colNames` argument to FALSE. To specify that the first column contain row names, you can set the `rowNames` argument to TRUE (it is not by default).

A second way of importing data directly from an Excel worksheet is by using a function called `odbcConnectExcel()` in the RODBC package. This method treats Excel as a database using Open Database Connectivity (ODBC) technology. A current advantage of using this approach is that you can access data held in Excel 2007 format using another function in RODBC called `odbcConnectExcel2007()`. To connect to an Excel workbook using `odbcConnectExcel()`, first load the RODBC library:

```
> library(RODBC)
```

You then open a connection to the workbook using the `odbcConnectExcel()` command:

```
> connection<-odbcConnectExcel("example.xls")
```

We have created a connection called "connection" to the "example.xls" workbook, which is the target file (more technically speaking, it is the data source name, or dsn). Now you can view the worksheets contained within the workbook using the `sqlTables()` command from RODBC and parsing the name of the connection as an argument (Code 6.4).

```
> sqlTables(connection)
TABLE_CAT
1 C:\\Program Files\\R\\R-2.5.0\\book\\ example    <NA>
2 C:\\Program Files\\R\\R-2.5.0\\book\\ example    <NA>
3 C:\\Program Files\\R\\R-2.5.0\\book\\ example    <NA>
     TABLE_SCHEM   TABLE_NAME   TABLE_TYPE    REMARKS
1    <NA>          Sheet1$      SYSTEM TABLE   <NA>
2    <NA>          Sheet2$      SYSTEM TABLE   <NA>
3    <NA>          Sheet3$      SYSTEM TABLE   <NA>
```

Code 6.4.

We can view the contents of a worksheet by using a function called `sqlFetch()`, as shown in Code 6.5.

```
> sqlFetch(channel, "Sheet1")
   patientid grade nodesexam nodespos extent nodalstatus size pgr er
1    pid00001    3       32        3     10          6    60  2  2
2    pid00002    2       13        1     10          6    15  1  1
3    pid00003    3        8        0     10          0     8  1  1
4    pid00004    3       20        0     10          0    10  2  2
5    pid00005    2       16        8     10          6    15  2  1
6    pid00006    3       19        0     10          0    48  1  1
7    pid00007    3        3        0     10          0    32  2  2
8    pid00008    2       13        0     10          0    15  1  1
9    pid00009    3       21        0     10          0    22  1  1
10   pid00010    2       15        0     10          0     6  2  1
```

Code 6.5.

Alternatively, you can place the contents of the worksheet into a data frame object:

```
> worksheet<-sqlFetch(channel, "Sheet1")
```

The help file for the RODBC package contains information on a number of other useful functions for manipulating data in Excel workbooks. These functions include `sqlSave()` to create a new worksheet and `sqlDrop()` to clear the contents from a worksheet.

6.4 Using Files Created by Other Statistical Software Packages

The safest way to import data you have created and stored using another statistical package is to export it first as a text file and then import that into R. However, the foreign package has functions that allow you to import data produced by a number of statistical systems. Two of the most common "user-friendly" data analysis software packages are SPSS and Minitab, and we will deal only with those here.

To read in the example SPSS data file called "example.sav," we use the `read.spss()` function:

```
> spssdata<-read.spss("example.sav")
```

read.spss() loads the data into a list of variables (spssdata) by default. To load data into a data frame object, you need to set the to.data frame argument to TRUE. Code 6.6 shows this option and the first 10 rows of data from the data frame created.

```
> spssdata<-read.spss("example.sav", to.data frame=TRUE)
> spssdata[1:10,]
   PATIENTI GRADE NODESEXA NODESPOS TUMOUREX NODALSTA TUMOURSI PGR ER
1  pid00001    3      32        3       10        6       60   2  2
2  pid00002    2      13        1       10        6       15   1  1
3  pid00003    3       8        0       10        0        8   1  1
4  pid00004    3      20        0       10        0       10   2  2
5  pid00005    2      16        8       10        6       15   2  1
6  pid00006    3      19        0       10        0       48   1  1
7  pid00007    3       3        0       10        0       32   2  2
8  pid00008    2      13        0       10        0       15   1  1
9  pid00009    3      21        0       10        0       22   1  1
10 pid00010    2      15        0       10        0        6   2  1
```

Code 6.6.

Data from Minitab-format files may be imported in a similar way using the read.mtp() function:

```
> minitabdata<-read.mtp("example.mtp")
```

This creates a list (minitabdata) with a variable for each column of data. When importing a minitab-format file, be aware that non-numeric variables are not currently allowed. So the patientid variable, which is character mode, has been removed from the example dataset that we have imported from other formats elsewhere in this chapter.

6.5 Editing Data by Invoking a Text Editor

Once you have loaded multivariate data sets into R, you may need to change one or more values. R has a built-in function called edit() that allows you to open a data frame object in a basic spreadsheet-style editor. As an example, read in the "example.txt" dataset using the read.table function:

```
> data<-read.table("example.txt",header=T,row.names=1)
```

Then call the editor to open the data object for editing (Figure 6.1):

```
> data<-edit(data)
```

You can also edit the original data file directly by specifying the file argument:

```
> edit(file="example.txt")
```

6.6 Summary

This chapter has demonstrated the ease with which data files can be imported from a wide range of file types into R and exported back to specified formats. The ability to import biomedical data for analysis from heterogeneous sources and file formats saves

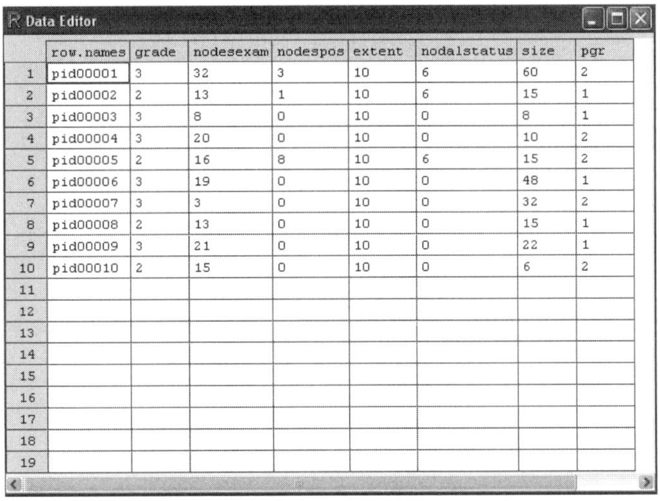

	row.names	grade	nodesexam	nodespos	extent	nodalstatus	size	pgr
1	pid00001	3	32	3	10	6	60	2
2	pid00002	2	13	1	10	6	15	1
3	pid00003	3	8	0	10	0	8	1
4	pid00004	3	20	0	10	0	10	2
5	pid00005	2	16	8	10	6	15	2
6	pid00006	3	19	0	10	0	48	1
7	pid00007	3	3	0	10	0	32	2
8	pid00008	2	13	0	10	0	15	1
9	pid00009	3	21	0	10	0	22	1
10	pid00010	2	15	0	10	0	6	2
11								
12								
13								
14								
15								
16								
17								
18								
19								

FIGURE 6.1 The R Editor.

time and reduces the need for multiple software tools used in preparing and processing data.

6.7 Questions

1. Why do you think a sound knowledge of data import and export is required when using R for biomedical data?
2. What type of R objects do you think can be assigned data read in from different data sources?
3. How can you read a tab-delimited text file into R so that the first row in the file is treated as column headers?
4. When we export data using `write.table()`, how are missing values treated?
5. What R function allows us to import Excel spreadsheets, and how can we specify the worksheet to import?
6. What is the advantage of using `scan()` over `read.table()`?
7. How can we simultaneously view and rapidly edit data that has been imported into R?

7 R, SQL, and Database Connectivity

7.1 Background

Storing a relatively small dataset in a text file or spreadsheet is practical in many situations, but multiple related or linked biomedical datasets in a project need managing in a more constructive way. In larger data analysis projects, it is more sensible to use a database management system (DBMS) and store your data in tables. Also, text files and spreadsheets aren't suited for very large datasets, whereas a database table could potentially store hundreds of thousands of rows (records).

The most common type of database in use today is the relational database based on the relational model. As we shall see in this chapter, the relational database model provides an efficient, quick, and cost-effective way to manage your biomedical data and seamlessly integrate the data with R for query and analysis. Relational DBMSs are often referred to by the acronym RDBMS. One of the most popular RDBMSs is the free, open-source MySQL software. Some excellent packages exist that allow you to connect to databases using R with little trouble. Querying and retrieving data from an RDBMS is done using the Structured Query Language (SQL). R can talk quite nicely to a range of RDBMSs using SQL, making it very flexible indeed.

The RODBC package provides functions that allow you to connect to a database source that supports the open database connectivity (ODBC) interface. For Windows users, these database systems include Microsoft Access, Microsoft SQL Server, MySQL, and PostgreSQL. Functions are provided that allow querying of a valid database using SQL with a return to R of data in a data frame object. A package called RMySQL is devoted solely to connecting R to data in MySQL databases (MySQL is also open-source). In the next chapter we will create a biomedical database in MySQL for the patient dataset using RMySQL.

Before we look at R's capabilities in communicating with databases, there is an introduction (or refresher for those who have already dipped their toes) to database systems. We will then follow this with a primer on SQL; database and SQL veterans can jump to Section 7.6.

To work through the chapter ensure that the "patient.mdb" file is in the Patient Records folder and that you point R to this working directory using the `setwd()` command or the "Change Dir" option in the RGUI File menu.

7.2 Overview of Database Systems

A database contains collections of data records. Databases can be a collection of one or more files made up of records, where the records contain data in one or more fields.

In the "example.txt" file that we used in the last chapter, we saw data for each patient (row) split into data values belonging to different variables (columns). If that exact same piece of data were stored in a database structure, each piece of data would be described as a field (column) belonging to a record (which is the equivalent to a row in the text file). We can think of the different fields as representing each variable in our dataset. Thus, a record is a collection of fields, and a file, in database terms, is a collection of records.

Types of database structures are usually described according to how the files and data are organized. The "example.txt" file is actually an example of perhaps the simplest type of database systems called the flat model. A flat model database is of a table format made up of rows and columns of particular data types. More complex database systems were developed to overcome the limitations of flat files, including hierarchical and network models. These systems were really for the hands of experts, and in the 1980s a new type of database concept became popular called the relational database model. The whole concept of a relational database makes life much easier in understanding how we can store our biomedical data. The literature abounds with solutions for storage and management of data using relational databases in both biology and medicine.

Relational databases use tables to store information as records and rows, and allow you to easily create, search, access, and amend tables in different ways. Relationships can then be defined between tables, allowing the user to combine data from several tables either for querying or reporting. Relationships can be of different types, such as a one-to-one where a record in one table is related to a record in a different table. A one-to-many relationship is one where one record in a table is related to many records in another table. Individual records in a table are identified by a unique primary key often in a code field. A relationship between two tables can be made using a foreign key. A foreign key field in one table can contain the primary key values of another table, thus linking all records in the first table to specific records in the second table.

Managing, querying, and reporting data from a database are done using a database management system (DBMS). The DBMS is the actual software you use, and some commercial and free/open-source DBMSs are shown in List 7.1.

List 7.1 Popular commercial and free/open-source database management systems.

- Oracle (commercial)
- IBM DB2 (commercial)
- Microsoft SQL Server (commercial)
- Ingres
- Microsoft Access (commercial)
- MySQL
- PostgresSQL
- SQLite

7.3 Creating Databases and Tables Using SQL

Structured Query Language, or SQL, is the standard language for relational database management systems according to the American National Standards Institute (ANSI).

Thankfully, SQL isn't difficult to grasp and you can accomplish much with just a few basic commands. Be aware, though, that despite SQL being declared a standard language, DBMS-specific extensions of the language do exist. The following examples work with MySQL, among others.

CREATE

The SQL statement for creating a new database (such as that in MySQL) is:

```
CREATE DATABASE newdatabase;
```

where "newdatabase" is the name of the database to create—it is that simple. Note that SQL statements end with a semicolon. To make the current database active, you provide the USE term:

```
USE newdatabase;
```

To create tables in the database, you also use the CREATE term:

```
CREATE TABLE newtable (field1, field2, field3);
```

This statement creates a table called "newtable" in "newdatabase" with three fields: "field1," "field2," and "field3." You can specify the data type and field length:

```
CREATE TABLE newtable (field1 CHAR(50));
```

SHOW

To view table names in an active MySQL database, you use the SHOW command:

```
SHOW TABLES;
```

The SHOW command can also be used to reveal other objects within the database, including the field names in a table:

```
SHOW COLUMNS FROM newtable;
```

7.4 Selecting Data with SQL

SELECT

The SELECT statement allows you to choose data from a database that match criteria you specify in the statement. The following line of code is a simple query of a table called "tablename" to return all fields for each record:

```
SELECT * FROM tablename;
```

The "*" character, when used with SELECT, is a wildcard to search all fields. If we wanted to be more selective and just select data from three fields called "field1," "field2," and "field3," we would type:

```
SELECT field1, field2, field3 FROM tablename;
```

This is fine if you want to retrieve all records from a table, but if you wish to return records given certain conditions, you can combine SELECT with the WHERE clause:

```
SELECT field1 FROM tablename WHERE field2 = 10;
```

The WHERE clause can be combined with a number of different operators and conditionals (Table 7.1). Furthermore, operators can be combined in a SELECT statement:

```
SELECT field1 FROM tablename WHERE (field2 <= 10) OR (field2 > 40);
SELECT field1 FROM tablename WHERE (field2 <= 10) AND (field2 > 40);
```

When using operators in this way, you must use parentheses to enclose the expression being evaluated. If a field contains a string, the relevant expression within the SELECT statement must be enclosed in single quotes:

```
SELECT field1 FROM tablename WHERE field4 = 'a_character';
```

LIKE and NOT LIKE

In dealing with strings, a pair of conditionals that give great power to the SELECT statement are LIKE and NOT LIKE. LIKE and NOT LIKE are pattern-matching operators and usually combined with the "_" and "%" wildcard characters to return records that are like or not like what you specify. As an example, we could use:

```
SELECT * FROM genes WHERE name LIKE 'epiderm%';
```

to return all records from a table called "genes" where the entry in the name field begins with "epiderm" (e.g., epidermal).

Alternatively, to search the same table for gene names that have a single letter after the term "factor," we would use the underscore character:

```
SELECT * FROM genes WHERE name LIKE 'factor_';
```

The NOT LIKE conditionals can be used in the same way.

ORDER BY

Queries can be ordered by selecting a field as an argument for the ORDER BY term in the SELECT statement:

```
SELECT field1, field2 FROM tablename ORDER BY field1;
```

Table 7.1   Common SQL Operators and Conditionals Used with the WHERE Expression.

=	equals
<	less than
>	greater than
<=	less than or equal to
>=	greater than or equal to
!=	not equal to
AND(&&)	two conditions are true
NOT(!)	condition is not true
OR(||)	one or the other condition is true
BETWEEN	within a specified range
NOT BETWEEN	not within the specified range
IS NULL	has no value
IS NOT NULL	has a value
LIKE	pattern matching by similarity

7.5 Inserting, Updating, and Deleting Data with SQL

INSERT

To add records to a table, use the INSERT command:

```
INSERT INTO newtable (field1, field2) VALUES ('value1', 'value2');
```

In this statement we have added a record into "newtable" with values for "field1" and "field2."

UPDATE

To amend data in records in MySQL, you first select the record you want:

```
SELECT patientid FROM patients WHERE forename = 'john' AND surname = 'smith';
```

This returns a primary key value for this record ("patientid" here is the primary key column). Let's say that the "patientid" value returned was 154. We can then apply the UPDATE command combined with SET to parse the new data to this record:

```
UPDATE patients SET age = 75 WHERE patientid = 175;
```

DELETE

To delete the record with patientid of 175 from the patients table, we can use the DELETE command:

```
DELETE FROM patients WHERE patientid = 175;
```

DROP

Finally, we can delete tables or entire databases using the DROP command:

```
DROP TABLE patients;
DROP DATABASE clinic;
```

7.6 Connecting R to Database Sources with RODBC

In Chapter 6 we used a function called odbcConnectExcel() to create a connection between R and a Microsoft Excel workbook. This was our first taste of using the RODBC package to create connections to databases that support an ODBC interface. ODBC stands for open database connectivity, which is a standardized application programming interface (API) for accessing data in many different types of DBMSs. ODBC is an interface that sits between front-end applications and databases without the need for either to be developed by the same vendor (i.e., it is product-neutral). ODBC works by inserting a piece of middleware called the database driver between the application and the DBMS. In Windows, these drivers should already be provided. RODBC can connect to databases in many different types of DBMSs. Although we'll only provide an example of connecting to a Microsoft Access database here, you can actually apply the same functions and procedures to other database formats. The best way to learn how to connect R to a database is to jump straight in with an example and try a few commands. First, install the RODBC library:

```
> install.packages("RODBC")
```

Now load the RODBC library:

```
> library(RODBC)
```

Now we'll try connecting to the "patient.mdb" database, which, as its .mdb file extension suggests, is in Microsoft Access format. There are two ways of doing this. The easiest way is to use the specialized functions `odbcConnectAccess()` or `odbcConnectAccess2007()`, but we'll first look at the generic way of connecting to a data source by creating a Data Source Name or DSN. Using this approach you can connect to different format database sources with the function `odbcDriverConnect()` and select the relevant driver. Begin by typing:

```
> odbcDriverConnect("")
```

By not specifying any arguments and just using two double quotes between parentheses you will invoke R to display the "Select Data Source" dialog box. The main box will display any DSNs already available, but might be blank. Therefore, you need to create a DSN for the "patient.mdb" file by clicking the "New" button adjacent to the "DSN Name" field. A "Create New Data Source" dialog box should appear with a list of drivers for different database formats. Scroll down and select "Microsoft Access Driver (*.mdb)" and click the "Next" button. Give the connection a name, such as "access_dsn," and click "Next" followed by "Finish." Another dialog box appears called "ODBC Microsoft Access Setup," which allows you to select your target database. If you click the "Select" button, you can navigate to the "patient.mdb" file in the working directory. Select the file and click "OK" twice to finish. Finally, back in the "Select Data Source" dialog box, select the DSN you have just created and click "OK." If all works well, R should display details for the connection (Code 7.1).

```
> odbcDriverConnect("")
RODB Connection 10
Details:
  case=nochange
  DBQ=C:\Program Files\R\R-2.5.0\book\patient.mdb
  DefaultDir=C:\Program Files\R\R-2.5.0\book
  Driver={Microsoft Access Driver (*.mdb)}
  DriverId=25
  FIL=MS Access
  FILEDSN=C:\Program Files\Common Files\ODBC\Data Sources\access.dsn
  MaxBufferSize=2048
  MaxScanRows=8
  PageTimeout=5
  SafeTransactions=0
  Threads=3
  UID=admin
  UserCommitSync=no
```

Code 7.1.

The easiest way to connect to an Access database such as "patient.mdb" is to parse the database name to the `odbcConnectAccess()` function, as shown in Code 7.2. You may be wondering why we went the long way round, using `odbcDriverConnect("")` to connect to "patients.mdb," when using `odbcConnectAccess()` is so much easier.

However, by working through the procedure using `odbcDriverConnect()`, you have gained the experience of creating a DSN for other DBMS formats.

```
> connection.access<-odbcConnectAccess("patient.mdb")
> connection.access
RODB Connection 11
Details:
  case=nochange
  DBQ=C:\Program Files\R\R-2.5.0\book\patient.mdb
  Driver={Microsoft Access Driver (*.mdb)}
  DriverId=25
  FIL=MS Access
  MaxBufferSize=2048
  PageTimeout=5
  UID=admin
```

Code 7.2.

In the "Select Data Source" dialog box, you may have noticed a tab called "Machine Data Source." This is another way to connect to an Access database by selecting "MS Access Database" and then finding your .mdb file to connect to without having to set up a DSN first. Available data source names on your computer can be accessed in this way and obtained using the `odbcDataSources()` function (Code 7.3).

```
> odbcDataSources()
MS Access Database
"Microsoft Access Driver (*.mdb)"
Excel Files
"Microsoft Excel Driver (*.xls)"
dBASE Files
"Microsoft dBase Driver (*.dbf)"
```

Code 7.3.

You may have a number of connections open to different databases simultaneously. To view the details on any connection, you can use the `odbcGetInfo()` function. Details for the connection we established to patients.mdb are shown in Code 7.4.

```
> connection.access<-odbcConnect("patient.mdb")
> odbcGetInfo(connection.access)
DBMS_Name        DBMS_Ver        Driver_ODBC_Ver    Data_Source_Name
"ACCESS"         "04.00.0000"    "03.51"            ""
Driver_Name      Driver_Ver      ODBC_Ver           Server_Name
"odbcjt32.dll"   "04.00.6304"    "03.52.0000"       "ACCESS"
```

Code 7.4.

To close a connection, simply call the `odbcClose()` function:

```
> odbcClose(connection.access)
```

Now that you have created a connection to "patients.mdb," let's quickly look at some of the useful functions in RODBC that allow you to retrieve and process data in the database. To view the tables, use the `sqlTables()` function. When applying this to "patient.mdb," the important details displayed are shown in Code 7.5. The five individual tables containing patient data (demographic, diagnosis, pathology, staging, treatment) are displayed after the system tables.

```
> connection.access<-odbcConnect("patient.mdb")
> sqlTables(connection.access)
                                            TABLE_CAT TABLE_SCHEM
1  C:\\Program Files\\R\\R-2.5.0\\book\\patient
             TABLE_NAME   TABLE_TYPE REMARKS
1  MSysAccessObjects SYSTEM TABLE    <NA>
2           MSysACEs SYSTEM TABLE    <NA>
3         MSysObjects SYSTEM TABLE   <NA>
4         MSysQueries SYSTEM TABLE   <NA>
5  MSysRelationships SYSTEM TABLE    <NA>
6         demographic       TABLE    <NA>
7           diagnosis       TABLE    <NA>
8           pathology       TABLE    <NA>
9             staging       TABLE    <NA>
10          treatment       TABLE    <NA>
```

Code 7.5.

RODBC has a number of functions for operational procedures. For instance, you can clear records from database tables without writing any SQL. As an example, to clear all records from the "demographic" table in "patients.mdb," type:

```
> sqlClear(connection.access, "demographic")
```

To remove the table, you would type:

```
> sqlDrop(connection.access, "demographic")
```

To retrieve the names of columns in a particular table, use the `sqlColumns()` function, which also returns data types (e.g., double, variable character) and the length of the field size (Code 7.6).

```
> connection.access<-odbcConnect("patient.mdb")
> sqlColumns(connection.access, "pathology")
   COLUMN_NAME DATA_TYPE TYPE_NAME COLUMN_SIZE
1    patientid        12   VARCHAR         255
2        grade         8    DOUBLE          53
3     nodesexam        8    DOUBLE          53
4      nodespos        8    DOUBLE          53
5        extent        8    DOUBLE          53
6    nodalstatus       8    DOUBLE          53
7          size        8    DOUBLE          53
8           pgr        8    DOUBLE          53
9            er        8    DOUBLE          53
```

Code 7.6.

To copy a table from one database to another, use the `sqlCopyTable()` function:

```
> sqlCopyTable(connection.a, table_1, table_2, connection.b)
```

To create a table in a database from a data frame object, use the `sqlSave()` command. The example in Code 7.7 creates a data frame of two variables. One is a variable of the first 10 patient IDs in the patient dataset, and the second is a list of random values assigned to those cases.

```
> scores<-rnorm(20)
> patientid<-c("pid00001", "pid00002", "pid00003", "pid00004",
"pid00005","pid00006","pid00007","pid00008","pid00009","pid00010")
> scores_table<-data.frame(patientid, scores)
> scores_table
   patientid      scores
1   pid00001   0.3577554
2   pid00002   0.5937775
3   pid00003   0.4121308
4   pid00004   1.3572288
5   pid00005   0.2945161
6   pid00006  -2.3109309
7   pid00007   1.7153430
8   pid00008  -1.4380264
9   pid00009  -1.8585721
10  pid00010   0.6860374
> connection.access<-odbcConnect("patient.mdb")
> sqlSave(connection.access, scores_table, rownames=FALSE)
```

Code 7.7.

The `sqlFetch()` function is used to read a database table into a data frame (Code 7.8).

```
> data<-sqlFetch(connection.access, "pathology")
> data[1:10,]
   patientid grade nodesexam nodespos extent nodalstatus size pgr er
1   pid00001     3        32        3     10           6   60   2  2
2   pid00002     2        13        1     10           6   15   1  1
3   pid00003     3         8        0     10           0    8   1  1
4   pid00004     3        20        0     10           0   10   2  2
5   pid00005     2        16        8     10           6   15   2  1
6   pid00006     3        19        0     10           0   48   1  1
7   pid00007     3         3        0     10           0   32   2  2
8   pid00008     2        13        0     10           0   15   1  1
9   pid00009     3        21        0     10           0   22   1  1
10  pid00010     2        15        0     10           0    6   2  1
```

Code 7.8.

So far we have used many functions to communicate with database tables, yet we have not had to construct a single SQL statement. That's the big advantage with using RODBC (and RMySQL, as we'll see in the next chapter). However, to work with your data at a finer level, you will need to parse SQL statements using a function called `sqlQuery()`. This involves the easy step of providing the text of a SQL statement as the query argument in `sqlQuery()`. The example in Code 7.9 first creates a character string called "query" containing a SQL statement to select data from rows for a series of demographic table columns in the "patient.mdb database" where values in the "ageatdiagnosis" column are greater than 80. The "query" object is then parsed as a SQL query argument to the `sqlQuery()` function.

```
> connection.access<-odbcConnect("patient.mdb")
> query<-"select patientid, dateofbirth, race, ageatdiagnosis, alivestatus,
survivaltime from demographic where ageatdiagnosis > 80 order by survivaltime"
> sqlQuery(connection.access, query)[1:13,]
    patientid dateofbirth race ageatdiagnosis alivestatus survivaltime
1     pid40833  1904-11-06    2             90           1            0
2     pid59466  1914-01-12    2             82           1            1
3     pid36134  1900-12-31    2             91           1            1
```

Code 7.9.

You could add the SQL statement character string directly to the `sqlQuery()` function like the example in Code 7.10.

```
> sqlQuery(connection.access, "select patientid, surname, dateofbirth
from demographic where surname like 'jones%' ")[1:10,]
    patientid surname dateofbirth
1     pid01277   JONES  1948-02-27
2     pid03358   JONES  1919-05-03
3     pid05560   JONES  1948-07-31
```

Code 7.10.

In reality, if you are familiar with SQL, you only need use `sqlQuery()` to manage your data. Another example is adding records to a table:

```
> sqlQuery(connection.access, "INSERT INTO scores_table
(patientid, scores) VALUES ('pid00011', '0.5')"
```

Yet another example is the following statement that updates a value in a field for a specific record:

```
> sqlQuery(connection.access, "UPDATE scores_table SET scores =
'1.2' WHERE patientid = 'pid00011'")
```

7.7 Summary

Storing and managing large biomedical datasets these days is usually done with the help of a relational database and DBMS. Communicating with relational databases from within R is straightforward, with support from some excellent packages. One of the advantages of using R to manage data in this way is that the user who is inexperienced with SQL still has full functionality by taking advantage of the built-in functions within the RODBC or MySQL package. On the other hand, SQL queries against a relational database can be made from within the environment in which you will analyze the data without having to open the DBMS.

7.8 Questions

1. Why do you think database structures might be the most efficient way of storing biomedical data?

2. What is an RDBMS?

3. Using SQL, how can we view tables in a MySQL database?

4. What is the SQL syntax for retrieving data from a table for two fields and ordering the data according to values in the second field?

5. How can we use SQL to delete tables from a database?

6. What is ODBC?

7. How does a connection facilitate communication between R and a database?

8. Using the RODBC package, how can we copy a table from one database to another?

8 Using R to Build a Biomedical Database in MySQL

8.1 Background

MySQL is a popular database management system (DBMS) that runs on many different platforms as a server, allowing multiuser access to multiple databases. MySQL has become an extremely popular choice for web applications, often in combination with PHP, Perl, Ruby, or Python. It is supported by excellent documentation, and many books and websites are devoted to learning and using MySQL. According to its developers, MySQL is the "World's most popular open-source database." MySQL is actually big business, with over 10 million copies of its software downloaded, and in January 2008, it was announced that the company behind MySQL (MySQL AB) was being acquired by Sun Microsystems.

MySQL comes as both commercial (MySQL Enterprise) and noncommercial versions, depending on whether you require additional services or simply want to maintain your own system. The MySQL Community Edition is freely available, and the stable release is version 5.0 at the time of writing. This chapter begins with a brief guide to obtaining, installing, and setting up MySQL 5.0 on Microsoft Windows XP. We will then use MySQL as our relational system to create a biomedical database using the "patient dataset." There is a lot more documentation and help available on this subject across the Internet. The steps that we will perform to create a MySQL database from the patient dataset are described in Table 8.1. A schema of the processes, datasets, and R functions used to extract and store data is shown in Figure 8.1. This chapter also provides the foundation for developing a platform for working with heterogeneous datasets in biomedical research studies, which we will explore in Chapter 9.

8.2 Downloading and Installing MySQL

To download the MySQL 5.0 release, navigate to http://dev.mysql.com/downloads/ and click the download link for MySQL Community Server. All you need to do then is find the link to your operating system and choose a download mirror (as a new user, you will have to register first). Assuming you have selected the option for Windows, you will have downloaded a zip file, which should contain a "Setup.exe" file. By clicking this, you will then enter the Setup Wizard and follow the steps for a typical installation, accepting all the default options. You must also select an administrative

Table 8.1　Steps and Key R Functions Involved in Creating a MySQL Patient Database.

Step	Description	R functions
Download and install MySQL, MySQL Query Browser, and the RMySQL package.	We will download and use the popular and free DBMS MySQL for storing, querying, and managing biomedical data via R.	
Create databases called "patient," "codes," and "studies."	We can quickly create databases for our patient data, in readiness for data import, using the MySQL Query Browser.	
Import patient data from different sources (file types).	The patient data exists in disparate data sources, and we use R to import and bring together the data into data frames.	`read.table()` `read.csv()` `read.xls()` `odbcConnectAccess()` `sqlQuery()` `cbind()`
Connect to MySQL.	Before we store the data in our patient database, we quickly create a connection from R to MySQL.	`dbConnect()`
Write tables in patient database.	Using the connection, we write the R data frames housing the patient data into tables that are created in the patient database.	`dbWriteTable()`
Manage, query, and retrieve data from R.	Data in tables can then be viewed, queried, modified, or deleted directly from R.	`dbListTables()` `dbReadTable()` `dbGetQuery()`
Retrieve data for analysis.	Data can easily be selected and retrieved into data frames for analysis.	`dbGetQuery()`

password. Changing these options is not a problem, as once installed you can amend settings using the MySQL Server Instance Config Wizard.

To use MySQL in Windows, you can either run the Command Line Client in MS-DOS or use two useful tools called MySQL Administrator and MySQL Query Browser. Both tools can be downloaded freely by navigating to the previously mentioned URL and selecting the GUI Tools option in the Main menu. After downloading the relevant "MySQL GUI Tools" file for your operating system, it should self-install.

You also need to ensure that the latest MySQL driver is installed on your PC to allow R to connect to a database. Use your browser's Back button to return to the MySQL "downloads" page and click the "Connectors" menu item. Then select the "Connector/ODBC" option, which will take you to a web page that has the download options for different operating systems. Follow the instructions for downloading the Windows driver, which will self-install.

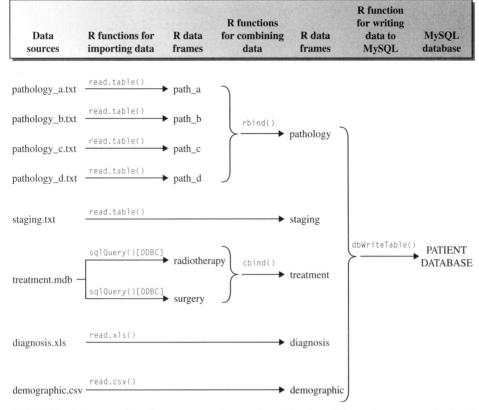

FIGURE 8.1 Schema detailing the extraction and processing of data from different data sources using key R functions followed by the storing of data in the MySQL patient database.

8.3 The RMySQL Package

The optimal way to manage databases from within R is to combine functions from both the RODBC and RMySQL packages. Whereas the functions in RODBC are universal for a variety of DBMSs, one or two functions in RMySQL perform faster, such as dbWriteTable() for writing data frames to tables in MySQL. First you need to install the RMySQL and DBI packages from CRAN:

```
>install.packages(c("RMySQL", "DBI"))
```

Before you load the libraries, you should search for the library file "libmysql.dll" in the bin folder of MySQL Server 5.0 in Program Files and copy it to the root folder of R. Similarly, copy the library file "RMySQL.dll" from the "C:\Program Files\R\R-2 .x.x\library\RMySQL\libs" folder, or wherever R resides in your folder system, to the R root folder. Once these files are in place, you can load the RMySQL library:

```
> library(RMySQL)
```

Loading RMySQL will also load the DBI library, which is necessary for full functionality.

To create a connection to a database, we need to use a function called dbConnect().We parse, by argument, the database name, username, and password (if these have been set). So to connect to "patient," where I have set up a username called "root" and a password called "a_password," we would type:

```
> connection.rmysql<-dbConnect(dbDriver("MySQL"),
dbname="patient", username="root", password="a_password")
```

Notice that I have called the connection object "connection.rmysql" to distinguish this from a connection I would set up for the RODBC package (e.g., "connection.rodbc").

8.4 Using MySQL to Create the "Patient Database"

The easiest way to start MySQL is using MySQL Administrator. When you open MySQL Administrator you will be asked to provide the name of the server host as well as a username and password that you provided during the install. On my PC (running Windows XP Professional), the server host is "localhost" and the username is "root." By clicking "OK," you are taken to the Server Information page. If MySQL is not running, the server status should read "Disconnected from Server." To begin MySQL, you need to click the "Service Control" icon in the menu and then click the "Start Service" button. MySQL can also be started using Windows Component Services.

With MySQL running you can create the patient database using the MySQL Query Browser. When you open MySQL Query Browser, again you will have to provide server name, username, and password information. You will also have to provide the name of a "Default-Schema," which at this point will be blank. Type in "patient" and then click the "OK" button. You should then see the Query Browser interface with a large window for displaying query results (ResultSet) and a list of database schemata in a window on the right that includes the "patient" schema. Below the schemata window is a menu for selecting help on SQL statement syntax. At the top of the interface is the query window for typing in SQL statements. You now have all the tools at your mercy for creating large, comprehensive biomedical databases!

To test your new skills further, create two more database schemas called "codes" and "studies" using the MySQL Query Browser. The patient database contains a whole host of values, many of which are merely codes for actual annotations, as was explained in Chapter 5. The "codes" database will contain a series of tables that allow us to look up what a code actually means. The "studies" database will be used to store datasets prepared for study analyses.

Once you've created a database schema using MySQL Query Browser, you can then switch to R to do pretty much everything else, from creating tables, records, and columns to retrieving data for analysis.

8.5 Importing Data into the MySQL Patient Database Using R

Before you can connect in R to the patient and code databases (or any database for that matter) you need to create data source names using the odbcDriverConnect() function in the RODBC package. To do this you can follow the Microsoft Access example in Section 7.6 by naming the new DSNs "mysql_patient" and "mysql_codes" and setting the targets as appropriate.

With all the groundwork done, we can now start to use our (currently empty) database. Our goal is to create a single data source to include heterogeneous biomedical data for more than 60,000 breast cancer patients. We've referred to this data as the "patient dataset," but as you can see from List 8.1, this data is actually a collection of datasets describing data that is demographic, pathological, treatment, and so on. Each dataset resides in a number of different files, which are also of different formats.

List 8.1 Data files needed to create "patient database."

- demographic.csv
- diagnosis.xls
- pathology_a.txt
- pathology_b.txt
- pathology_c.txt
- pathology_d.txt
- staging.txt
- treatment.mdb

The pathology data sits in a series of text files, where each file contains a subset of rows. The treatment data, held in an Access database, has columns of data split into two tables. Now we're going to use R to import each dataset into a data frame object and then save each data frame as an individual table in the MySQL patient database. We'll also need to bind together the four pathology datasets as a single data frame and then do the same for the treatment data. Every individual dataset has a first column called "patientid," and the values are identical across datasets. This column will be the one that provides a common ID for patients and links all tables together in the database.

In Chapter 6 we looked at ways of importing files into R, including text and spreadsheet formats. In Chapter 7, we also saw how we can import tables from database tables held in different DBMSs. By looking at the file formats in List 8.1, we can see that the datasets we need to import to create our patient database are held in text, Microsoft Excel, and Access formats. We know that we can easily import datasets from text files into data frames using the `read.table()` function. Similarly, we can also read in data from Excel workbooks using the `read.xls()` function in the xlsReadWrite package. To read in data from Access tables, we can create a connection using `odbcConnect()` and read in the table using `sqlFetch()` from the RODBC package. You could, of course, also use RODBC to connect to the Excel workbook, but `read.xls()` does involve fewer steps.

Let's start by creating data frames from the "staging" and "demographic" datasets in text (tab-delimited and CSV) files and the "diagnosis" dataset in the Excel workbook (Code 8.1).

```
> library(xlsReadWrite)
> staging<-read.table("staging.txt", header=T)
> demographic<-read.csv("demographic.csv", header=T)
> diagnosis<-read.xls("diagnosis.xls")
```

Code 8.1.

The next step is to read in the four text files representing the "pathology" dataset into individual data frames and then combine the data into a single data frame object called "pathology" using the rbind() function (Code 8.2).

```
> path_a<-read.table("pathology_a.txt", header=T)
> path_b<-read.table("pathology_b.txt", header=T)
> path_c<-read.table("pathology_c.txt", header=T)
> path_d<-read.table("pathology_d.txt", header=T)
> pathology<-rbind(path_a, path_b, path_c, path_d)
```

Code 8.2.

Now we need to connect to the "treatment.mdb" database to read in data from the "radiotherapy" and "surgery" tables and create a single data frame called "treatment" (Code 8.3). We create a data frame called "radiotherapy" by parsing a SQL statement in sqlQuery() to select the "patientid" and "radiotherapy" columns from the radiotherapy table. We then create a second data frame called "surgery" by parsing a SQL statement to select the "surgery" column from the "surgery" table. Both tables in "treatment.mdb" contain an identical column called "patientid," which is why we only select this once when querying the "radiotherapy" table. Finally, we create the "treatment" data frame using the cbind() function.

```
> library(RODBC)
> connection.access<-odbcConnectAccess("treatment.mdb")
> radiotherapy<-sqlQuery(connection.access, "SELECT patientid, radiotherapy
FROM radiotherapy")
> surgery<-sqlQuery(connection.access, "SELECT surgery FROM surgery")
> treatment<-cbind(radiotherapy, surgery)
```

Code 8.3.

We now have all data stored in a series of five data frames. Next we need to set up a connection in R to the patient database so we can write the tables using the dbWriteTable() function from the MySQL package:

```
> connection.rmysql<-dbConnect(dbDriver("MySQL"),
dbname="patient", username="root", password="a_password")
```

Once the connection is created, you can start to create tables in the empty MySQL patient database and parsing to dbWriteTable() each data frame as the "dat" argument (Code 8.4).

```
> dbWriteTable(connection.rmysql, "diagnosis", diagnosis, row.names=FALSE)
> dbWriteTable(connection.rmysql, "demographic", demographic, row.names=FALSE)
> dbWriteTable(connection.rmysql, "pathology", pathology, row.names=FALSE)
> dbWriteTable(connection.rmysql, "staging", staging, row.names=FALSE)
> dbWriteTable(connection.rmysql, "treatment", treatment, row.names=FALSE)
```

Code 8.4.

If all is well, R should return a TRUE for each case.

Last, we turn to the codes database we created alongside the patient database. Many of the columns in the patient database tables contain values that are just codes for what the actual variable means. In the "Patient Dataset" folder are a series of text files that contain the codes and annotation for each of these variables (List 8.2).

List 8.2. Data files needed to create "codes database."

- alivestatus.txt
- er.txt
- extent.txt
- grade.txt
- histology.txt
- maritalstatus.txt
- nodalstatus.txt
- pgr.txt
- primary.txt
- race.txt
- radiotherapy.txt
- stage.txt
- surgery.txt

For each variable we will create a table of two columns—one for the code and one for the annotation. As we create the table, in the same line of R code, we will also import the value codes and annotation from the relevant text file (Code 8.5).

```
> connection.mysql2<- dbConnect(dbDriver("MySQL"), dbname="codes",
username="root", password="a_password")
> dbWriteTable(connection.mysql2, "er", read.table("er.txt", header=T),
row.names=FALSE)
> dbWriteTable(connection.mysql2, "alivestatus", read.table("alivestatus.txt",
header=T), row.names=FALSE)
> dbWriteTable(connection.mysql2, "er", read.table("er.txt", header=T),
row.names=FALSE)
> dbWriteTable(connection.mysql2, "extent", read.table("extent.txt", header=T),
row.names=FALSE)
> dbWriteTable(connection.mysql2, "grade", read.table("grade.txt", header=T),
row.names=FALSE)
> dbWriteTable(connection.mysql2, "histology", read.table("histology.txt",
header=T), row.names=FALSE)
```

```
> dbWriteTable(connection.mysql2, "maritalstatus",
read.table("maritalstatus.txt", header=T), row.names=FALSE)
> dbWriteTable(connection.mysql2, "nodalstatus", read.table("nodalstatus.txt",
header=T), row.names=FALSE)
> dbWriteTable(connection.mysql2, "pgr", read.table("pgr.txt", header=T),
row.names=FALSE)
> dbWriteTable(connection.mysql2, "primarytumor", read.table("primary.txt",
header=T), row.names=FALSE)
> dbWriteTable(connection.mysql2, "race", read.table("race.txt", header=T),
row.names=FALSE)
> dbWriteTable(connection.mysql2, "radiotherapy",
read.table("radiotherapy.txt", header=T), row.names=FALSE)
> dbWriteTable(connection.mysql2, "stage", read.table("stage.txt", header=T),
row.names=FALSE)
> dbWriteTable(connection.mysql2, "surgery", read.table("surgery.txt",
header=T), row.names=FALSE)
```

Code 8.5.

8.6 Working with a MySQL Database

Like RODBC, the RMySQL package has many functions to manage and query your data. To list the tables in the patient database, we use the dbListTables() function (parsing the name for the connection):

```
> dbListTables(connection.rmysql)
```

To read a table, we use the dbReadTable() function, which returns the data as a data frame (as demonstrated for the "demographic" table in Code 8.6).

```
> dbReadTable(connection.rmysql, "diagnosis")[1:10,]
   patientid yearofdiagnosis histology primary__1 numberofprimaries
1  pid00001             1993      8500        502                 1
2  pid00002             1994      8500        504                 2
3  pid00003             1997      8500        508                 2
4  pid00004             1992      8500        501                 2
5  pid00005             1996      8541        501                 1
6  pid00006             1994      8510        504                 1
7  pid00007             1995      8500        502                 1
8  pid00008             1995      8500        504                 2
9  pid00009             1991      8500        502                 1
10 pid00010             1995      8500        504                 2
```

Code 8.6.

Most importantly, we can query our database by parsing SQL statements using dbGetQuery(). For instance, to update a value in the "ageatdiagnosis" column of the demographic table for a patient, you could type:

```
> dbGetQuery(connection.rmysql, "UPDATE demographic SET
ageatdiagnosis = '74' WHERE patientid = 'pid00001'")
```

Data is not always stored as you might expect in databases. For example, when saving data to a table from a data frame, sometimes the fields in the last column will end in a pilcrow character "¶". This can happen if the last value is a character. If you then try to retrieve the data into an R data frame by query from the database, this character will be read as "\r" and appear at the end of the last value in each row. To avoid such a problem, you can easily remove the pilcrow character from the database using a SQL query in the dbGetQuery() function:

```
> dbGetQuery(connection.rmysql, "UPDATE codes.alivestatus SET
annotation =(SELECT TRIM('\r' From annotation))")
```

The SQL Trim function removes white space or specified characters such as "\r" from each end of the value in a field.

8.7 Creating Datasets for Analysis

The ability to combine data from different tables to create new datasets prior to analysis is critical. In this section, as in Chapter 9, we will see how this can be achieved using the dbGetQuery() function in our quest to create heterogeneous biomedical datasets ready for analysis.

In Chapters 14 and 15 we will attempt to generate models for survival in breast cancer using multivariate statistical and machine learning methods. The dataset for these analyses needs to contain a number of variables from different tables within the patient dataset, as shown in Table 8.2.

If you ensure that each table within the patient database is ordered according to the "patientid" field, it is simple to retrieve this data without having to match by key (Code 8.7).

```
> library(RMySQL)
> connection<-dbConnect(dbDriver("MySQL"), dbname="patient", username="root",
password="a_password")

# demographic table
> ageatdiagnosis<-dbGetQuery(connection, "SELECT ageatdiagnosis FROM
demographic")
> alivestatus<-dbGetQuery(connection, "SELECT alivestatus FROM demographic")
> survivaltime<-dbGetQuery(connection, "SELECT survivaltime FROM demographic")
# diagnosis table
> yearofdiagnosis<-dbGetQuery(connection, "SELECT yearofdiagnosis FROM
diagnosis")

# pathology table
> grade<-dbGetQuery(connection, "SELECT grade FROM pathology")
> nodesexam<-dbGetQuery(connection, "SELECT nodesexam FROM pathology")
> nodespos<-dbGetQuery(connection, "SELECT nodespos FROM pathology")
> size<-dbGetQuery(connection, "SELECT size FROM pathology")

# treatment table
> surgery<-dbGetQuery(connection, "SELECT surgery FROM treatment")
> radiotherapy<-dbGetQuery(connection, "SELECT radiotherapy FROM treatment")
```

Code 8.7.

After retrieving data from a database, it is wise to ensure that each variable (i.e., object within the data frame) is of the right type (Code 8.8). Within the data frame, the

| Table 8.2 | Variables from the Patient Dataset Required for Analysis in Chapter 14. |

Table	Field
demographic	ageatdiagnosis
	alivestatus
	survivaltime
diagnosis	yearofdiagnosis
pathology	size
	grade
	nodesexam
	nodespos
treatment	surgery
	radiotherapy

objects that are called alivestatus, grade, surgery, and radiotherapy must be factors, whereas the other objects are simply numeric.

```
> ageatdiagnosis[,c(1)]<-as.numeric(ageatdiagnosis[,1])
> alivestatus[,c(1)]<-as.factor(alivestatus[,1])
> survivaltime[,c(1)]<-as.numeric(survivaltime[,1])
> yearofdiagnosis[,c(1)]<-as.numeric(yearofdiagnosis[,1])
> grade[,c(1)]<-as.factor(grade[,1])
> nodesexam[,c(1)]<-as.numeric(nodesexam[,1])
> nodespos[,c(1)]<-as.numeric(nodespos[,1])
> size[,c(1)]<-as.numeric(size[,1])
> surgery[,c(1)]<-as.factor(surgery[,1])
> radiotherapy[,c(1)]<-as.factor(radiotherapy[,1])
```

Code 8.8.

Now we can create a data frame using cbind():

```
> data<-data.frame(cbind(ageatdiagnosis, yearofdiagnosis, grade,
nodesexam, nodespos, size, surgery, radiotherapy, survivaltime,
alivestatus))
```

The summary() function can return information about the distribution of each object and counts for levels in each factor (Code 8.9). This is also a good way to ensure that objects are of the right type.

```
> summary(data)
 ageatdiagnosis    yearofdiagnosis grade          nodesexam
 Min.   : 19.00   Min.   :1990    1: 9112   Min.   : 1.00
 1st Qu.: 47.00   1st Qu.:1993    2:26489   1st Qu.:11.00
 Median : 58.00   Median :1994    3:24566   Median :15.00
 Mean   : 57.81   Mean   :1994    4: 2451   Mean   :15.39
 3rd Qu.: 68.00   3rd Qu.:1996              3rd Qu.:19.00
 Max.   :106.00   Max.   :1997              Max.   :90.00

    nodespos          size         surgery      radiotherapy
 Min.   : 0.000   Min.   : 1.00   50  :30578   1:29042
 1st Qu.: 0.000   1st Qu.: 11.00  20  :27218   3:33576
```

```
Median : 0.000   Median : 17.00   58    : 3143
Mean   : 1.565   Mean   : 20.86   10    :  767
3rd Qu.: 1.000   3rd Qu.: 25.00   40    :  467
Max.   :97.000   Max.   :200.00   60    :  133
                                  (Other): 312
 survivaltime     alivestatus
Min.   :  0.00   0:52822
1st Qu.: 68.00   1: 9796
Median : 87.00
Mean   : 88.24
3rd Qu.:111.00
Max.   :155.00
```

Code 8.9.

Save the data frame to a text file in your working directory:

```
> write.table(data.all, "data.all.txt")
```

We will use this dataset later in the book. We will also require a dataset containing patients with 10-year survival data. Notice that the year of diagnosis ranges from 1990 to 1997. The survival data for all of the patients, however, was recorded up until 2002. To create a dataset that contains only those patients who have 10-year survival data, we return rows where the yearofdiagnosis object has a value less than 1993:

```
> data.10yr<-subset(data, yearofdiagnosis<1993)
```

Thus, we have retrieved patients diagnosed between 1990 and 1992 inclusive into a data frame called data.10yr. If we apply nrow() to data.10yr, we can determine the number of patients retrieved (Code 8.10).

```
> nrow(data.10yr)
[1] 15194
```

Code 8.10.

Finally, we can save the data frame as a text file for future use using the write.table() function:

```
> write.table(data.10yr, "data.10yr.txt")
```

8.8 Summary

In this chapter we have learned how we can combine two of the most powerful free and open-source data management and analysis platforms available. It is not unreasonable to suggest that the combination of R and a DBMS such as MySQL can be visualized as a complete data storage and analysis "system" for biomedical datasets. As you will see in later chapters, this system can be extended further by combining R with other freely available tools, such as Waikato Environment for Knowledge Analysis (WEKA), to provide an extremely powerful open-source data mining platform for Windows that costs nothing.

8.9 Questions

1. Why do R and MySQL complement each other when creating a biomedical data management and analysis platform?

2. What is a database schema?

3. Why is the SQL Trim function useful when dealing with biomedical data?

4. What is the RMySQL function for parsing a SQL query to a database, and what do the two main arguments represent?

5. After importing data from a MySQL database, what should one always check, and amend if necessary, before binding objects into a data frame?

6. What function allows us to specify certain rows of a data frame given defined criteria?

9 Creating Heterogeneous Datasets for Analysis in R

9.1 Background

Translational medicine is now an often-used phrase to describe efforts to link basic research to patient care. When talking of translational medicine or research, one often hears the phrase "bench to bedside," referring to the process of taking discoveries in the laboratory to improve diagnosis, prognosis, and treatment of patients. Use of terms such as translational medicine and biomedicine simply reflect the fact that medical research has become truly interdisciplinary. Clinicians require biomedical researchers to discover new biomarkers, such as novel drug targets, whereas a biologist might require access to human tissue and anonymous patient data prior to searching for such molecules.

Modern medical research doesn't just involve the integration of personnel from different disciplines. It also involves the mixing of data, either generated for the study or by the study. The biomarker discovery pipeline may involve the aggregation and processing of many tissue samples from hundreds of patients using a variety of high-throughput technologies. Just think of some of the data stored in a hospital information system that could be used in the design of a disease study when selecting suitable cases. In the researcher's laboratory, a technique will provide "results" data, such as differentially expressed genes, a new protein structure, or a new mutation.

What this all means is that one study alone could generate huge datasets that are heterogeneous in nature. What do we mean when we say that "the data is heterogeneous"? We could actually mean a number of different things. We may mean that the data resides in disparate sources, perhaps over many countries, as in the case of a clinical trial. We could mean that different datasets are kept in a range of differing file formats. Heterogeneous could also refer to the fact that the variables in a dataset are of different modes, such as numerical or categorical.

In preceding chapters we encountered the patient dataset, which itself is a good example of a biomedical, heterogeneous dataset with files containing demographic, treatment, pathological, and biological data. Indeed, the patient dataset satisfies our multiple definitions of "heterogeneous" in that all the data originated from different sources (many hospitals and departments across the United States), was obtained in various file formats such as text or database, and is a mixed bag of continuous and categorical data.

It is important to stress that the integration, organization, and processing of heterogeneous data for a study can be a huge challenge. When reading the findings of a biomedical study, it is often not realized that perhaps the most time-consuming and arduous part of the process was the preparation of data for analysis. As a biologist, physician, statistician, or informaticist involved in analysis of study data, you need to consider beforehand the structure of the data, how the data will be integrated from different sources, and how the data will be stored. In Chapter 8 we saw how we could integrate individual related datasets into a single database structure using R. Thus, we unified the source of all the patient data into a manageable system. Good planning and data organization in this way really does make an analysis easier. Many of the remaining chapters are devoted to the analysis of biomedical data with R. But with a little thought and practice, much of the data preprocessing can also be performed using R.

In this chapter we will look at some of the tasks involved in processing data as part of a hypothetical biomedical study. The scenario depicts typical research that would involve players from a number of disciplines and a lot of data generation. Using our patient database, we will first see how we can select patients and relevant data for the study, which aims to evaluate a series of protein biomarkers by immunohistochemistry. We will then look at how we can combine novel results data from the study with existing data to create a heterogeneous dataset ready for statistical analysis in subsequent chapters.

9.1.1 An Example Biomarker Study

A medical oncologist has an interest in the growing evidence of age-related biologic differences in breast cancer tumors. She would like to carry out a small study to evaluate whether any age-associated patterns of differences exist for the levels of two cell-cycle protein markers called "proteinA" and "proteinB" in elderly breast cancer patients. It is known that both proteins are differentially expressed in breast cancer. If age-associated patterns do exist in older patients, this would indicate the need for further research into how elderly patients should be treated according to the biology of their tumor. She is particularly concerned that grade three, node-negative (G3N0) breast cancer patients with invasive ductal carcinomas receive the same chemotherapy regimen, regardless of their age, yet the tolerability and effectiveness to chemotherapeutic agents is reduced in some of these patients over the age of 70. Furthermore, she would like to investigate the expression of these two biomarkers alongside the estrogen (ER), progesterone (PgR), and human epidermal growth factor 2 (Her2) receptors that already are routine biomarkers for treatment determination.

She decides to investigate the expression of these biomarkers in two diverse age groups of G3N0 patients. The first group will contain patients who were over the age of 70 when diagnosed, whereas the second group will contain patients under the age of 45 at diagnosis.

She collaborates with a pathologist who specializes in breast cancer biomarker research. He has already created and maintains the "patient database" using MySQL that we covered in Chapter 7. He considers this database invaluable for assisting in the selection of suitable cases for research studies and providing additional patient data when queried. This database system is also useful for storing the results of the studies he participates in. The pathologist has access to archived tumor tissue from all of the patients in the database.

It is agreed that available funds would allow for a tissue microarray (TMA) study with 21 patients in each group. The pathologist advised the oncologist that ER and

Table 9.1	Steps and Key R Functions Involved in Selecting Patients for Experimental Studies and Storing Results Datasets.

Step	Description	R functions
Connect to "patient" database.	Via the RMySQL package, we can create a connection to the patient database to retrieve data.	`dbConnect()`
Retrieve selected data from patients.	Using SQL, we can return data for all patients to a data frame by selecting the required fields.	`dbGetQuery()`
Select and create random patient lists for study.	From our data frame, we can then create, using random selection, two lists of 21 patients under the age of 45 and 21 over the age of 70.	`sample()`
De-identify patient lists.	Each patient list then must have any patient identifying data replaced with an encrypted ID number (hash string).	`digest()` `dbWriteTable()`
Create dataset from research data.	The patient lists can then be used to identify available tissue for experimental research. The results of the research must then be stored as a dataset.	`c()` `data.frame()`
Combine research data with patient data.	The research dataset can then be merged with clinical and pathological data retrieved for each patient from the patient database.	`dbGetQuery()` `paste()` `order()` `cbind()` `merge()`
Store study datasets.	The heterogeneous datasets created within the study can then be stored, ready for further analysis.	`dbWriteTable()` `write.table()`

PgR data for each patient selected will already be available in his database, saving the need (and money) to repeat the analysis. Thus, laboratory work would involve generating immunohistochemistry score data for proteinA, proteinB, and Her2.

As soon as all the required data is collected, the pathologist will provide the oncologist researcher with a tabular dataset saved as a text file. As we work through this example, you can assume the role of the pathologist as you use the patient database to query, process, and store the data for the study. Table 9.1 details the steps involved, as well as key R functions used, as we select patients for study, add experimental results, and subsequently store the study datasets.

9.2 Selecting Patients for a Study

With the study design agreed upon, the next step is to select suitable patients whose archived tissue will provide samples. This can be done by searching the patient database from within R using the RMySQL package, as demonstrated in Chapter 8. Our task (as the pathologist) is to search the database for 21 G3N0 patients over the age of 70 and another 21 patients under the age of 45.

Let's start by returning two lists of G3N0 patients for each age category to data frames called "over70.list" and "under45.list" (Code 9.1). To generate each list, we create a connection to the MySQL patient database and supply relevant SQL statements to the RMySQL dbGetQuery() function. We use a SQL SELECT statement to return the patient ID numbers from the demographic table, where exact criteria are met for age at diagnosis in this table, as well as grade and the number of nodes positive in the pathology table. You will see in the code that the patient's ID value is returned from the demographic table (demographic.patientid) only if the "ageatdiagnosis" column in the demographic table (demographic.ageatdiagnosis) is greater than "70," the value in the "grade" column of the pathology table (pathology.grade) is equal to "3," and the value in the "nodespos" column, also in the pathology table, is equal to "0" (i.e., node-negative). We must also state in our query that, in order to search all specified values for the same patients in both tables, the "demographic.patientid" value must equal the "pathology.patientid."

```
> library(RMySQL)
> connection<-dbConnect(dbDriver("MySQL"), dbname="patient", username="root",
password="a_password")
> over70.list<-dbGetQuery(connection, "SELECT demographic.patientid FROM
demographic, pathology WHERE demographic.patientid = pathology.patientid AND
demographic.ageatdiagnosis > 70 AND pathology.grade = 3 AND nodespos = 0")
> under45.list<-dbGetQuery(connection, "SELECT demographic.patientid FROM
demographic, pathology WHERE demographic.patientid = pathology.patientid AND
demographic.ageatdiagnosis < 45 AND pathology.grade = 3 AND nodespos = 0")
> over70.list[1:10,1]
 [1] "pid00003" "pid00007" "pid00014" "pid00039" "pid00042" "pid00056"
 [7] "pid00082" "pid00084" "pid00105" "pid00125"
> under45.list[1:10,1]
 [1] "pid00234" "pid00338" "pid00471" "pid00476" "pid00483" "pid00502"
 [7] "pid00503" "pid00507" "pid00510" "pid00522"
> nrow(over70.list)
[1] 2414
> nrow(under45.list)
[1] 3016
```

Code 9.1.

The SQL for returning the younger patients list is identical, barring the "demographic.ageatdiagnosis" conditional. For the novice, querying multiple tables in this way may be a little daunting at first, but by studying the SQL query you can see that statements are built by combining conditions using the AND operator. Note, too, that columns in each table are referred to by the table, a full stop, and then the column name (e.g., pathology.grade).

The last four lines in Code 9.1 show how we use the nrow() function to return the number of patient IDs for both groups. The next step is to select 21 patients for each group from this total. You decide that to avoid potential bias, you will select both lists randomly. An easy method of doing this is to use the sample() function to return 21 random numbers and specify a range between 1 and the list length as well as the total random numbers to return (Code 9.2). So as to return the actual patient IDs, the sample() function is used within the "over70.list" and "under45.list" data frames. Remember that different random numbers would be generated every time, so the output in Code 9.2

would differ from yours. To follow the example all the way through, you may want to create the "over70.list" and "under45.list" data frames using the output in Code 9.2.

```
> over70.list<-over70.list[sample(1:nrow(under45.list), 21),1]
> over70.list
 [1] "pid02257" "pid26942" "pid44825" "pid09937" "pid59896" "pid61901"
"pid00214" "pid39456" "pid25678" "pid34710" "pid06430" "pid39664" "pid58096"
"pid10839" "pid61845" "pid14996" "pid46298" "pid03236" "pid53040" "pid28927"
"pid20582"
> under45.list<-under45.list[sample(1:nrow(over70.list), 21),1]
> under45.list
 [1] "pid46093" "pid27764" "pid03636" "pid41576" "pid60741" "pid37429"
"pid18634" "pid05574" "pid33710" "pid57505" "pid27830" "pid04672" "pid44062"
"pid27291" "pid57687" "pid01527" "pid25439" "pid51423" "pid53558" "pid27926"
"pid37549"
```

Code 9.2.

9.3 De-identifying the Patient List

The hypothetical patient ID numbers in the dataset represent what would be a hospital number in real life. As a pathologist, you are able to view such confidential information but cannot supply a list of these patient ID numbers to a researcher. There will always be a "guardian" of health data whose role is to ensure safe-keeping of patient records. In our example, the pathologist is the guardian of data in the database he has created. In the United States, he must comply by the federal regulations, HIPAA[3] and the Common Rule,[4] which protect private medical data. Under these regulations, the sharing of private data can only occur if the patient has given consent. However, in a project such as this, a patient's data may be used for research purposes if all identifiable information has been removed. Thus, in a study you must first de-identify the lists holding patient data such as those created previously.

You may be wondering why, after selecting tissue samples, you have to provide any form of case reference to the researcher. There are, in fact, a number of reasons. We are dealing here with patients being retrospectively selected for a study, and it is quite possible that at some point, the researchers may want to carry out further experiments with the same tissue. If the cases provided by you were no longer identifiable, the original tissue samples could not be matched for future studies. If the researchers required further data in the current study, such as tumor extent or survival time, that hadn't originally been passed on, you would need some way to identify those original patients.

You need to create a reference to each case, which essentially becomes a "case reference label" but retains the anonymity of the patient. You could create a random case label for each patient and store these alongside the patient's ID number so that you can refer to it in the future. To safeguard against any unauthorized person viewing this data, all this information would have to be stored within the patient database and not externally. However, there is the problem that if you (the pathologist) were to retire, move on, or the database get destroyed (without a backup), the researcher would not have a means of accessing data for the patient cohort in the future. For example, assume that you had assigned labels such as "case1," "case2," "case3," etc. to the 21 patients selected for the under-45 breast cancer group. If the researcher were to return to the pathology department 10 years later, long after the pathologist collaborator had retired, how could a new pathologist match up these codes to patient records?

You could instead use one of the popular one-way hash algorithms to create a string case reference label, which cannot be decrypted to reveal the identity of the original patient ID. In addition, this type of algorithm applied to a patient ID would always produce the same case label, as the output hash string depends on the input string. By using a one-way hashing procedure, you could, in theory, provide a list of unique case labels or hash strings to individual patient samples and not even have to store these strings since they can be regenerated by reapplying the algorithm. In other words, so long as a hash string for a patient exists in a study list, the patient can be identified at a later date by somebody with authorized access to the patient database.

Some of the more common one-way hashing algorithms are the MD5 and SHA-1 algorithms, and both are implemented in the `digest()` function in the package called "digest." To install the digest package, type:

```
> install.packages("digest")
```

To create a one-way hash string, you just need to parse the target R object as an argument (Code 9.3). In this example we are parsing the patient ID value held as the first element in the "over70.list" data frame. The `digest()` function has applied the default hashing algorithm called MD5 that has returned a 32-character string.

```
> library(digest)
> digest(over70.list[1])
[1] "afd29bd1a493305e666596cf98d2eb6d"
```

Code 9.3.

To change the algorithm you can specify the `algo` argument, which can be "sha1" for SHA-1 as well as "md5." Application of the SHA-1 algorithm to the first element in the "over70.list" object returns a 40-character string (Code 9.4).

```
> digest(over70.list[1], algo = "sha1")
[1] "b3a7e9cbb010965513eb4e9124e62a94b1c453d2"
```

Code 9.4.

To facilitate research projects, the pathologist should create a new table in the patient database to house individual hash strings for each patient ID number. The first step to do this using R is to retrieve all the patient ID numbers from one of the tables in the patient database and store these numbers in an array (patientid.numbers):

```
> patientid.numbers<-dbGetQuery(connection, "SELECT patientid FROM demographic")
```

The next step is to create a second array of the same size as the "patientid.numbers" array. This array will house the hash numbers for each patient ID number we generate using the digest() function:

```
> hash.strings<-dim(nrow(patientid.numbers))
```

With both arrays created, we can create the hash numbers:

```
> for(i in 1:nrow(patientid.numbers)){hash.strings[i]<-
digest(patientid.numbers[i,1])}
```

We use here a FOR LOOP statement to iterate through the digest() function by the number of elements in the "patientid.numbers" array (which equals the number of patient IDs). The digest() function then creates all of the hash numbers and places them into the "hash.strings" array. The FOR LOOP statement is one of a number of control statements that we will look at more closely in Chapter 12.

The whole hashing procedure will probably take a few minutes since there are more than 60,000 patient IDs to iterate through. Once complete you can check the hash strings that were produced by selecting a small subset (Code 9.5).

```
> hash.strings[1:10]
 [1] "26d3e097821a58c1ee43fdd530fa667e"
 [2] "07f88205a632a1c740af699f30105537"
 [3] "f12b4f3422aed6cfbb33e43d0212ab32"
 [4] "d0cab857320abf6c0fd6b9e6f8415333"
 [5] "262287e20dec3db3b6b25fbb3d6117a9"
 [6] "d6be450eb07183e329782fd42bce098a"
 [7] "297dd1c1b97d9aeaac376456b118ffe5"
 [8] "f5d4900b5604ef4463e1ec52e9823eb9"
 [9] "3d42d5cd618da6e696a42e5370b744c0"
[10] "ce9847cb7e9d21d9ac468e76a7141956"
```

Code 9.5.

Now we can create a new table in the patient database called "hashcodes" using the dbWriteTable() function. We just need to parse the "patientid.numbers" and "hash.strings" as a data frame object for the value argument:

```
> dbWriteTable(connection, "hashcodes",
data.frame(patientid.numbers, hash.strings), row.names=FALSE)
```

9.4 Creating Hash Strings for Patient Lists in a Study

There is no need for you to search the "hashstrings" table every time you wish to retrieve hash strings for a list of patient IDs such as in our example study. The "hashstring" table merely acts as a reference for you to identify cases of a study at a later stage if need be. To generate hash strings as case reference numbers for this study and future ones, all you have to do is generate the strings using the digest() function. The hashing algorithms will always produce the same string for a given input.

We can apply this procedure first to the patient ID numbers held in the "over70.list" object (Code 9.6). Initially, we need to create a one-dimensional array of

21 elements called "over70.hash." This array will hold the hash strings. Using the FOR LOOP, we then apply the `digest()` function 21 times to create the hash strings. We then repeat the procedure for the patient ID numbers held in the "under45.list."

```
> over70.hash<-dim(21)
> for(i in 1:21){ over70.hash[i]<-digest(over70.list[i])}
> over70.hash
 [1] "afd29bd1a493305e666596cf98d2eb6d"
 [2] "c49757a6fd4f178ccf05271c37839989"
 [3] "089bf2f200acc2e1a807f86e243e244b"
 [4] "4ab32d1a86d6b9ff1a39d95c2e6d8c01"
 [5] "f946d1c3960f0de968cb68ab68e7ddad"
 [6] "8f757751fbf43de2253b5bc794c80c13"
 [7] "6be9ebab67aa834ad709e00342e1670e"
 [8] "edacff44b17a7ac55ddd248292a46890"
 [9] "037adfe485ecf86e0c6fd1f0c227ebac"
[10]  "e9655fd9174d63327a06d08a2ad4e782"
[11] "c3b4fd2083befd4c8481247ce1a542a2"
[12] "8da2886ad40b0a96d6c7ef54ff9ffc1c"
[13] "02fdeff7e48627a6c276d7b38fac5143"
[14]  "5c0af2b80323b428cf01df55dd605980"
[15] "db2ca77f89e6880f55225ee5b4cd2059"
[16]  "ff75e9c081a112449d69942a1ec8470c"
[17] "f86962d0781cba4310a4a60157a53435"
[18]  "9c790cf2f97f788fb1e5a3dbb15805a5"
[19] "8edd650ef769ab2064126a60bbcab56a"
[20]  "23d7c6bf79341f05a2b462b153bbc393"
[21] "a7a815d3c4a7171b08257fa4a5c80d64"

> under45.hash<-dim(21)
> for(i in 1:21){ under45.hash[i]<-digest(under45.list[i])}
> under45.hash
 [1] "a6baf0062a71a9c21392f73079a6076b"
 [2] "cfb19203c4dc16d88ecf6ccf2ec99ed0"
 [3] "9dfd10cb6ae5ca1c56467b2fe27ad659"
 [4] "8e61448206f14e437f4bc1e31ede4e92"
 [5] "e540f0e1e2e88ed07f44d0bf9960c077"
 [6] "1ec5198bdd64e130d5a6ca39811039cf"
 [7] "c5cc4df40504b69f6082c94606596630"
 [8] "9028e0766b4c689423f55b37e2aae595"
 [9] "ed61084a95d96da25d50fd82fecc3f60"
[10] "9dd44ccb48a1ec8d7a25acefb5e9a895"
[11] "3daeb41c8910f3f3d8e4eda6e237b48d"
[12] "2275fb5df821d0abc234fc3d88610679"
[13] "51a7bbe92e44b7a27a03369a76a79e00"
[14] "b79b8ec7784d0d34b0ac2eef5b26464a"
[15] "fbb5bd86ea0df2b035eafb6d15e882c2"
[16] "77764243684965da53bdd3cf7d1be453"
[17] "230077d3bd3342fb15c0d438baec4d1b"
[18] "04d01d8ca9039632b2239fbafbe37b57"
[19] "910a440071ad3b6fc4a24a754252a7bb"
[20] "085fedf676902da772bf25c3502923cd"
[21] "6cf0f2d9228f507d5157fa1cfd272aec"
```

Code 9.6.

We now have case reference labels that can be passed to the researcher for the tissue microarray (TMA) breast cancer study, and samples can be identified at any later time.

9.5 Creating a Dataset for Study Results

With 42 cases identified for the TMA study, the pathologist can provide laboratory staff with tissue samples. Once the experimental procedure is complete, images are returned to him, and he then generates a series of scores for each biomarker across all patients. He records scores for proteinA and proteinB as the proportion of cells stained positive (0%–100%). The original scores for ER and PgR for these patients that were already in the patient database were generated using the Allred system (described previously) and converted into scores of 0 (ER- or PgR-negative), 1 (marginal), and 2 (ER- or PgR-positive). Scores for Her2 range from 0 to 3, where scores of 2 and 3 are considered positive. The experimental design dictated that three cores per tumor be embedded into the TMA, and so the pathologist records the highest score for a marker into the data set.

Having recorded all of the scores on a piece of paper (using any other means would be too simple!), the pathologist then types the numbers into vectors directly in R, ready to be transferred to the patient database (Code 9.7).

```
proteinA.over70<-c(8,2,28,3,28,8,8,34,8,0,8,40,15,5,10,34,13,36,38,38,34)
proteinA.under45<-c(0,28,0,30,39,31,70,25,17,3,18,1,30,35,31,28,58,0,70,13,17)
proteinB.over70<-c(7,1,38,1,16,11,18,29,47,7,83,6,30,13,32,89,20,15,61,36,62)
proteinB.under45<-c(16,63,0,39,86,94,81,84,21,5,96,7,80,93,2,91,91,0,92,31,87)
her2.over70<-c(0,1,0,0,0,2,0,2,0,0,0,3,0,0,1,0,0,0,0,0,1)
her2.under45<-c(3,0,2,0,0,0,0,0,3,2,3,3,0,0,3,0,0,0,0,3,1)
```

Code 9.7.

We can at this point combine each vector into two data frames called "under45.scores" and "over70.scores" that will eventually form the structure of the data file passed back to the researcher (Code 9.8). We will also add the hash reference number so that each row can be matched when we add data at a later stage.

```
> over70.scores<-data.frame(over70.hash,proteinA.over70, proteinB.over70,
her2.over70)
> under45.scores<-data.frame(under45.hash,proteinA.under45, proteinB.under45,
her2.under45)
```

Code 9.8.

It was agreed that the pathologist would also provide data for tumor size, ER, and PgR status in each patient. Therefore, you need to query the patient database to retrieve this information. To do this you must create a SQL query to parse to the patient database in MySQL using the dbGetQuery() function. In MySQL, you can retrieve data for specific patients by matching the hash string value to the patientid value for each patient in the "hashcodes" table. These patientid values can then be used to retrieve the ER and PgR values as well as the tumor size for each patient from the "pathology" table. Let's look at the SQL statement to retrieve this data for the first patient in the over 70 group (Code 9.9).

```
> dbGetQuery(connection,"SELECT pathology.er, pathology.pgr, pathology.size
FROM pathology, hashcodes WHERE pathology.patientid = hashcodes.patientid AND
hashcodes.hash_strings = 'a6baf0062a71a9c21392f73079a6076b'")

  er pgr size
   2   2   16
```

Code 9.9.

The values of the ER and PgR fields are a bit meaningless to a researcher without a description for each code. Therefore, you need to retrieve the code annotation from the "codes" database for both these variables. This involves a slight amendment to the SQL statement by inserting SQL JOINS between the "pathology" table and er and pgr tables in the "codes" database (Code 9.10).

```
> dbGetQuery(connection,"SELECT pathology.er, pathology.pgr, pathology.size,
codes.er.er_annot, codes.pgr.pgr_annot
FROM (patient.pathology JOIN codes.er ON pathology.er = codes.er.code JOIN
codes.pgr ON pathology.pgr = codes.pgr.code), hashcodes
WHERE pathology.patientid = hashcodes.patientid AND hashcodes.hash_strings =
'a6baf0062a71a9c21392f73079a6076b'")

  er pgr size er_annot  pgr_annot
   2   2   16 Negative  Negative
```

Code 9.10.

Note that the expression containing the table JOINS is encompassed by brackets. However, we need to return data for multiple patients, so let's look at the same statement amended for two patients (Code 9.11).

```
> dbGetQuery(connection,"SELECT pathology.er, pathology.pgr, pathology.size,
codes.er.er_annot, codes.pgr.pgr_annot FROM (patient.pathology JOIN codes.er ON
pathology.er = codes.er.code JOIN codes.pgr ON pathology.pgr = codes.pgr.code),
hashcodes
WHERE (pathology.patientid = hashcodes.patientid) AND hashcodes.hash_strings =
'a6baf0062a71a9c21392f73079a6076b'
OR  (pathology.patientid = hashcodes.patientid) AND hashcodes.hash_strings =
'cfb19203c4dc16d88ecf6ccf2ec99ed0'")

  er pgr size er_annot  pgr_annot
   1   2   30 Positive  Negative
   2   2   16 Negative  Negative
```

Code 9.11.

By studying the statement, you should see that the piece of code specifically retrieving data for the two patients follows the WHERE clause and code specific for an individual patient is separated by the OR operator. Thus, to retrieve ER, PgR, and tumor size data for the 21 "over 70" patients, we need to add to the SQL statement this same code for each patient, separated by an OR operator, and which differs only by the hash string.

To do this in R we can create a character string object, which we'll call "over70.sql" to hold the SQL code and then iteratively add the code that is specific for retrieving data from each patient. It is much simpler to consider the SQL statement we need to construct in sections. The elements of text that make up the SQL statement can be added to the "over70.sql" string by using the `paste()` function. If you examine the code in Code 9.10 and Code 9.11, you can see that there is a single piece of text used, regardless of the number of patients being queried, that starts with the SELECT clause and ends with the WHERE clause. This tells MySQL exactly which tables and columns to look in for the data given the specified arguments. We can begin constructing the "over70.sql" string by pasting in this initial text (Code 9.12).

```
> over70.sql<-paste("SELECT hashcodes.hash_strings, pathology.er,
pathology.pgr, pathology.size, codes.er.er_annot, codes.pgr.pgr_annot FROM
(patient.pathology JOIN codes.er ON pathology.er = codes.er.code JOIN codes.pgr
ON pathology.pgr = codes.pgr.code), hashcodes WHERE")
```

Code 9.12.

We've modified the query this time to also return the hash string for each patient. We can then use a FOR LOOP to iteratively add the code that follows the WHERE clause for each patient to yield an almost complete SQL statement:

```
> for (i in 1:20){over70.sql<-
paste(over70.sql,"(pathology.patientid = hashcodes.patientid) AND
hashcodes.hash_strings = '",over70.hash[i],"'OR",sep="")}
```

This line of code constructs a piece of text that includes the value in the "over70.hash" array (given the loop number), flanked on either side by the texts "(pathology.patientid = hashcodes.patientid) AND hashcodes.hash_strings = '" and "'OR". We also parse a blank character to the `sep` (separator) argument of the `paste()` function so as not to include any spaces in the text that makes up the "hashcodes.patientid" value. The statement is not quite complete since we have only added SQL code for 20 patients (the FOR LOOP was set from 1 to 20). The reason for this is that the end of each patient-specific expression has the OR operator, and if this appeared after the last patient-specific line, the SQL syntax would be incorrect, causing MySQL to throw an error. Therefore, we can just add a final line, again using the `paste()` function, and parse text to include the last element in the "over70.list":

```
> over70.sql<-paste(over70.sql,"(pathology.patientid =
hashcodes.patientid) AND hashcodes.hash_strings =
'",over70.hash[length(over70.hash)],"'",sep="")
```

The last element in the "over70.hash" array is specified by using the `length()` function, which returns the array length as an integer, which in this case is "21." If you view the "over70.sql" string, you should immediately see the advantage of using R to create repetitive expressions when building long and cumbersome SQL statements (Code 9.13).

```
> over70.sql
[1] "SELECT hashcodes.hash_strings, pathology.er, pathology.pgr,
pathology.size, codes.er.er_annot, codes.pgr.pgr_annot FROM (patient.pathology
JOIN codes.er ON pathology.er = codes.er.code JOIN codes.pgr ON pathology.pgr =
codes.pgr.code), hashcodes WHERE(pathology.patientid = hashcodes.patientid) AND
hashcodes.hash_strings =
'afd29bd1a493305e666596cf98d2eb6d'OR(pathology.patientid = hashcodes.patientid)
AND hashcodes.hash_strings =
'c49757a6fd4f178ccf05271c37839989'OR(pathology.patientid = hashcodes.patientid)
AND hashcodes.hash_strings =
'089bf2f200acc2e1a807f86e243e244b'OR(pathology.patientid = hashcodes.patientid)
AND hashcodes.hash_strings =
'4ab32d1a86d6b9ff1a39d95c2e6d8c01'OR(pathology.patientid = hashcodes.patientid)
AND hashcodes.hash_strings =
'f946d1c3960f0de968cb68ab68e7ddad'OR(pathology.patientid = hashcodes.patientid)
AND hashcodes.hash_strings =
'8f757751fbf43de2253b5bc794c80c13'OR(pathology.patientid = hashcodes.patientid)
AND hashcodes.hash_strings =
'6be9ebab67aa834ad709e00342e1670e'OR(pathology.patientid = hashcodes.patientid)
AND hashcodes.hash_strings =
'edacff44b17a7ac55ddd248292a46890'OR(pathology.patientid = hashcodes.patientid)
AND hashcodes.hash_strings =
'037adfe485ecf86e0c6fd1f0c227ebac'OR(pathology.patientid = hashcodes.patientid)
AND hashcodes.hash_strings =
'e9655fd9174d63327a06d08a2ad4e782'OR(pathology.patientid = hashcodes.patientid)
AND hashcodes.hash_strings =
'c3b4fd2083befd4c8481247ce1a542a2'OR(pathology.patientid = hashcodes.patientid)
AND hashcodes.hash_strings =
'8da2886ad40b0a96d6c7ef54ff9ffc1c'OR(pathology.patientid = hashcodes.patientid)
AND hashcodes.hash_strings =
'02fdeff7e48627a6c276d7b38fac5143'OR(pathology.patientid = hashcodes.patientid)
AND hashcodes.hash_strings =
'5c0af2b80323b428cf01df55dd605980'OR(pathology.patientid = hashcodes.patientid)
AND hashcodes.hash_strings =
'db2ca77f89e6880f55225ee5b4cd2059'OR(pathology.patientid = hashcodes.patientid)
AND hashcodes.hash_strings =
'ff75e9c081a112449d69942a1ec8470c'OR(pathology.patientid = hashcodes.patientid)
AND hashcodes.hash_strings =
'f86962d0781cba4310a4a60157a53435'OR(pathology.patientid = hashcodes.patientid)
AND hashcodes.hash_strings =
'9c790cf2f97f788fb1e5a3dbb15805a5'OR(pathology.patientid = hashcodes.patientid)
AND hashcodes.hash_strings =
'8edd650ef769ab2064126a60bbcab56a'OR(pathology.patientid = hashcodes.patientid)
AND hashcodes.hash_strings =
'23d7c6bf79341f05a2b462b153bbc393'OR(pathology.patientid = hashcodes.patientid)
AND hashcodes.hash_strings = 'a7a815d3c4a7171b08257fa4a5c80d64'"
```

Code 9.13.

The "over70.sql" string can then be parsed as the statement argument in `dbGetQuery()`, which returns data to an object that we call "over70.data" (Code 9.14).

```
> over70.data<-dbGetQuery(connection,over70.sql)
> over70.data
                    hash_strings er pgr size    er_annot   pgr_annot
1  6be9ebab67aa834ad709e00342e1670e 2  2   30    Negative    Negative
2  afd29bd1a493305e666596cf98d2eb6d 1  1   22    Positive    Positive
3  9c790cf2f97f788fb1e5a3dbb15805a5 2  2   15    Negative    Negative
4  c3b4fd2083befd4c8481247ce1a542a2 2  2   19    Negative    Negative
5  4ab32d1a86d6b9ff1a39d95c2e6d8c01 3  2   35  Borderline    Negative
6  5c0af2b80323b428cf01df55dd605980 2  2   11    Negative    Negative
7  ff75e9c081a112449d69942a1ec8470c 2  2   22    Negative    Negative
8  a7a815d3c4a7171b08257fa4a5c80d64 2  2   38    Negative    Negative
9  037adfe485ecf86e0c6fd1f0c227ebac 2  2   16    Negative    Negative
10 c49757a6fd4f178ccf05271c37839989 1  2   34    Positive    Negative
11 23d7c6bf79341f05a2b462b153bbc393 2  1   20    Negative    Positive
12 e9655fd9174d63327a06d08a2ad4e782 1  1   22    Positive    Positive
13 edacff44b17a7ac55ddd248292a46890 1  2   25    Positive    Negative
14 8da2886ad40b0a96d6c7ef54ff9ffc1c 3  2   25  Borderline    Negative
15 089bf2f200acc2e1a807f86e243e244b 2  2   18    Negative    Negative
16 f86962d0781cba4310a4a60157a53435 2  2   20    Negative    Negative
17 8edd650ef769ab2064126a60bbcab56a 2  2   40    Negative    Negative
18 02fdeff7e48627a6c276d7b38fac5143 2  2   30    Negative    Negative
19 f946d1c3960f0de968cb68ab68e7ddad 2  2   21    Negative    Negative
20 db2ca77f89e6880f55225ee5b4cd2059 1  1   16    Positive    Positive
21 8f757751fbf43de2253b5bc794c80c13 1  3   10    Positive  Borderline
```

Code 9.14.

Next you can combine the TMA scores data frame with the data frame that you have just created using the patient database. Each of these data frames contains a column with the same hash reference strings. If you take a look at these data frames, though, you will probably find that the order of cases, identifiable by the hash strings, is different. Thus, to avoid combining data from the wrong patients, you need to do a quick order on both data frames using the columns containing the hash strings as references (Code 9.15).

```
> over70.data<-over70.data[order(over70.data[,1]),]
> over70.scores<-over70.scores[order(over70.scores[,1]),]
```

Code 9.15.

Now we can combine the columns from the two data frames to create a single data frame for the over-70 age group:

```
> over70.dataset<-cbind(over70.data, over70.scores[2:4])
```

The same procedure should then be used for the under-45 patient group to create a data frame called "under45.dataset." You are now just two steps away from creating a complete dataset for the study. Since we are going to combine the 42 patients from both groups into a single dataset, we should add a label column to the two age-related data frames first so that patients in each group can be distinguished (Code 9.16).

```
> over70.dataset<-cbind(over70.dataset, label=rep("over70",21))
> under45.dataset<-cbind(under45.dataset, label=rep("under45",21))
```

Code 9.16.

We need to tidy up and standardize the column names between datasets so that they match when merged. We can change the column names for both datasets using the names() function and passing a vector of new labels (Code 9.17).

```
> names(over70.dataset)<-c("hash_strings", "er", "pgr", "size",
"er_annotation", "pgr_annotation", "proteinA", "proteinB", "her2", "label")
> names(under45.dataset)<-c("hash_strings", "er", "pgr", "size",
"er_annotation", "pgr_annotation", "proteinA", "proteinB", "her2", "label")
```

Code 9.17.

All that's left to do is combine both datasets into one using the merge() function:

```
> breast.dataset<-merge(under45.dataset, over70.dataset, all=TRUE)
```

The "breast.dataset" object now contains the complete dataset and is ready to be given to the researcher. The dataset is shown in Code 9.18.

```
> breast.dataset
                   hash_strings er pgr size er_annotation
1   02fdeff7e48627a6c276d7b38fac5143  2  2   30      Negative
2   037adfe485ecf86e0c6fd1f0c227ebac  2  2   16      Negative
3   04d01d8ca9039632b2239fbafbe37b57  2  2    8      Negative
4   085fedf676902da772bf25c3502923cd  2  2   25      Negative
5   089bf2f200acc2e1a807f86e243e244b  2  2   18      Negative
6   1ec5198bdd64e130d5a6ca39811039cf  2  2   16      Negative
7   2275fb5df821d0abc234fc3d88610679  1  2   23      Positive
8   230077d3bd3342fb15c0d438baec4d1b  2  2    8      Negative
9   23d7c6bf79341f05a2b462b153bbc393  2  1   20      Negative
10  3daeb41c8910f3f3d8e4eda6e237b48d  2  2   20      Negative
11  4ab32d1a86d6b9ff1a39d95c2e6d8c01  3  2   35      Borderline
12  51a7bbe92e44b7a27a03369a76a79e00  2  2   20      Negative
13  5c0af2b80323b428cf01df55dd605980  2  2   11      Negative
14  6be9ebab67aa834ad709e00342e1670e  2  2   30      Negative
15  6cf0f2d9228f507d5157fa1cfd272aec  2  2   24      Negative
16  77764243684965da53bdd3cf7d1be453  2  2   15      Negative
17  8da2886ad40b0a96d6c7ef54ff9ffc1c  3  2   25      Borderline
18  8e61448206f14e437f4bc1e31ede4e92  1  1   18      Positive
19  8edd650ef769ab2064126a60bbcab56a  2  2   40      Negative
20  8f757751fbf43de2253b5bc794c80c13  1  3   10      Positive
21  9028e0766b4c689423f55b37e2aae595  2  2   23      Negative
22  910a440071ad3b6fc4a24a754252a7bb  2  2   18      Negative
23  9c790cf2f97f788fb1e5a3dbb15805a5  2  2   15      Negative
24  9dd44ccb48a1ec8d7a25acefb5e9a895  1  1   15      Positive
25  9dfd10cb6ae5ca1c56467b2fe27ad659  2  2   12      Negative
26  a6baf0062a71a9c21392f73079a6076b  2  2   16      Negative
27  a7a815d3c4a7171b08257fa4a5c80d64  2  2   38      Negative
28  afd29bd1a493305e666596cf98d2eb6d  1  1   22      Positive
29  b79b8ec7784d0d34b0ac2eef5b26464a  1  2   25      Positive
30  c3b4fd2083befd4c8481247ce1a542a2  2  2   19      Negative
31  c49757a6fd4f178ccf05271c37839989  1  2   34      Positive
32  c5cc4df40504b69f6082c94606596630  2  2    2      Negative
```

```
33 cfb19203c4dc16d88ecf6ccf2ec99ed0  1  2  30      Positive
34 db2ca77f89e6880f55225ee5b4cd2059  1  1  16      Positive
35 e540f0e1e2e88ed07f44d0bf9960c077  2  2  17      Negative
36 e9655fd9174d63327a06d08a2ad4e782  1  1  22      Positive
37 ed61084a95d96da25d50fd82fecc3f60  3  2  25      Borderline
38 odacff11b17a7ac66ddd248292a46890  1  2  25      Positive
39 f86962d0781cba4310a4a60157a53435  2  2  20      Negative
40 f946d1c3960f0de968cb68ab68e7ddad  2  2  21      Negative
41 fbb5bd86ea0df2b035eafb6d15e882c2  1  3  17      Positive
42 ff75e9c081a112449d69942a1ec8470c  2  2  22      Negative
   pgr_annotation proteinA proteinB her2    label
1        Negative       15       30    0   over70
2        Negative        8       47    0   over70
3        Negative        0        0    0  under45
4        Negative       13       31    3  under45
5        Negative       28       38    0   over70
6        Negative       31       94    0  under45
7        Negative        1        7    3  under45
8        Negative       58       91    0  under45
9        Positive       38       36    0   over70
10       Negative       18       96    3  under45
11       Negative        3        1    0   over70
12       Negative       30       80    0  under45
13       Negative        5       13    0   over70
14       Negative        8       18    0   over70
15       Negative       17       87    1  under45
16       Negative       28       91    0  under45
17       Negative       40        6    3   over70
18       Positive       30       39    0  under45
19       Negative       38       61    0   over70
20      Borderline       8       11    2   over70
21       Negative       25       84    0  under45
22       Negative       70       92    0  under45
23       Negative       36       15    0   over70
24       Positive        3        5    2  under45
25       Negative        0        0    2  under45
26       Negative        0       16    3  under45
27       Negative       34       62    1   over70
28       Positive        8        7    0   over70
29       Negative       35       93    0  under45
30       Negative        8       83    0   over70
31       Negative        2        1    1   over70
32       Negative       70       81    0  under45
33       Negative       28       63    0  under45
34       Positive       10       32    1   over70
35       Negative       39       86    0  under45
36       Positive        0        7    0   over70
37       Negative       17       21    3  under45
38       Negative       34       29    2   over70
39       Negative       13       20    0   over70
40       Negative       28       16    0   over70
41      Borderline      31        2    3  under45
42       Negative       34       89    0   over70
```

Code 9.18.

9.6 Managing a Database to Contain Study Datasets

As the pathologist, you would like to store all datasets prepared for a study in a single database for easy reference. In Section 8.4, you created a MySQL database called "studies" in addition to the "patient" and "codes" databases. This database is currently empty, but before we store the "breast.dataset" data frame as a table, the pathologist would like to create a table that contains the details of each study that he participates in. You can then update this table every time you create a new study dataset. You decide that you need to include columns for the dataset name, the researcher, the creation date for the dataset, and the study type.

To create the data frame object, called "details," that will be passed as a table to MySQL, you just need to provide the necessary column names and values for this study as the first row:

```
> details<-data.frame(dataset="breast_TMA_age",researcher="PD
Lewis",date_created="15.02.08",study_type="TMA")
```

Then after setting up a connection to the "studies" database, you can save the "details" data frame as a table using dbWriteTable() (Code 9.19).

```
> connection.mysql2<-dbConnect(dbDriver("MySQL"), dbname="studies",
username="root", password="a_password")
> dbWriteTable(connection.mysql2, "details", details, row.names=FALSE,
overwrite=TRUE)
```

Code 9.19.

The "breast.dataset" data frame should also be saved to a table called "breast_tma_age":

```
> dbWriteTable(connection.mysql2, "breast_tma_age",
breast.dataset, row.names=FALSE, overwrite=TRUE)
```

We will be using the breast TMA dataset you have just created in subsequent chapters, and you could also save the data to a favored working directory as the text file "breast_tma_age.txt":

```
> write.table(breast.dataset, "breast_tma_age.txt",
row.names=FALSE)
```

This will give you the option of either loading the dataset directly from a working directory using the read.table() function or setting up a connection to your database.

9.7 Summary

This chapter described an example of how to select suitable cases for a study and create and store a heterogeneous biomedical dataset of patient data and experimental results. Although the example described a patho-biological study, the basics and concepts described apply to many types of biomedical research projects. The patient database does not have to be of the same format as the one described in this book or one specifically created by a pathologist. It could be a database within an existing hospital information system.

Perhaps the most important message is that R does not just have to be your data analysis platform. The example is given purely to highlight the capabilities of R within the biomedical research setting. R can be your data management system linked to your favorite DBMS. The example study used MySQL as the DBMS, but other DBMSs could be used.

9.8 Questions

1. What do we mean by a "heterogeneous biomedical dataset"?

2. What are the benefits of using R to manage your heterogeneous biomedical data?

3. How can we select random patients (rows) from a data frame?

4. What is so important about having a "patientid" field in each table of the patient database?

5. Why is it so important to de-identify data from patients?

6. What are the advantages of creating hash strings using the `digest()` function?

10 | Descriptive Statistics in R

10.1 Background

The most basic forms of statistics you can apply to your data are descriptive. Descriptive statistics differ from inferential statistics, which we will explore in the next chapter. When we talk about our data, we often use descriptive statistical values without even realizing that we are using "statistics" to do it. Descriptive statistics are usually generated before a data analysis to give us a feel for the data and provide us with ways to summarize our data. The descriptive statistical values produced can also assist us in learning about the underlying distribution of our data and help us decide exactly what statistical tests we can apply. The values can be taken as parameters for a statistical test and further used to calculate any error in the results.

A number of descriptive statistics are familiar to most people, including mean, median, standard deviation, and confidence intervals. Either singularly or collectively, they provide you with a summary of the data. More often than not, it is easiest to summarize your data graphically, and a variety of plots may be used in conjunction with descriptive statistics to help you get to know your data.

R, being a statistical environment, has a tremendous range of methods for summarizing your data, as you might expect. Descriptive or summary statistics can be obtained using functions held in the "base" package, which is installed by default with the software. The graphic capabilities of R are highly impressive, and as we saw in Chapter 4, basic plots are available using the "graphics" package, which is also installed by default.

In this chapter we will explore the different functions that R possesses for obtaining summaries of the data, both as output to the console and graphically as plots. We'll begin with a discussion on how information can be gathered on the different types of data you are likely to encounter in biomedical datasets. The types of descriptive statistics are split into three key areas that reflect different qualities of the data. Location statistics include values estimating "location" or central tendency, such as the mean. Scale statistics measure the variation or dispersion or spread of the data in a sample. Distribution statistics measure whether the shape of the data differs from a normal distribution and is often aided by graphic techniques.

10.2 Knowing the Data Types and Storage Modes

In Chapter 9 we created a biomedical dataset that included scores for a tissue microarray study in 21 patients over the age of 70 and 21 patients under the age of 45.

The dataset also included other biomedical variables as well as text labels describing scoring systems. The types of variables we would expect in biomedical datasets and their storage modes were introduced in Chapter 3.

Let us now focus on the different types of variables present in the breast cancer tissue microarray dataset. The dataset can be fetched into a data frame object (we'll call "breast.dataset") from the "breast_tma_age" table in the MySQL "studies" database we created in Chapter 8, loaded from the "breast_tma_age.txt" created in the same chapter using read.table(), or loaded from the "Tissue Microarray Datasets" folder that accompanies this book that you should have downloaded to the R root folder. To use the latter option, change the R working directory to this folder and retrieve the names of the variables using the names() function (Code 10.1).

```
> breast.dataset<-read.table("breast_tma_age.txt", header=T)
> names(breast.dataset)
 [1] "hash_strings"   "er"             "pgr"            "size"
 [5] "er_annotation"  "pgr_annotation" "proteinA"       "proteinB"
 [9] "her2"           "label"
```

Code 10.1.

Having created the dataset ourselves, we already know that the variables are quantitative and qualitative and of different data types, both numeric and categorical. The numeric integer variables (proteinA, proteinB, age) are numeric in the sense that they can take on any number. To be more precise, the proteinA and proteinB variables are interval because they take on an integer value within the range 0 to 100. The age variable is actually discrete, as a value must be an integer within the age range specified by the sampled population. The categorical variables (er, pgr, her2) are ordinal because they range from zero to a maximum score. The remaining variables are simply text labels (hash_strings, er_annotation, pgr_annotation, label) and can be treated as such.

When we load in a dataset like this into R, we need to make sure that the variables are assigned as they ought to be in relation to their type and the mode in which they are stored. In our example, the continuous variables in the dataset should be assigned as numeric vectors (Table 10.1), the categorical (ordinal) variables should be factors,

Table 10.1   The Required Types for the breast.dataset Variables.

hash_strings	character
er	factor
pgr	factor
size	integer
er_annotation	character
pgr_annotation	character
proteinA	integer
proteinB	integer
her2	factor
label	character

and the text label variables should be character vectors. When we created the dataset, we didn't concern ourselves with assigning a type or storage mode to each vector within the data frame. In later chapters we are to going to analyze this data using functions that will process data according to the modes, hence the reason to ensure prior to this that each variable is of the right type.

The type and storage mode of each vector can be found using the typeof() and mode() functions, respectively. Regardless of what each vector storage mode is in the data frame after being read in from a file, they can be changed easily using the as() function (Code 10.2).

```
> breast.dataset[,c(1)]<-as.character(breast.dataset[,1])
> breast.dataset[,c(2)]<-as.factor(breast.dataset[,2])
> breast.dataset[,c(3)]<-as.factor(breast.dataset[,3])
> breast.dataset[,c(4)]<-as.numeric(breast.dataset[,4])
> breast.dataset[,c(5)]<-as.character(breast.dataset[,5])
> breast.dataset[,c(6)]<-as.character(breast.dataset[,6])
> breast.dataset[,c(7)]<-as.numeric(breast.dataset[,7])
> breast.dataset[,c(8)]<-as.numeric(breast.dataset[,8])
> breast.dataset[,c(9)]<-as.factor(breast.dataset[,9])
> breast.dataset[,c(10)]<-as.character(breast.dataset[,10])
```

Code 10.2.

In similar fashion, using the as() function, you can ensure that each variable is of the right data type including, among others, integer, double, and character:

```
> breast.dataset[,c(8)]<-as.integer(breast.dataset[,8])
```

The summary() command can be used to view the information concerning variable vectors in the data frame (Code 10.3). Numeric vectors, such as size, display some descriptive statistics, such as the median and quartiles, which we will look at later in the chapter. The categorical factor variables list the different values and their frequencies.

```
> summary(breast.dataset)
hash_strings        er        pgr      size
Length:42           1:12      1: 6     Min.   : 2.00
Class :character    2:27      2:34     1st Qu.:16.00
Mode  :character    3: 3      3: 2     Median :20.00
                                       Mean   :20.62
                                       3rd Qu.:25.00
                                       Max.   :40.00


er_annotation       pgr_annotation         proteinA
Length:42           Length:42              Min.   : 0.00
Class :character    Class :character       1st Qu.: 8.00
Mode  :character    Mode  :character       Median :21.50
                                           Mean   :22.43
                                           3rd Qu.:34.00
                                           Max.   :70.00


proteinB            her2  label
Min.   : 0.00       0:27  Length:42
```

```
1st Qu.:11.50     1: 4   Class:character
Median :31.50     2: 4   Mode:character
Mean   :42.40     3: 7
3rd Qu.:82.50
Max.   :96.00
```

Code 10.3.

10.3 Summarizing and Displaying Categorical Data

Individual vectors with categorical data can be summarized using the `table()` function (Code 10.4).

```
> table(breast.dataset$er)

 1  2  3
12 27  3
```

Code 10.4.

The primary function of `table()` is to produce a contingency or frequency table of counts for each level of specified factors. Thus, one could produce a contingency table to display the counts at levels of two categorical vectors, such as for patients according to their ER and PgR status (Code 10.5).

```
> table(breast.dataset[,c(2,3)])
    pgr
er   1  2  3
  1  5  5  2
  2  1 26  0
  3  0  3  0
```

Code 10.5.

Contingency tables for more than two variables can be displayed using the `table()` function, but a better display format can be obtained using the `ftable()` function (Code 10.6).

```
> ftable(breast.dataset[,c(2,3,9)])
        her2  0  1  2  3
er pgr
1  1          3  1  1  0
   2          2  1  1  1
   3          0  0  1  1
2  1          1  0  0  0
   2         20  2  1  3
   3          0  0  0  0
3  1          0  0  0  0
   2          1  0  0  2
   3          0  0  0  0
```

Code 10.6.

Pie chart to display ER category proportions

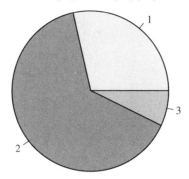

FIGURE 10.1 Generating a pie chart with `pie()`.

Categorical data can be displayed graphically in a number of ways. A pie chart is a commonly used circular chart where the sectors are proportional to the frequency or percent of cases in a group. To generate a pie chart (Figure 10.1) to display the proportions of each ER category in the "breast.dataset" object, you can use the `pie()` function:

```
> pie(table(breast.dataset[,2]))
```

The drawback in using pie charts is that plotting anything more than a handful of categories means it is difficult to visualize actual differences in the sizes of segment areas. Bar or column charts are a preferable way to compare two or more values within a variable. We can generate a bar chart for the ER variable (Figure 10.2) using the `barplot()` function:

```
> barplot(table(breast.dataset$er), main="Bar chart to display ER categories")
```

We can also produce a bar chart to visualize frequencies of the values of one categorical variable given the value in another categorical variable (Figure 10.3). This is

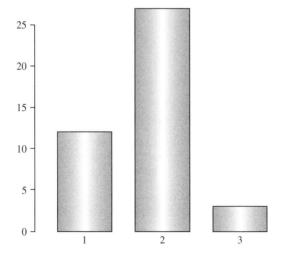

FIGURE 10.2 Generating a bar chart using `barplot()`.

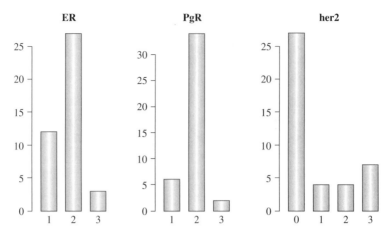

FIGURE 10.3   Producing multiple bar charts.

another way of visualizing a contingency table, such as that shown in Code 10.6. The code shown in Code 10.7 produces such a graphic for the ER and PgR data, and the bar chart is shown in Figure 10.4.

```
> data<-table(breast.dataset[,c(2,3)])
> chart<-barplot(height=data, beside = TRUE, main="ER", legend.text=c("PgR =
1","PgR = 2","PgR = 3"))
```

Code 10.7.

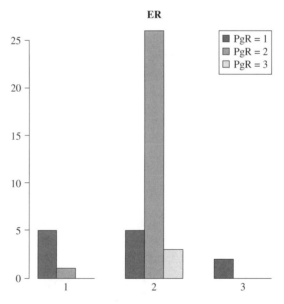

FIGURE 10.4   Changing the color format of a bar chart.

10.4 Summarizing Numeric Data Distributions

There are more ways of describing numeric data than there are for categorical data. There are a number of different aspects and properties about numeric data that can be summarized in tabular fashion or graphically. We can apply functions in R to feedback information on the spread of data, as well as measurements about the distribution that can be subject to statistical testing.

We have already used the `summary()` function to reveal properties of each variable in the breast.dataset data frame (Code 10.3). The output didn't tell us much about each of the categorical variables aside from giving the counts for each category. The output did reveal some measurements for the numeric variables. For example, the minimum and maximum values in the range for tumor size are 1 and 40 with an (arithmetic) mean value of 20.62. The summary also gives measurements for the first, second (median), and third quartiles, which are 16, 20, and 25, respectively. These measurements can also be obtained individually for a vector by applying the `summary()` function to retrieve all of the `mean()`, `median()`, and `quantile()` functions where you must also specify the three quartile points for the latter (Code 10.8).

```
> summary(breast.dataset$size)
   Min. 1st Qu. Median   Mean 3rd Qu.   Max.
   2.00  16.00  20.00  20.62  25.00  40.00
> mean(breast.dataset$size)
[1] 20.61905
> median(breast.dataset$size)
[1] 20
> quantile(breast.dataset$size, probs = 1:3/4)
25% 50% 75%
 16  20  25
> min(breast.dataset$size)
[1] 2
> max(breast.dataset$size)
[1] 40
```

Code 10.8.

There are occasions when you may want to calculate the geometric mean as opposed to the arithmetic mean, particularly if you suspect the data distribution to be highly skewed. There is no base function in R to calculate the geometric mean, but it can be returned using the expression in Code 10.9. In the example, the arithmetic mean is calculated for log values of the tumor size data and the exponent is returned.

```
> exp(mean(log(breast.dataset$size)))
[1] 18.69630
```

Code 10.9.

You may also need to calculate a weighted arithmetic mean where each value in the vector to be measured must have a particular weight attached. To do this you can use the `weighted.mean()` function parsing the measurable vector and a vector of weights:

```
> weighted.mean(measurable.variable, weights.vector)
```

10.5 Visualizing Numeric Data Distributions

The frequency distribution of numeric data can also be displayed using a bar chart. The vertical bar chart graphic of tumor size in Figure 10.5 was produced using the `barchart()` function and ordering the values according to size:

```
> barplot(breast.dataset[order(breast.dataset$size),4], horiz=T,
main="Barchart of tumor size")
```

An alternative and more visually appealing plot type than the bar chart is the dot plot (Figure 10.6), which can be obtained using the `dotchart()` function:

```
> dotchart(breast.dataset[order(breast.dataset$size),4], main="Dot
plot of tumor size")
```

Another way to visualize the distribution of the data is to generate a stem-and-leaf plot in the R console (Code 10.10).

```
> stem(breast.dataset[,4])

  The decimal point is 1 digit(s) to the right of the |

  0 | 2
  0 | 88
  1 | 012
  1 | 5556666778889
  2 | 00001222334
  2 | 55555
  3 | 0004
  3 | 58
  4 | 0
```

Code 10.10.

The plot reveals that in the tumor size variable, there is one value of 2; two values of 8; one value each of 10, 11, and 12; four values of 15; and so on.

A more graphic type of plot is the histogram (Figure 10.7), which can be generated using the `hist()` function:

```
> hist(breast.dataset[order(breast.dataset$size),4],
main="Histogram of tumor size", xlab="tumor size")
```

The stem-and-leaf plot and histogram give visual insight into the "spread" of the data. Another graphic that displays information to help interpret the distribution of data is the box-and-whisker plot, generated using the `boxplot()` function. In fact, the box-and-whisker plot is essentially a graphic representation of the information displayed by the `summary()` function. The "box" in the plot shows the interquartile range between first and third quartiles and encloses the median. The two "whiskers" protrude from the box to the maximum and minimum values of the data.

Barchart of tumor size

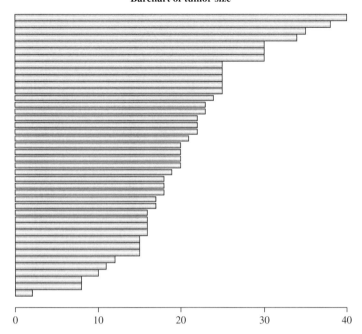

FIGURE 10.5   Horizontal bar chart.

Dot plot of tumor size

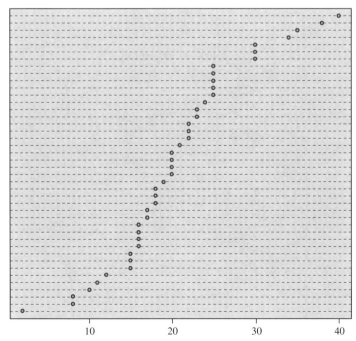

FIGURE 10.6   Dot plot for tumor size.

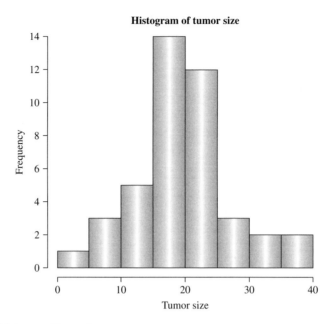

FIGURE 10.7 Histogram of tumor size.

The example in Figure 10.8 shows two adjacent box-and-whisker plots for the tumor size variable, but where the cases have been split into two groups according to the label variable, (i.e., whether the patient was under the age of 45 or over the age of 70). Code 10.11 shows how this was achieved. First, the breast.dataset data frame was ordered according to the label variable, and the tumor size data was passed into two vec-

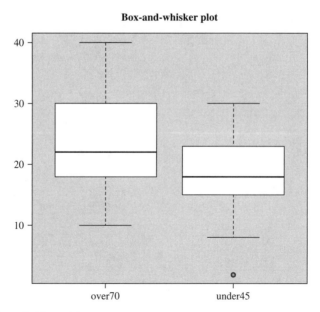

FIGURE 10.8 Box and whisper plots.

tors (over70 and under45) according to age category. These vectors were then parsed as a data frame to the `boxplot()` function.

```
> breast.dataset<-breast.dataset[order(breast.dataset$label),]
> over70<-breast.dataset[1:21,4]
> under45<-breast.dataset[22:42,4]
> boxplot(as.data.frame(cbind(over70,under45)),main="Box-and-whisker plot")
```

Code 10.11.

This example reveals that the range of tumor size values for patients over the age of 70 in this particular cohort is greater than that for patients under the age of 45.

10.6 The Normal Distribution

The last couple of sections have looked at ways in which the distribution of your data can be assessed. Knowledge of the distribution of your data dictates what statistical tests may be applied. The most important member of the family of continuous probability distributions is the standard normal distribution (also called the Gaussian distribution). The normal distribution is important, as many variables may be distributed approximately in a normal way. Robust statistical analysis procedures can be applied if the data distribution is or approximately is normal.

Most people will be familiar with the characteristic "bell shape" curve produced when plotting the distribution of normal data. The variance, or "spread," of data around the arithmetic mean in a normal distribution can be measured to give an idea of how varied the values are within a sample range. In R, the variance can be obtained using the `var()` function (Code 10.12). Similarly, to calculate the standard deviation, you can use the `sd()` function. In fact, a variable conforming to a normal distribution can be completely described by the mean and standard deviations.

```
> var(breast.dataset$size)
[1] 64.09524
> sd(breast.dataset$size)
[1] 8.00595
```

Code 10.12.

So how do we know if our data approximates a normal distribution? We can make a visual assessment using stem-and-leaf or histogram plots of the data and look for a bell-shape curve. We could also produce a boxplot to see if the median line approximately cuts the rectangular box in half and the whiskers are roughly of the same length. We can generate a curve for our tumor size vector using the `density()` function and plotting the resulting kernel density function (Figure 10.9):

```
> plot(density(breast.dataset$size),main="Density estimate of
data")
```

Density estimate of data

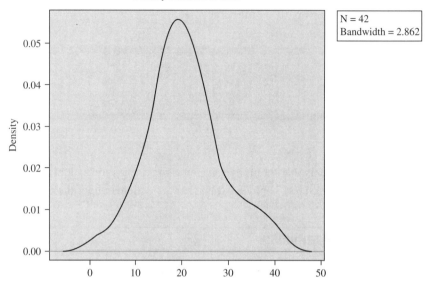

FIGURE 10.9   Plotting a density estimate.

We can inspect the shape of the curve not just for a bell shape, but also for kurtosis and skewness. Better still, we can produce a normal plot, which plots the cumulative frequency distribution of the data (Code 10.13).

```
> plot(ecdf(breast.dataset$size),horizontal=F, col.hor="white", main =
"Cumulative frequency distribution", xlab="Tumor size", ylab="Standard Normal
deviate" )
> range<-seq(1,40,0.01)
> distribution<-pnorm(range, mean=mean(breast.dataset$size),
sd=sqrt(var(breast.dataset$size)))
> lines(range, distribution)
```

Code 10.13.

The first line of code calculates the empirical cumulative frequency distribution for the tumor size vector using the `ecdf()` function and plots the generated function using `plot()`. The next line creates a vector of a sequence of numbers with a range between 1 and 40 in increments of 0.01. This vector, called range, is then used to produce a normal distribution function via `pnorm()`, with a mean and standard variation equivalent to that for the tumor size data. This distribution function is stored in an object called "distribution," which is then parsed to the `lines()` function and the distribution plotted. The plot is shown in Figure 10.10, where the x-axis represents the cumulative frequency distribution of the data and the y-axis the same frequency but that expected of a normal distribution. If the data distribution of tumor size is indeed normal, the points should fall close to the line for the normal distribution function. Substantial deviation of points from the line could signify deviation from normality.

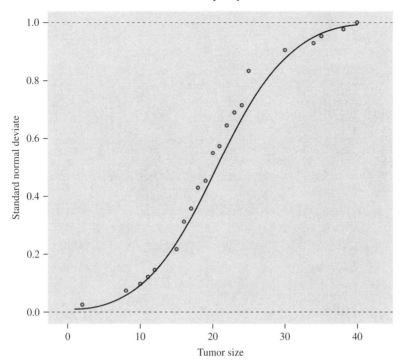

Cumulative frequency distribution

FIGURE 10.10   Plotting a cumulative frequency distribution.

Another useful plot that allows us to compare the distribution in our data to a normal distribution is a quantile-quantile plot, or Q-Q plot (Figure 10.11). We can parse the tumor-size vector to a function called `qqnorm()`, which generates what is essentially a scatterplot of the observed values versus the expected values of a normal distribution (Code 10.14). A line can then be superimposed on the plot using `line()` to show where the points should lie if the data was normally distributed.

```
> qqnorm(breast.dataset$size)
> qqline(breast.dataset$size)
```

Code 10.14.

Using graphic procedures to assess normality of data is, of course, subjective. To make a more objective assessment of your data, you could use one of two tests called Kolmogorov-Smirnov and Shapiro-Wilk, provided as functions by R. These are goodness-of-fit tests that test whether the data distribution differs from normality. A p-value is generated where a significant result suggests that you should reject the null hypothesis that the data follows a normal distribution. Be careful in your interpretation of the result, though, as a nonsignificant result does not guarantee that the data is normally distributed. The code for applying these two tests to the tumor size data is shown in Code 10.15.

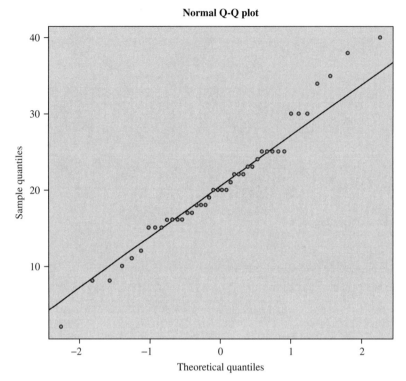

FIGURE 10.11   Normal quantile-quantile (Q-Q) plot.

```
> ks.test(breast.dataset$size, "pnorm", mean=mean(breast.dataset$size),
sd=sqrt(var(breast.dataset$size)))

        One-sample Kolmogorov-Smirnov test

data:  breast.dataset$size
D = 0.1254, p-value = 0.5231
alternative hypothesis: two-sided

# Shapiro-Wilk Test
> shapiro.test(breast.dataset$size)

        Shapiro-Wilk normality test

data:  breast.dataset$size
W = 0.9755, p-value = 0.4963
```

Code 10.15.

The p-values suggest that the tumor size data distribution does not differ from normality, which we might have assumed from the density, normal, and Q-Q plots.

Finally, if we suspect our data to be not normal, we could consider transforming data using one of a number of methods. By applying a mathematical transformation of each value in a data vector, it could have the effect of producing a new distribution that approximates normality. Some of the more common transformation functions are

Table 10.2	Transformation Functions.
logarithmic	z=log(y), z=log2(y), z=log10(y)
square root	z=sqrt(y)
reciprocal	z=1/y
square	z=y^2

shown in Table 10.2 and can be quite easily implemented in R using functions such as `log()`, `log2()`, `log10()`, `sqrt()`, etc.

10.7 Summary

Before we consider carrying out inferential statistical analyses of our data, we need to summarize our data with measurements. We can summarize data in tabular format or use graphic techniques such as bar charts. R facilitates simple generation of these estimates with built-in functions and provides plots with excellent graphic tools. These measurements reveal the distribution of our data and allow us to determine what statistical tests can be applied to test for similarity or difference between variables.

10.8 Questions

1. What are the three key areas of descriptive statistics?

2. What do we mean by the "distribution" of the data?

3. What is the difference between quantitative and qualitative data, and how does this difference relate to R object types?

4. What is the most common nongraphic way of displaying categorical data, and what R function may be used to achieve this?

5. How might we graphically display the distribution of numeric data in R?

6. What does the output from the `ecdf()` function tell us about our data?

7. How can we graphically compare the distribution of our data to a normal distribution in R?

8. How can we test that our data is normal?

11 R and Basic Inferential Statistical Analysis

11.1 Background

We began the last chapter by declaring that statistical analysis of our data can be divided into two key domains: descriptive statistics and inferential statistics. In the last chapter, we saw how we can use R to provide descriptive statistics to summarize our data and provide measurements and information regarding the distribution of the data. In this chapter, we explore R's capabilities in using basic statistics to make inferences about our data.

In Chapter 10 we looked at how we can produce a number of measurements from our data sample, including the mean and standard deviation. We also stated that these measurements can be subject to statistical testing. Inferential statistics is, if you like, the next step of our analysis where we can take such measurements and estimate how good they really are in representing the true population values (such as an estimated mean versus the population mean) and test hypotheses about our experimental data. This latter phase of a statistical analysis is perhaps what most people consider "statistics" to be—applying a statistical test to some data and testing the null hypothesis.

The chapter begins by demonstrating how we can estimate the error of parameter measurements from our sample data and calculate confidence intervals. The rest of the chapter then concentrates on using R to apply basic statistical methods to test for differences between samples. We will also look at using R to measure associations between two variables. Before you begin the chapter, download the following packages either by using the Menu option in RGui or typing:

```
>install.packages(c("irr", "epitools"))
```

11.2 Estimating Error and Confidence Intervals of a Measurement

In Chapter 10 we saw how we can create statistics for a single sample of values from a defined population. Using an example of tumor size from a population of 42 breast cancer patients, we were able to calculate the mean, median, standard deviation, and so on. What if we were to create a second and third set of samples from different patients using the same clinical criteria and record the mean each time? We would very likely see a different mean each time.

In practice, we only tend to generate one sample of data for a variable of interest. We take a value like the mean and use this as an estimate of the true population mean. But how good is our estimate of the true population mean? We may, after all, use this measurement for statistical analysis, which might influence clinical decision making.

By knowing the standard deviation of our sample, we can estimate the standard error (SE) of the mean using the equation:

$$SE = s/\sqrt{n}$$

The SE of the mean can be calculated in R by encoding this equation, and we can apply the code to the breast cancer tumor size variable (Code 11.1). First, though, load in the breast cancer TMA dataset by changing the working directory to the appropriate folder.

```
> breast.dataset<-read.table("breast_tma_age.txt", header=T)
> sd(breast.dataset$size)/sqrt(length(breast.dataset$size))
[1] 1.235345
```

Code 11.1.

A small SE indicates more precision than a larger one. You may be wondering how useful the SE actually is—after all, we don't really know what constitutes a large or small value. The usefulness of the SE (of the mean) comes in providing us with an interval estimate for the mean. We can return a confidence interval (CI) for the mean by calculating confidence limits each side of the parameter. The CI for the mean is the estimated range in which the mean is expected to fall. We tend to calculate 95% CI values, which means that if we repeatedly sampled the population and calculated a CI each time, 95% of the CI values would contain the estimated mean. The CI for the mean can be calculated from the normal distribution, and the R code for this, using tumor size data as an example, is shown in Code 11.2

```
> mean<-mean(breast.dataset$size)
> SD<-sd(breast.dataset$size)
> N<-length(breast.dataset$size)
> error <- qnorm(0.975)*SD/sqrt(N)
> left.limit<-mean-error
> right.limit<-mean+error
> mean
[1] 20.61905
> left.limit
[1] 18.19782
> right.limit
[1] 23.04028
```

Code 11.2.

So for the tumor size data, we estimate that the population mean falls between the values of 20.61905 and 23.04028.

11.3 One-Group Tests: The *t*-test

In the previous section, we estimated an interval for the sample mean of the breast cancer tumor size data and estimated that it lay between the values of 18.2 and 23.0. If we already knew what the population mean was for tumor size in breast cancer cases, we could directly compare this to our sample mean and test the null hypothesis that no size difference exists.

Because the breast cancer TMA dataset originates from patients in the breast cancer dataset, we can easily obtain the population mean. If you have created a MySQL database or other relational database, as described in Chapter 8, you could obtain the population mean by setting up a connection and parsing the following SQL statement using `dbGetQuery()`:

```
"Select AVG(pathology.size) From pathology"
```

The population mean for tumor size in the entire breast cancer dataset is 20.8632. As we saw in Code 11.2, the mean for our patients in our TMA sample dataset was 20.61905. However, the TMA study was based around two groups of patients that differed according to age range, where 21 patients were under the age of 45 and 21 were over the age of 70. Let's look at the mean values for each age group by applying the creating vectors for each mean for the over-75 group (over75) and under-45 group (under45) (Code 11.3).

```
> breast.dataset<-breast.dataset[order(breast.dataset$label),]
> over70.size<-breast.dataset$size[1:21]
> under45.size<-breast.dataset$size[22:42]
> over70.mean<-mean(over70.data)
> under45.mean<-mean(under45.data)
> over70.mean
[1] 23.28571
> under45.mean
[1] 17.95238
```

Code 11.3.

We can see that both mean values fall, unsurprisingly, above and below the population mean, but we should ask how representative of the population mean these values are. We can test the null hypothesis that there is no difference between each sample mean and population mean using a one-sample *t*-test option in the `t.test()` function (Code 11.4).

```
> t.test(over70.size, mu=20.8632)

        One Sample t-test

data:  over70.size
t = 1.3155, df = 20, p-value = 0.2032
alternative hypothesis: true mean is not equal to 20.8632
95 percent confidence interval:
 19.44439 27.12704
sample estimates:
mean of x
 23.28571

> t.test(under45.size, mu=20.8632)

        One Sample t-test
```

```
data:  under45.size
t = -1.983, df = 20, p-value = 0.06127
alternative hypothesis: true mean is not equal to 20.8632
95 percent confidence interval:
 14.89045 21.01431
sample estimates:
mean of x
 17.95238
```

Code 11.4.

As we only parse one vector to t.test() in each case, R recognizes that we want to perform a one-sample *t*-test; note also that we provide the population mean value as the mu parameter. The output provides a range of values that include the *t*-statistic (*t*), the degrees of freedom (df), and the P Value along with confidence intervals for the sample means. We can see from the results that there was no significant difference between either group mean or the population mean at the 5% level, but we could conclude that the under-45 patient group shows a trend toward having a smaller tumor size than the population average. We do need to be careful in our interpretation, though, as this sample does come from the population that provided the population mean.

11.4 One-Group Tests: The Wilcoxon Signed-Rank Test

In the previous section, we compared the tumor size sample mean to the population mean. We could also apply a one-sample Wilcoxon signed-rank test to compare the median value in our samples to the population median. The function to perform this test is wilcoxon.test() and takes the numeric vector and a parameter mu. The population median value for the breast cancer dataset is 17, and we can parse this value to mu along with each age-related dataset (Code 11.5).

```
> wilcox.test(over70.size, mu=17)

        Wilcoxon signed rank test with continuity correction

data:  over70.size
V = 197.5, p-value = 0.004578
alternative hypothesis: true location is not equal to 17

> wilcox.test(under45.size, mu=17)

        Wilcoxon signed rank test with continuity correction

data:  under45.size
V = 113, p-value = 0.4805
alternative hypothesis: true location is not equal to 17 1 sample t test;
t.test
One sample sign
One sample wilcoxon
```

Code 11.5.

The results tell us that the median value of tumor size in the under-45 age group does not differ significantly from the population median, but the over-70 group median is significantly different from that of the population.

11.5 Two-Group Tests: The *t*-test

A one-sample *t*-test told us that the mean tumor size in each of the age-related breast cancer patient groups does not differ significantly from the population mean. Given the difference in group means, however, we are left wondering whether tumor size may differ according to age.

With the knowledge that tumor size values in both groups follow a normal distribution, we would be inclined to carry out a two-sample *t*-test to test for difference between the means (23.29 versus 17.95). The variables we are testing are unpaired, which leaves us with two choices of *t*-test, depending on whether we assume that the variance in each group is equal or unequal. If we make the assumption that the variances in each group do differ, we can apply a two-sample *t*-test, commonly known as Welch's *t*-test, with the default option of unequal variance (Code 11.6).

```
> t.test(over70.size, under45.data)

        Welch Two Sample t-test

data:  over70.size and under45.size
t = 2.2647, df = 38.106, p-value = 0.0293
alternative hypothesis: true difference in means is not equal to 0
95 percent confidence interval:
  0.5664138 10.1002529
sample estimates:
mean of x mean of y
 23.28571  17.95238
```

Code 11.6.

The P Value indicates that there is a significant difference at the 5% level in mean tumor size between the two age categories. We may, however, be mistaken in assuming that the variance of the two groups differs. When applying Welch's *t*-test, there is a loss of power compared to using a *t*-test with equal variance assumed. If the group variances are equal, we can apply a second option of *t*-test where we assume equal variance. First, we can determine the standard deviation for each group using the sd() function (Code 11.7).

```
> over70.SD<-sd(over70.size)
> under45.SD<-sd(under45.size)
> over70.SD
[1] 8.438856
> under45.SD
[1] 6.726635
```

Code 11.7.

Is the variation significantly different between both age groups? With the assumption that our data follows a normal distribution satisfied, we can test the equality of the variance using the *F*-test (Code 11.8).

```
> var.test(over70.size, under45.size)

        F test to compare two variances

data:  over70.size and under45.size
F = 1.5739, num df = 20, denom df = 20, p-value = 0.3186
alternative hypothesis: true ratio of variances is not equal to 1
95 percent confidence interval:
 0.6386241 3.8788005
sample estimates:
ratio of variances
          1.573879
```

Code 11.8.

At the 5% level, there is no significant difference between the age group variances and we can now justly perform a *t*-test assuming equal variance (Code 11.9). This time, the function call differs by stating the `var.equal` parameter and setting it to TRUE.

```
> t.test(over70.size, under45.size, var.equal = TRUE)

        Two Sample t-test

data:  over70.size and under45.size
t = 2.2647, df = 40, p-value = 0.02902
alternative hypothesis: true difference in means is not equal to 0
95 percent confidence interval:
  0.5737952 10.0928715
sample estimates:
mean of x mean of y
 23.28571   17.95238
```

Code 11.9.

If you compare the output with that for Welch's *t*-test in Code 11.6, you will notice that the results differ very little, but you will have more confidence in the results knowing that you have applied the appropriate *t*-test.

There is a third common variant of the *t*-test that is applied to paired samples. Let's look at an example for a group of 20 male patients who have been diagnosed with high blood pressure and drink more than 30 units of alcohol per week. Each patient is asked to reduce his alcohol intake to no more than two units per day for a period of 12 months. Measurements are taken for systolic blood pressure (mmHg) at the start and end of the trial. The measurements are entered into two vectors, called "pretrial" and "posttrial" (Code 11.10). We then apply the `t.test()` function, setting the parameter called `paired` to TRUE.

```
> pretrial<-
c(147,155,140,165,151,160,158,144,143,162,154,171,148,168,141,148,170,156,148,1
61)
> posttrial<-
c(134,145,141,155,151,148,148,128,137,160,140,160,155,145,138,147,164,158,144,1
45)
> t.test(pretrial, posttrial, paired=TRUE)


        Paired t-test

data:  pretrial and posttrial
t = 4.4289, df = 19, p-value = 0.0002881
alternative hypothesis: true difference in means is not equal to 0
95 percent confidence interval:
  3.876539 10.823461
sample estimates:
mean of the differences
               7.35
```

Code 11.10.

The P Value returned shows that the reduction of systolic blood pressure after reduced alcohol intake for 12 months is highly significant.

11.6 Two-Group Tests: The Mann-Whitney *U*-Test

If the assumption that the data is normally distributed is not satisfied, we need to use a nonparametric test for comparing two samples. The Mann-Whitney *U*-test is considered the nonparametric equivalent of the unpaired *t*-test and is based on ranking the values. The Mann-Whitney *U*-test can be called in R using the Wilcox.test() function. We have already met this function previously (Section 11.4) for applying the Wilcoxon signed-rank test to one sample. To obtain the Mann-Whitney *U*-test from Wilcox.test(), you just have to pass two vectors. As an example, we will apply the Mann-Whitney *U*-test to the proteinB data in the "breast.dataset" data frame. The proteinB data represents immunohistochemistry scores for an antibody, and it is expected that the data would be skewed toward lower scores and thus not be normally distributed. Code 11.11 shows the creation of two vectors, called over70.proteinB and under45.proteinB, to hold the scores for the two age groups. Both vectors are then fed as input to the Wilcox.test() function.

```
> over70.proteinB<-breast.dataset$proteinB[1:21]
> under45.proteinB<-breast.dataset$proteinB[22:42]
> wilcox.test(over70.proteinB, under45.proteinB, conf.int = TRUE)


        Wilcoxon rank sum test with continuity correction

data:  over70.proteinB and under45.proteinB
W = 141.5, p-value = 0.04823
alternative hypothesis: true location shift is not equal to 0
95 percent confidence interval:
 -5.800006e+01 -4.744962e-05
```

```
sample estimates:
difference in location
             -27.12772
```

Code 11.11.

Interestingly, the P Value tells us that there may be a significant difference for proteinB scores between the two age groups.

11.7 More Than Two Groups: ANOVA and Kruskal-Wallis Rank Sum Tests

If we have data for a factor containing more than two independent groups that follow a normal distribution, we can use a one-way analysis of variance (ANOVA) to test for differences between group means. ANOVA can best be explained by example. In Section 11.5 we saw that there was a significant difference in tumor size between the over-70 and under-45 age groups. These group ages fall at both ends of the entire age range for breast cancer patients. It would be of interest to see if tumor size of patients whose ages lie in the mid range also differ from any of these two groups.

The file "breast_age_size.txt" contains tumor size data for three groups of patients: under the age of 45, between the age of 55 and 60, and over the age of 70. The dataset also contains scores for estrogen receptor (ER) status (which we shall ignore for now). Each age group contains data for 100 patients. We can load the dataset into a data frame called size.data using `read.table()` and determine the variable names using `names()` (Code 11.12).

```
> size.data<-read.table("breast_age_groups.txt", header=T)
> names(size.data)
[1] "size" "er"    "age"
```

Code 11.12.

We need to ensure that both `age` and `er` variables are classed as factors. The first two lines of code in Code 11.13 demonstrate how this can be enforced, and the `summary()` function then shows us the numbers and names of levels in both these groups.

```
> size.data$er<-as.factor(size.data$er)
> size.data$age<-as.factor(size.data$age)
> summary(size.data)
      size         er            age
 Min.   : 1.00   1:210   midage :100
 1st Qu.:12.00   2: 85   over70 :100
 Median :18.00   3:  5   under45:100
 Mean   :19.85
 3rd Qu.:25.00
 Max.   :50.00
```

Code 11.13.

In relation to tumor size and age, to carry out a one-way ANOVA, we can think of our data as essentially a single variable (tumor size) defined by levels of a factor (age). When we say we have three levels for age, we mean we have three groups, each with 100 cases.

The code for applying an ANOVA to the age groups is shown in Code 11.14.

```
> anova(lm(size.data$size ~ size.data$age))
Analysis of Variance Table

Response: size.data$size
                Df Sum Sq Mean Sq F value    Pr(>F)
size.data$age    2   2714    1357  12.551 5.851e-06 ***
Residuals      297  32111     108
---
Signif. codes:  0 '***' 0.001 '**' 0.01 '*' 0.05 '.' 0.1 ' ' 1
```

Code 11.14.

In the first line of code, we create a linear model object, using the lm() function, by specifying a formula symbolically, where the size.data$size vector is the response variable and size.data$age is a term. When we specify formulas in this way, the response variable always comes first, followed by a tilde character ("~"), followed by the terms or groups. The model is itself parsed to the anova() function and the ANOVA output table displayed. The table shows the within and between group variation, the *F*-statistic, and corresponding P Value. The *F*-statistic tells us that the ratio of between-group to within-group variance is high and that this difference is highly significant.

The nonparametric equivalent to a one-way ANOVA is the Kruskal-Wallis test, which is an extension of the Wilcoxon test and thus based on ranking. If our data was not normally distributed, we could have applied this test as shown in Code 11.15.

```
> kruskal.test(list(size.data$under45, size.data$middleage, size.data$over70))

        Kruskal-Wallis rank sum test

data:  list(size.data$under45, size.data$middleage, size.data$over70)
Kruskal-Wallis chi-squared = 17.6784, df = 2, p-value = 0.0001449
```

Code 11.15.

If there are two factors to consider, a two-way ANOVA may be used. We performed a one-way ANOVA previously to assess tumor size according to the single factor: age. The dataset also contained a third variable called er, and we saw in Code 11.13 that er also contained three levels. We can ask the question, "Is there a difference in tumor size according to age group and ER status?" The null hypothesis would be that there is no difference. To extend the ANOVA to two factors in this way is simple (Code 11.16). We just add a term to the formula in the lm() function, in this case, size.data$er. Note that terms are separated by an asterisk.

```
> anova(lm(size.data$size ~ size.data$age * size.data$er))
Analysis of Variance Table

Response: size.data$size
                           Df  Sum Sq Mean Sq F value   Pr(>F)
size.data$age               2  2714.0  1357.0 12.7259 5.023e-06 ***
size.data$er                2   852.9   426.4  3.9991   0.01935 *
size.data$age:size.data$er  3   121.2    40.4  0.3789   0.76826
Residuals                 292 31136.9   106.6
---
Signif. codes:  0 '***' 0.001 '**' 0.01 '*' 0.05 '.' 0.1 ' ' 1
```

Code 11.16.

A glance at the ANOVA table shows that a two-way ANOVA compares the response variable within individual groups and the interaction of both groups together. There is a significant difference between groups of ER status for tumor size (as there was for age groupings), but not when considering interaction between age and ER.

11.8 Categorical Data Analysis: The Chi Square Test and Fisher's Exact Test

Let us say we have two independent groups of patients that display a certain characteristic. We may be interested in whether the proportion of patients who display the characteristic in each group is the same. For our example, let's ask if the proportion of ER score categories is the same in the young and old breast cancer age groups held in size.data. The first step is to create tables of counts of ER category for each age group and place these counts in the vectors under45.er and over70.er (Code 11.17).

```
> under45.er<-table(size.data$er[1:100])
 1  2  3
57 39  4

> over70.er<-table(size.data$er[201:300])
 1  2  3
84 16  0
```

Code 11.17.

We can then use the matrix() function to create a two-by-three contingency table (Code 11.18). The counts are parsed as a vector to matrix(), and the nrow parameter is set to 2, specifying the number of rows in the table. Take note of the order of the counts in the vector.

```
> con.table<-matrix(c(57,84,39,16,4,0), nrow=2)
> con.table
     [,1] [,2] [,3]
[1,]   57   39    4
[2,]   84   16    0
```

Code 11.18.

The table object called `con.table` can be parsed to the function called `chi.test()` to perform Pearson's chi-square test (Code 11.19).

```
> chisq.test(con.table)

        Pearson's Chi-squared test

data:  con.table
X-squared = 18.7884, df = 2, p-value = 8.32e-05
```

Code 11.19.

The resulting P Value shows that the difference between ER status categories in patients over 70 and under 45 years of age is highly significant.

For contingency tables with lower counts, it is more pertinent to use Fisher's exact test. As a rough rule of thumb, you should use Fisher's exact test if the total count number is less than 50 or if some cells in the contingency table have counts less than 5. Let's return to the TMA breast cancer dataset held in the "breast.data" data frame. We'll ask the question whether her2 categories differ between patients under the age of 45 and over the age of 70. Code 11.20 shows the code for creating a matrix (as we did previously for the chi-square test), and we can see that a number of cells in the contingency table have counts less than 5.

```
> x<-table(breast.dataset$her2[1:21])
> y<-table(breast.dataset$her2[22:42])
> x

 0  1  2  3
15  3  2  1
> y

 0  1  2  3
12  1  2  6
> con.table<-matrix(c(15,12,3,1,2,2,1,6), nrow=2)
> con.table
      [,1] [,2] [,3] [,4]
[1,]   15    3    2    1
[2,]   12    1    2    6
```

Code 11.20.

To carry out Fisher's exact test, we just parse the `con.table` matrix to the function `fisher.test()` (Code 11.21).

```
> fisher.test(c)

        Fisher's Exact Test for Count Data

data:  c
p-value = 0.245
alternative hypothesis: two.sided
```

Code 11.21.

The result suggests that there is no difference in her2 category counts between the two age groups.

11.9 Testing for Association Between Variables: Correlation

To measure the degree of association between two groups, we use correlation analysis. To measure association between two variables in this way requires that the variables contain values measured from the same subjects. If measurements are numerical and we assume there to be a possible linear relationship between variables, we can calculate the Pearson product-moment correlation coefficient, better known as the correlation coefficient (r). If we were to plot the measurements in both variables as a scatterplot and draw a straight line through the points to represent the linear relationship, the value of r tells us how close the points are to the line. Values of r fall within the range -1 through zero to 1. Values close to -1 indicate a strong negative correlation between variables, values close to 1 suggest strong positive correlation, and values close to zero no correlation.

For our example, we will create a hypothetical dataset that includes measurements for age, body mass index (BMI), and systolic blood pressure (mmHg) in males who are categorized as overweight. Measurements are taken for 20 cases and placed into vectors called age, bmi, and systolic (Code 11.22).

```
> age<-c(18,20,21,24,25,28,29,32,33,36,37,38,40,42,44,45,47,47,49,50)
> bmi<-c(23,26,24,25,27,24,30,28,31,30,33,29,32,36,30,38,33,35,34,36)
> systolic<-c(128,132,129,135,130,132,140,137,135,139,140,136,144,
148,136,140,160 ,137,148,151)
```

Code 11.22.

We can create a series of scatterplots (Figure 11.1) to visualize the association of variable pairs in each case (Code 11.23).

```
> par(mfrow = c(1, 3))
> plot(age, bmi)
> plot(age, systolic)
> plot(bmi, systolic)
```

Code 11.23.

The plots reveal possible positive association between all pairs of variables, and we can test the degree of association and significance between each pair using the cor.test() function (Code 11.24).

```
> cor.test(age, bmi)

        Pearson's product-moment correlation

data:  age and bmi
t = 8.1758, df = 18, p-value = 1.796e-07
```

```
alternative hypothesis: true correlation is not equal to 0
95 percent confidence interval:
 0.7329964 0.9550135
sample estimates:
      cor
0.8876074

> cor.test(age, systolic)

        Pearson's product-moment correlation

data:  age and systolic
t = 5.2374, df = 18, p-value = 5.57e-05
alternative hypothesis: true correlation is not equal to 0
95 percent confidence interval:
 0.5098205 0.9075069
sample estimates:
      cor
0.777037

> cor.test(bmi, systolic)

        Pearson's product-moment correlation

data:  bmi and systolic
t = 4.4769, df = 18, p-value = 0.0002915
alternative hypothesis: true correlation is not equal to 0
95 percent confidence interval:
 0.4173837 0.8843188
sample estimates:
      cor
0.725841
```

Code 11.24.

The *r* (cor) values for all comparisons signify strong correlation between each variable. Furthermore, cor.test() calculates a *t*-statistic and P Value, and the correlation for each comparison in this case is significant.

If either of the two variables for analysis does not follow a normal distribution or is ordinal, is small in size, or the relationship is nonlinear, we can calculate the non-parametric equivalent to *r*, called Spearman's rank correlation coefficient. To do this, we still use the cor.test() function but parse the string "spearman" to the method parameter. The default for the method parameter is "pearson," hence our not needing

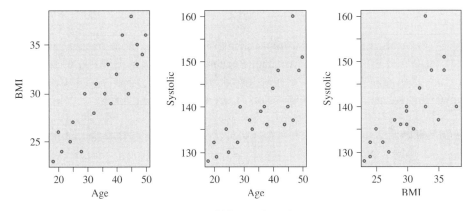

FIGURE 11.1 Scatterplots of pairs of variables: BMI, systolic, and age.

to state this previously. We can apply this test to the ER and PgR data of the TMA age study held in the "breast.dataset" data frame to determine any association in score patterns of patients under the age of 45 (Code 11.25).

```
> cor.test(x,y, method="spearman")

        Spearman's rank correlation rho

data:  x and y
S = 765.83, p-value = 0.02020
alternative hypothesis: true rho is not equal to 0
sample estimates:
      rho
0.5027078
```

Code 11.25.

The correlation coefficient using this test is called "rho," and there is a significant association between ER and PgR status in the under-45 breast cancer cases.

11.10 Describing the Relationship Between Variables: Regression

Correlation can tell you if there is an association between two variables, but it doesn't provide us with information about the relationship or allow us to predict a value of one variable given a value in the other. Regression analysis provides us with the means to do this. In its simplest form, linear regression allows us to model the relationship between a response variable (usually referred to as "Y") and a predictor variable ("X").

In the previous section we created variables for BMI and systolic blood pressure in overweight males and discovered a positive correlation between them. Given this result, we may want to generate a model or linear equation to predict BMI in overweight males given their age (remember that the data was made up). In more technical terms, we can perform a regression of age on BMI where age is the predictor variable and BMI is the response variable. Linear regression fits a straight line through the data using least squares to yield, if there is only a single predictor, an equation of the form:

$$Y = aX + b$$

We have already seen earlier in the chapter how a linear model is generated using lm() and used in analysis of variance. Because the linear model also underlies regression, we again use the lm() function to create a regression model. Code 11.26 shows how we can perform a regression of age on BMI using lm().

```
> reg.model<-lm(bmi~age)
> summary(reg.model)

Call:
lm(formula = bmi ~ age)
```

```
Residuals:
     Min      1Q  Median      3Q     Max
-3.5591 -1.0839 -0.2557  1.6568  4.0570

Coefficients:
            Estimate Std. Error t value Pr(>|t|)
(Intercept) 16.66766    1.72066   9.687 1.45e-08 ***
age          0.38390    0.04695   8.176 1.80e-07 ***
---
Signif. codes:  0 '***' 0.001 '**' 0.01 '*' 0.05 '.' 0.1 ' ' 1

Residual standard error: 2.103 on 18 degrees of freedom
Multiple R-Squared: 0.7878,     Adjusted R-squared: 0.7761
F-statistic: 66.84 on 1 and 18 DF,  p-value: 1.796e-07
```

Code 11.26.

This call to `lm()` produces a linear model stored in an object we have called `reg.model`. After creating the linear model, we can display a considerable amount of information about the model using `summary()`. The first section shows statistics about the residuals. The most important information (that which you will probably be interested in) then follows, showing the regression coefficients, including the intercept and slope and corresponding standard errors, *t*-statistics, and P Values. The third section provides statistics on the variance and residuals.

We can use the linear model object produced to predict BMI scores for new cases. Code 11.27 displays the fitted BMI scores alongside the BMI scores used to create the model, as well as the residual differences in our model. To create this display, we parse a vector for the original ages (`age`) and the original BMI scores (`bmi`) as the first two objects to a data frame. We then parse a function call to `predict()` itself, taking the linear model as input to return the predicted BMI scores. Finally, we parse a function call to `resid()`, again taking the model as input to return the residual scores for the model.

```
> data.frame(age, obs.BMI=bmi, predict(reg.model), residual= resid(reg.model))
   age obs.BMI predict.model.     residual
1   18      23       23.57779  -0.57778886
2   20      26       24.34558   1.65441855
3   21      24       24.72948  -0.72947775
4   24      25       25.88117  -0.88116665
5   25      27       26.26506   0.73493706
6   28      24       27.41675  -3.41675184
7   29      30       27.80065   2.19935186
8   32      28       28.95234  -0.95233703
9   33      31       29.33623   1.66376667
10  36      30       30.48792  -0.48792222
11  37      33       30.87182   2.12818148
12  38      29       31.25571  -2.25571482
13  40      32       32.02351  -0.02350742
14  42      36       32.79130   3.20869999
15  44      30       33.55909  -3.55909261
16  45      38       33.94299   4.05701109
17  47      33       34.71078  -1.71078150
18  47      35       34.71078   0.28921850
19  49      34       35.47857  -1.47857410
20  50      36       35.86247   0.13752960
```

Code 11.27.

If we have more than two predictor variables, we can perform multiple regression in much the same way. As an example, we can use the same dataset to perform a regression of age and BMI scores on systolic blood pressure (Code 11.28).

```
> multiplereg.model<-lm(systolic~bmi+age)
> summary(multiplereg.model)

Call:
lm(formula = systolic ~ bmi + age)

Residuals:
    Min      1Q  Median      3Q     Max
-9.0965 -2.8733 -0.1602  2.4787 14.5210

Coefficients:
            Estimate Std. Error t value Pr(>|t|)
(Intercept) 112.2326    10.8460  10.348 9.36e-09 ***
bmi           0.3087     0.5961   0.518   0.6112
age           0.4906     0.2578   1.903   0.0741 .
---
Signif. codes:  0 '***' 0.001 '**' 0.01 '*' 0.05 '.' 0.1 ' ' 1

Residual standard error: 5.318 on 17 degrees of freedom
Multiple R-Squared: 0.6099,    Adjusted R-squared: 0.5641
F-statistic: 13.29 on 2 and 17 DF,  p-value: 0.0003347
```

Code 11.28.

11.11 Estimating Risk: Odds Ratio and Relative Risk

In epidemiological and clinical research we are often concerned with the risk of a person having a disease given an exposure to a particular etiological factor. In studies evaluating risk, cases are often categorized according to one of four groups (or more), depending on disease status and exposure. A classic example would be evaluating the risk of a person having a certain cancer type if they had or had not been exposed to a particular mutagen in the workplace.

There are two statistics that can be evaluated when estimating risk that often confuse people. The first of these is the odds ratio, and the second is relative risk. Both of these statistics can be calculated in standard ways or robustly estimated using specialized functions in R.

For many medical scenarios, an odds ratio is the ratio of the probability of outcome (disease/death) given an exposure to the probability of outcome without exposure. We can calculate an odds ratio for males who will die in middle age because they smoke relative to those who will die but don't smoke. First, let's find the odds of dying in middle age if we are male and smoke. A Norwegian study in 2006 found that about 40% of male smokers died in middle age (between 40 and 70 years old).[5] The proportion of people dying in this age category who do not smoke is about 14%. Therefore, we can calculate that the odds of dying of smoking-related disease in middle age are:

$$\frac{\text{frequency of smokers who die}}{\text{frequency of nonsmokers who die}} = \frac{0.40}{0.14} = 2.857$$

Similarly, the odds of not dying in middle age for smokers can also be calculated in the same way:

$$\frac{\text{frequency of smokers who live}}{\text{frequency of nonsmokers who live}} = \frac{0.60}{0.86} = 0.698$$

Now that we've calculated the odds for the exposed and unexposed groups of cases, we can calculate the required odds ratio, which is:

2.857 / 0.698 = 4.093

This is not the same as saying that people who smoke have a 4.093 greater probability of dying from disease in middle age than those who don't. There are advantages to using the odds ratio in a study, but a more intuitive measurement that provides probabilistic information is relative risk. The relative risk for the previous example would be calculated as:

$$\frac{\text{smokers who die / (smokers who die + smokers who live)}}{\text{nonsmokers who die / (nonsmokers who die + nonsmokers who live)}}$$

Therefore, for relative risk, we compare the probability of death in smokers to the probability of death in nonsmokers in this age group, which is the risk ratio.

Using actual figures for the previous example, we will apply functions from the epitools library called `oddsratio()` and `riskratio()` to estimate both the odds ratio and relative risk. We will create a dataset for 1000 people, of which 250 are smokers (25% estimate for the population), and base the frequencies of mortality in middle age on the figures we used previously. In our four groups of people, there are 100 smokers who die in middle age (0.4 × 250), 150 smokers who live (0.6 × 250), 105 nonsmokers who die (0.14 × 750), and 645 nonsmokers who live (0.86 × 750). We can put this information into a matrix and apply labels accordingly (Code 11.29).

```
> data<-matrix(c(100,105,150,645),2,2)
> dimnames(data)<-list(Smoker=c("yes","no"),Death=c("yes","no"))
> data
       Death
Smoker yes  no
   yes 100 150
   no  105 645
```

Code 11.29.

The input format of the contingency table to functions in epitools needs consideration. For `oddsratio()`, the 2 × 2 table in our example should be in the format shown in Code 11.30. For the `riskratio()` function, both the columns and rows should be reversed, as we will see.

To estimate an odds ratio, you simply need to parse the matrix called `data` to `oddsratio()` and select one of four robust methods: median-unbiased ("midp"), conditional maximum likelihood ("fisher"), unconditional maximum likelihood estimation ("wald"), or small sample adjustment ("small"). The example in Code 11.30 shows an odds ratio estimated using maximum likelihood (wald).

```
> library(epitools)
> oddsratio(data, method="wald")
$data
        Death
Smoker  yes  no Total
  yes   100 150   250
  no    105 645   750
  Total 205 795  1000

$measure
        odds ratio with 95% C.I.
Smoker estimate     lower     upper
  yes  1.000000        NA        NA
  no   4.095238 2.954653 5.676123

$p.value
        two-sided
Smoker midp.exact fisher.exact    chi.square
  yes         NA           NA            NA
  no           0 5.020675e-17 1.156091e-18

$correction
[1] FALSE

attr(,"method")
[1] "Unconditional MLE & normal approximation (Wald) CI"
```

Code 11.30.

The output provides the contingency table with odds ratio and lower/upper confidence intervals. Note that if the lower confidence interval was close to or less than one, we would have little confidence in the result being significant. A P Value is also provided for the Fisher's exact test applied to the contingency table, which again shows the result to be highly significant.

To estimate relative risk, we use the riskratio() function. The rows and columns of the data matrix must be reversed if they are in the format shown in Code 11.29. To do this, the matrix is parsed and the rev parameter is set to "both" (i.e., both columns and rows are reversed). We can also select one of two methods for risk ratio estimation: unconditional maximum likelihood estimation ("wald") or small sample adjustment ("small"). The example in Code 11.31 actually shows no method selected, meaning the default option of "wald" is automatically chosen.

```
> riskratio(data, rev="both")
$data
        Death
Smoker   no yes Total
  no    645 105   750
  yes   150 100   250
  Total 795 205  1000

$measure
        risk ratio with 95% C.I.
```

```
Smoker estimate     lower    upper
   no  1.000000        NA       NA
  yes  2.857143  2.262214  3.60853

$p.value
      two-sided
Smoker midp.exact  fisher.exact   chi.square
   no          NA            NA           NA
  yes           0  5.020675e-17  1.156091e-18

$correction
[1] FALSE

attr(,"method")
[1] "Unconditional MLE & normal approximation (Wald) CI"
```

Code 11.31.

The results assume the same format as that for the `oddsratio()` function, and here we note a significant estimated odds ratio of 2.857.

11.12 Measuring Agreement: Cohen's Kappa Test

Biomedical data can often take the form of sets of human-generated scores, such as results from pathology or psychiatric experiments. These scores can reveal a degree of subjectivity, which could affect the overall conclusions drawn from an experiment or test. In some situations there will be more than one person generating scores, and there may be a requirement to make an assessment about each scorer's (rater's) ability to provide reliable results. Alternatively, there may be a need to simply obtain a consensus result from a number of raters and provide a measurement of how well each rater agrees.

If the data is categorical or ordinal, one can generate Cohen's kappa coefficient as a statistical measure of inter-rater reliability. In R we can generate Cohen's kappa coefficient for two raters using the `kappa2()` function in the `irr` package. This value falls in the range of 0 to 1, with a score of 1 being complete agreement between both raters. `kappa2()` also provides a kappa statistic and P Value for the kappa coefficient.

As an example, we will create a dataset of three vectors of categorical scores (1, 2, or 3) for nuclear staining of an antibody for a protein that is a marker for tumor proliferation. The scores are generated by two experienced pathologists and a student who has no previous experience in scoring for this antibody aside from reading a training manual. The three raters each score 20 cases and record their scores as vectors (Code 11.32). The vectors are then put into a data frame called scores.

```
> library(irr)
> pathologist1<-c(1,1,2,1,1,3,2,1,1,1,2,1,1,2,2,2,1,1,1,3)
> pathologist2<-c(1,1,2,1,1,1,2,1,1,1,2,1,1,2,2,2,1,1,1,3)
> student<-c(1,2,1,1,2,1,3,2,1,1,2,1,1,1,2,2,1,1,1,2)
> scores<-data.frame(pathologist1, pathologist2, student)
```

Code 11.32.

We can now generate kappa coefficients, statistics, and P Values for raters on a pairwise basis by parsing the data frame to `kappa2()` and specifying the two relevant columns each time (Code 11.33).

```
> kappa2(scores[,c(1,2)])
 Cohen's Kappa for 2 Raters (Weights: unweighted)

 Subjects = 20
   Raters = 2
    Kappa = 0.903

        z = 4.93
  p-value = 8.25e-07
> kappa2(scores[,c(1,3)])
 Cohen's Kappa for 2 Raters (Weights: unweighted)

 Subjects = 20
   Raters = 2
    Kappa = 0.245

        z = 1.33
  p-value = 0.182
> kappa2(scores[,c(2,3)])
 Cohen's Kappa for 2 Raters (Weights: unweighted)

 Subjects = 20
   Raters = 2
    Kappa = 0.303

        z = 1.57
  p-value = 0.117
Kappa; kappa2
```

Code 11.33.

The results show that the two pathologists have a kappa coefficient close to 1, which is highly significant (P = 8.25E-07), indicating almost complete agreement. The student's attempt, however, was not particularly good, where there was little agreement with either pathologist, producing kappa coefficients of less than three in both cases and the results being nonsignificant.

11.13 Summary

This chapter has provided an overview of some of the more basic inferential statistical approaches used in medicine. It can often be confusing as to which is the most appropriate test to apply to one's data, and the reader is strongly advised to "know" the data by type and distribution using descriptive statistics first. This chapter hopefully provided a springboard for readers to explore a number of statistical approaches that can be applied to their data. R is a statistical environment, and there are a number of excellent introductory and specialized texts for statistical analysis using this software and a wealth of quality online tutorials.

11.14 Questions

1. How does inferential statistics differ from descriptive statistics?

2. Why is it important to know the variance and standard error of a statistic for sample data? How do we calculate these values in R?

3. What is the reason we apply a Wilcoxon signed-rank test to a single sample of data as opposed to a one-sample *t*-test?

4. In addition to the sample data as a vector, what other parameter must we parse to t.test() for a single-sample *t*-test, and what does the value of the parameter represent?

5. What are the differences between the varieties of two-sample *t*-tests, and how can we tell R how to apply the correct test using t.test()?

6. What is the R function for a nonparametric equivalent to a two-sample *t*-test?

7. Under what circumstances should you use fisher.test() with your categorical data?

8. Under what circumstances should you apply Spearman's correlation to your data as opposed to Pearson's correlation, and how is this achieved in R?

9. What is special about the first argument of the lm() function when performing linear regression on your data?

10. Why are the prerequisites the same in performing a chi-square test and calculating an odds ratio?

12 Writing Functions in R

12.1 Background

Given the enormous power of R as a statistical environment and the wealth of packages available, you may be wondering why you would have to write your own functions. You may, of course, never have to, depending on your requirements. In preceding chapters you have already experienced using many functions for all types of problems, such as `mean()`, `t.test()`, `dbGetQuery()`, and `library()`. You also know that to get a result back from a function, you will usually have to "parse" one or more objects as arguments to it.

To communicate with R, you use code of a defined syntax. R may be a statistical environment, but in essence, it is a complete programming environment. When you call a function, you're just writing a line of code in the console, e.g., "data<-mean(x)." When you call a function, you are simply accessing code already written for you but available as a function object with which you can communicate. Because a function is made up of code that is executed, we can think of it as a program. Indeed, we can use programming control statements in our user-defined functions, as we shall see.

Use of R generally involves typing in and evaluating series of statements or function calls at the command line. You often find that you use the same sets of statements over and over again, and it would be of great benefit to automate the input of repetitive code by placing the code in a function. A function might be a complete algorithm for statistical analysis, a series of commands or other function calls for setting and displaying options in a graphic, or even just a few lines of code that allow you to automate a process. All of the functions you have encountered so far have been built into the standard packages that come with R or additional packages that you have downloaded. When you create your own user-defined functions, you will not have to create packages for them, but just save them as files that can be loaded whenever required.

The fact that you can piece together many lines of code in a function means that you can iteratively loop through code using control statements. Other control statements allow you to create conditional structures. For those who are familiar with a programming language, such as C, control statements should be well known. The inexperienced can be reassured that the concepts and use of control statements are easy to grasp.

In this chapter we will look at how we can create our own functions and introduce some control statements with straightforward R programming syntax.

12.2 The Architecture of a Function

We can divide a function into three parts: the function *name*, the list of *arguments* for passing information to the code on which the function acts, and the lines of code that actually perform the function, called the *expression*. The following is the general syntax for a function in R:

```
name<-function(argument1, argument2…){expression}
```

The function is then called using the defined name. Let's have a look at the function called `var()` in the standard stats package for calculating variation in a numerical vector. To view the code for a function, you can type the function name (e.g., `var`) at the R command line and it should display the function definition and lines of code (Code 12.1).

```
> var
function (x, y = NULL, na.rm = FALSE, use)
{
    if (missing(use))
        use <- if (na.rm)
            "complete.obs"
        else "all.obs"
    na.method <- pmatch(use, c("all.obs", "complete.obs",
"pairwise.complete.obs"))
    if (is.data.frame(x))
        x <- as.matrix(x)
    else stopifnot(is.atomic(x))
    if (is.data.frame(y))
        y <- as.matrix(y)
    else stopifnot(is.atomic(y))
    .Internal(cov(x, y, na.method, FALSE))
}
<environment: namespace:stats>
```

Code 12.1.

Before we examine the `var()` function, look at the last line in Code 12.1 that reads "`<environment: namespace:stats>`." The "namespace" for a function informs us of the package in which the function resides. This is important when you have packages loaded that have the same function name. A common example is a function in a package called "plot" that has the same name as the `plot()` function in the standard graphics package. R will give one function priority over another with the same name, but any function can be called using the name preceded by the namespace and two colons:

```
stats::mean(x)
```

After we type "var" in the console, the first line we see displayed is the function definition. For `var()`, we see four arguments: `x`, `y`, `na.rm`, and `use`. If we type "?var" at the command line to call up the HTML help file for `var()`, we can find the description for each of these arguments.

Some arguments have a default value specified, such as "y = NULL." Now look at the code that is contained within the curly brackets (called braces). We won't detail what each line of code does, but notice how the arguments are used by different commands and other function calls within var(). You'll also notice commands sandwiched between statements, including "if" and "else," which form control statements for processing objects, either parsed as arguments to the function or created by code within the function. In the next few sections we'll briefly introduce control statements and how they can be used to control program flow.

12.3 Control Statements: *if-else, ifelse*

The if-else statement is used to evaluate a condition. If that condition is TRUE, a statement is evaluated. If the condition is FALSE, a different condition is evaluated. A simple example is shown in Code 12.2. We create a vector called a to hold the value of 20. The second line then evaluates the condition "a>5." If the condition is TRUE, the value of a is printed to the console. Otherwise, the string "a is less than 5" is printed. Because the value of a is 20 initially, "a" is printed. When we then assign the value of 2 to a and evaluate the condition using the same if-else statements, we see "a is less than 5" printed.

```
> a<-20
> if(a>5) print(a) else print("a is less than 5")
[1] 20
> a<-2
> if(a>5) print(a) else print("a is less than 5")
[1] "a is less than 5"
```

Code 12.2.

If-else statements can also be nested, as shown in Code 12.3. Here, given the nature of the nested statements, we put the lines of code into a function called f. The only argument to the function is the vector called a, and a is then evaluated by the nested if-then statements. The first if statement checks if a is less than five and if it is prints "a." If not, a is evaluated by the next if statement to see if its value is less than 10. If the result is FALSE, the string "a is greater than or equal to 10" gets printed.

```
> f<-function(a){
    if(a<5) print("a is less than 5")
    else
    if(a<10) print("a is greater than 5 but less than 10")
    else print("a is greater than or equal to 10")
  }
> f(2)
[1] "a is less than 5"
> f(7)
[1] "a is greater than 5 but less than 10"
> f(20)
[1] " a is greater than or equal to 10"
```

Code 12.3.

One drawback of using if-else is that the condition to be evaluated can only be a vector of length one. To evaluate a vector of greater than length one, you can use ifelse (Code 12.4). The vector a contains eight values from one to eight. The ifelse statement has three arguments: the condition, the condition if TRUE, and the condition if FALSE. The ifelse statement evaluates whether a is less than or equal to five and if TRUE, prints the number; if FALSE, it prints "greater than 5."

```
> a<-c(1:8)
> ifelse(a <= 5, a, "greater than 5")
[1] "1"  "2"  "3"  "4"  "5"  "greater than 5"  "greater than 5"
[8] "greater than 5"
```

Code 12.4.

12.4 Loop Statements: *for, while, repeat*

If you wish to apply an operation to a vector with more than one element, often, depending on the function, R will automatically evaluate all values in the vector for you. An example is the evaluation of the square root of a vector of length eight shown in Code 12.5.

```
> a<-c(1:8)
> sqrt(a)
[1] 1.000000 1.414214 1.732051 2.000000 2.236068 2.449490 2.645751 2.828427
```

Code 12.5.

However, there are many situations where you would have to use a "loop" construct to evaluate multiple functions or statements. The example shown in Code 12.3 is a case in point, where a vector of more than one element being parsed to the function f would result in only the first element being evaluated. (Try it.) Say we had a vector of eight integers, from one to eight, in which we wished to print the square root of numbers between three and six inclusive. We could use a nested if-else construct like that of Code 12.3, but rather than keep typing out the lines of code for each integer, we could wrap the code in a for loop (Code 12.6).

```
> f<-function(a){
   for(i in 1:length(a))
   {
     if(a[i]<3) print("NA")
     else
       if(a[i]<7) print(sqrt(a[i]))
     else print("NA")
   }
  }
> f(a)
[1] "NA"
```

```
[1] "NA"
[1] 1.732051
[1] 2
[1] 2.236068
[1] 2.449490
[1] "NA"
[1] "NA"
```

Code 12.6.

The for loop works by looping through and evaluating the lines of code held between the curly brackets (in practice, you only need to use curly brackets to contain code in a for loop if more code follows the construct). The syntax for the expression in for is simple. The number of iterations in the for loop must be specified. In the example, it is the length of vector a, which is 8. As R works through the loop, the number of the current iteration is held in an integer vector called i. This is useful in our example, as the i vector allows us to specify elements in the vector we wish to evaluate.

Another loop function that is similar to for is while. The while loop allows you to evaluate the expression provided until the condition is no longer TRUE (Code 12.7).

```
> a<-1
> while(a<4)
  {
    print(sqrt(a))
    a<-a+1
  }
[1] 1
[1] 1.414214
[1] 1.732051
```

Code 12.7.

Another similar looping function is repeat (Code 12.8). You don't provide an expression with repeat that will iterate through the lines of code indefinitely. Therefore, you should provide a break in the code when a certain condition is met.

```
> a<-1
> repeat
  {
    print(sqrt(a))
    if(a>5) break
    a<-a+1
  }
[1] 1
[1] 1.414214
[1] 1.732051
[1] 2
[1] 2.236068
```

Code 12.8.

12.5 Saving and Loading Functions

Aside from typing a function at the R console, you can write the lines of code using a text editor, such as Notepad, or more conveniently using the built-in text editor in R. To use the built-in editor, select "New script" from the File menu in R. The text editor window should open, and you can write your code. To execute the code, right-click the Editor window and select the "Run line or selection" option (or press the ctrl and R keys simultaneously). Whether you have created a function or just typed a few lines of code, you have created an R script. To save the script, click the File menu, select the "Save as…" option, and choose a file name and destination. To load a script into R, you can either use the "Open script…" option in the File menu or use the source() function directly from the console:

```
> f<-source('script.r')
```

Ensure that the correct working directory that contains the script is preselected before attempting to load it.

12.6 Summary

The ability to write your own functions and use control statements gives the same power to R as a programming language. User-defined functions allow the automation of repetitive code and the ability to develop your own algorithms. Control statements allow conditions to be set in code and iterative looping through lines of code.

12.7 Questions

1. Would you define R functions as simple or complex structures?

2. What are the three fundamental components of an R function?

3. How is code within an R function enclosed?

4. What R code might you use to print the word "True" to the screen if a random number between 1 and 10 is less than or equal to 5, or "False" if the value is greater than 5?

5. What are the differences between the for, while, and repeat loop statements?

13 | Multivariate Analysis in R

13.1 Background

Datasets that contain more than two dependent variables are multivariate. Modern technologies in biomedicine often produce multivariate datasets. In previous chapters we have seen a good example of a multivariate dataset for tissue microarray containing data for multiple proteins scored for more than 40 patients. Multivariate datasets are also multidimensional in structure, where each dependent variable represents one dimension. To analyze multivariate datasets, we need methods that can deal with this multidimensionality. There are many such methods available and they are widely applied. The literature abounds with publications about gene expression microarray data where the authors simply would not have generated meaningful patterns of knowledge from their data without multivariate analytical methods. Many textbooks on multivariate analysis are available; the one by Hair is an excellent introduction for the novice.[6]

Applying multivariate analysis methods to a dataset is all about gathering information concerning the underlying structure of the variables. By "structure," we mean relationships or similarities between variables. The structure of a multivariate dataset is either known or unknown prior to analysis, and this knowledge (or lack of) dictates what multivariate analysis methods are used and in what context.

Consider a gene expression microarray dataset where there are expression values for thousands of genes for 50 patients with a particular disease. The investigators may know little about the disease in terms of the biology and thus very little about the underlying structure of the data. They may wish to apply multivariate analysis methods to give insight into how many potential groupings of patients may exist based purely on the gene expression values. The results could show potential subgroups of patients that differ biologically. The investigators would essentially be "exploring" their data to determine these underlying groups. Thus, one use of multivariate analysis is to provide knowledge about the previously unknown structure of the data, which allows you to generate hypotheses for further testing. Following on from this, a second use of multivariate analysis is to apply a priori knowledge of the structure to a dataset to classify new objects (patients, genes, protein scores, etc.) into groups.

Multivariate analysis methods that are widely used in this context tend to fall into one of two domains: cluster analysis and data reduction. Cluster analysis methods attempt to split a whole dataset of objects into distinct subsets or groups (clusters). The degree of similarity between objects in the same cluster is much stronger than between objects of different clusters. Cluster analysis methods can be hierarchical or partitional. Data reduction methods are applied to multivariate datasets to reduce the dimensionality of the data, making the structure easier to understand. Dimension reduction is achieved by combining similar variables in a dataset, resulting in fewer variables that are independent of each other. In a less technical sense, you can think of these explanatory variables as groups of variables that give insight into the underlying structure of the data.

In this chapter we will explore the use of both cluster analysis and data reduction methods. For simplicity, we will apply methods to the breast cancer age-associated TMA dataset, whereby each method will determine underlying structure and ultimately the number of patient subgroups that differ by biology. This dataset is a small but typical biomedical multivariate dataset, and the methods described can easily be applied in many other domains.

13.2 Planning a Multivariate Analysis of the TMA Dataset

In Chapter 9 we assumed the role of a medical oncologist who had an interest in evaluating whether any age-associated patterns of differences exist between young and elderly breast cancer patients for the levels of two cell-cycle protein markers called "proteinA" and "proteinB" (see Section 9.2). In that chapter we created a truly multivariate dataset that included protein expression scores for 42 grade 3, node-negative breast cancer patients. Twenty-one patients were under the age of 45 and the remaining patients were over the age of 70. You should have saved the dataset as "breast_tma_age.txt" or, alternatively, use the file of the same name that accompanies this book.

As the medical oncologist, you could now explore the variables in this dataset using multivariate analysis to try to identify different biological age-related patient subgroups. In Chapter 11 we saw that there may be a significant difference for proteinB scoring intensities between old and young patients; however, this gave no insight into how many patient groups exist. In addition, there is no information regarding potential differences for proteinA or any of the other variables, such as tumor size and her2. Thus, you could take a purely exploratory approach to the multivariate analysis, where no assumptions are made about the structure of the data.

Before deciding on any methodological strategies, let's remind ourselves of the variables in the dataset available for analysis. Load the "breast_tma_age.txt" file into an object called `data`, ensuring first that you have set the working directory to the folder where the file resides:

```
> data<-read.table("breast_tma_age.txt", header=T)
```

To list the variables, we use the `names()` function (Code 13.1).

```
> names(data)
 [1] "hash_strings"  "er"             "pgr"        "size"
 [5] "er_annotation" "pgr_annotation" "proteinA"   "proteinB"
 [9] "her2"          "label"
```

Code 13.1.

Because the dataset contains two age-related groups of patients, it would be useful to order the data accordingly by group:

```
> data<-data[order(data$label),]
```

At this point you would need to give careful consideration as to what variables to include in the exploratory analysis. The variables proteinA and proteinB are of key biological interest, so will definitely be included in the study. In addition to proteinA and proteinB, our dataset contains variables for intensity scores on the estrogen receptor protein (ER), progesterone (PgR) receptor protein, and the her2 protein, as well as data for tumor size. Although we are primarily interested in any association between age and proteinA and proteinB, we should not rule out the possibility that one or more of the other variables may have an association.

For the sake of simplicity, we will just look at proteinA and proteinB data along with tumor size. All of these variables are of a numeric type. As we will see in later sections, the variable types dictate how we can combine them for multivariate analyses and what methods we can apply. It is far more straightforward to perform a multivariate analysis on numerical type datasets, although we can apply certain methods to mixed numerical and categorical data. An example will be given toward the end of the chapter where we will include the her2 variable in the dataset.

In readiness for the sections that follow, create a data frame for the numerical variables proteinA, proteinB, and tumor size:

```
> ndata<-data[,c(4,7,8)]
```

The proteinA and proteinB variable scores are on a scale from 0 to 100, but the scale for tumor size differs in that the range is smaller. Before we analyze the data, we need to make a decision about whether we should standardize the data. Standardizing the data would have the effect of making the scores of each variable unit less, thus making scores comparable. We can do this using the scale() function:

```
> ndata.sc<-scale(ndata)
```

The scale() function first subtracts the column mean from each value in that column. This is achieved when a parameter called center is set to TRUE, which it is by default. Second, the values in each centered column are scaled by dividing the values in the column by their root mean square. These options can be switched off by setting to FALSE.

We also need to set up the character vector to hold abbreviated labels to use for the cases (patients), as these will be useful for space saving in plots. Although the case labels in the dataset are relatively short ("over70" and "under45"), we can condense the labels to signify a case as old ("O") or young ("Y"). We can also give the label a unique number according to the position of the case in the ordered dataset. We create a vector called labels and use the paste() function to supply the label characters and then assign these labels to the row names of the ndata data frame using row.names() (Code 13.2).

```
> labels<-c(paste("O", c(1:21), sep=""), paste("Y", c(22:42), sep=""))
> labels
 [1] "O1"  "O2"  "O3"  "O4"  "O5"  "O6"  "O7"  "O8"  "O9"  "O10" "O11" "O12"
"O13" "O14" "O15" "O16" "O17" "O18" "O19" "O20" "O21" [22] "Y22" "Y23" "Y24"
"Y25" "Y26" "Y27" "Y28" "Y29" "Y30" "Y31" "Y32" "Y33" "Y34" "Y35" "Y36" "Y37"
"Y38" "Y39" "Y40" "Y41" "Y42"
> row.names(ndata)<-labels
```

Code 13.2.

Now we are ready to think about what analysis methods to apply. There are no hard and fast rules about choosing a method, and there is no harm in applying as many as you want. After all, our requirement is to explore the data and gather as much information about the structure as possible. For the analysis of this numerical dataset, the first step in our strategy will involve use of hierarchical cluster analysis (HCA) to obtain a feel for the number of potential patient groups that exist in the data. We will then explore the data structure further by using data reduction methods, including principal components analysis. These methods will provide useful visualization aids for structure determination but won't confirm the number of patient subgroups. Subgroup number confirmation will finally be obtained using partition clustering methods. We will take a similar approach for the mixed type dataset, but will not be able to use the range of methods available for numerical data.

Many packages are available in R that allow you to perform multivariate analyses. We will stick to the mainstream packages available, as we will be applying widely used analysis methods. The reader is encouraged, however, to explore the use of other clustering methods available in different packages. The packages used in this chapter are shown in List 13.1.

List 13.1 Packages with multivariate analysis functions used in this chapter.

- stats
- cluster
- randomForest
- RColorBrewer
- vegan
- labdsv

Although the stats package comes preinstalled, you will need to install the other three:

```
>install.packages(c("stats","cluster","randomForest",
"RColorBrewer", "vegan","labdsv"))
```

One final note on the use of terminology in multivariate data analysis. For the rest of this chapter, we will use the term "object" to refer to the variables we are clustering. Thus, if we are clustering patients, then these are our objects.

13.3 Hierarchical Agglomerative Cluster Analysis

HCA methods can be "agglomerative" or "divisive." Agglomerative HCA works by treating each case as a cluster and then building ever-larger clusters and subclusters according to how similar different cases are. The algorithm is a two-step process. First, a similarity (or dissimilarity) matrix is built between all cases, where the values in the similarity matrix correspond to distance measures between a pair of cases. Distances can be some mathematical measurement, such as Euclidean or correlation. These measurements are then used to gradually build clusters of cases. The clusters can be visualized as a hierarchical tree structure or dendrogram. This second step uses one of a number of tree building algorithms. If two cases are linked together in a cluster, they are more similar to each other than they are to cases in other clusters. The main steps involved in HCA are summarized in List 13.2.

List 13.2 Main steps involved in the HCA process.

- Calculate a distance matrix for all pairs of cases.
- Apply a tree building algorithm to distance matrix.
- Visualize trees for objects with a heatmap, if required.
- Estimate numbers of clusters (groups) from trees.

We stated previously that we would only use HCA to give us a "feel" for the data structure, which might give the impression that we don't possess a lot of confidence in the method. HCA is useful as a first analysis method, but there are a number of problems that should be considered. In applying HCA, you can use a number of types of distance measures and one of many tree building procedures. Often, though, you will get slightly different results in the way cases will group. A second problem is that HCA will build a tree from any dataset. You could give HCA a dataset of completely unrelated data and it would still give you a tree!

Before you start to apply HCA to your data, you need to think about what distance measure and tree building algorithms you will use. Unfortunately, people often try one combination and then another until they achieve the sort of result they want without any consideration of the data types. The following sections will outline some of the options available in the more commonly used R functions (although be aware that there are other specialized packages for cluster analysis).

13.3.1 Create a Distance Matrix

As we shall see, the chosen distance metric will depend on the type of data. It is no use applying Euclidean distance or correlation to nominal data, as the distance would be meaningless. You can usually pick and choose distance functions and tree building functions from different packages, depending on your needs. For instance, you can use distance measure functions from the cluster package and tree building methods from the stats package. The most commonly used distance metric function in R is called `dist()` and can be found in the base stats package. The actual metrics available from the `dist()` function for creating a distance matrix are shown in List 13.3.

List 13.3 Distance metric methods available in `dist()`.

- euclidean

- maximum

- manhattan

- canberra

- binary

- minkowski

Distances are calculated between rows of the matrix. The chosen distance metric can be parsed to `dist()` by specifying the method parameter. Create a distance matrix called `ndist` for the numerical dataset where the required distance metric is euclidean (this is actually the default):

```
> ndist<-dist(ndata.sc, method = "euclidean")
```

We have chosen to use the standardized dataset, as we are unsure what effects the raw scores would have when calculating distance. In other words, we are avoiding creating a distance matrix that is influenced by one object having more weight.

13.3.2 Create a Hierarchical Tree

Once you have a distance matrix object, you can provide this as input to the tree building function. The stats package also contains the most common R function for tree building, called `hclust()`. The methods available in `hclust()` are shown in List 13.4.

List 13.4 HCA tree building methods in `hclust()`.

- ward

- single

- complete

- average

- mcquitty

- median

- centroid

These methods generally differ according to how they link clusters together. For example, Ward's method attempts to create compact spherical clusters (type "?hclust" for further details). The average linkage method will link together clusters with the shortest average difference between cases in each cluster. The method of choice can be parsed to `hclust()` by specifying the method parameter (in this case, `average`) along with the distance matrix object (`ndist`):

```
> ntree<-hclust(ndist, method="average")
```

You can now visualize the tree using the `plot()` function:

```
> plot(ntree, hang=-1, cex=0.5, xlab="Cases")
```

We parse the `hclust` tree object (`ntree`) and set parameters to improve the look and feel of the plot. The `hang` parameter set to -1 aligns the labels in a neat straight line along the bottom of the plot. Label character size is set using `cex`.

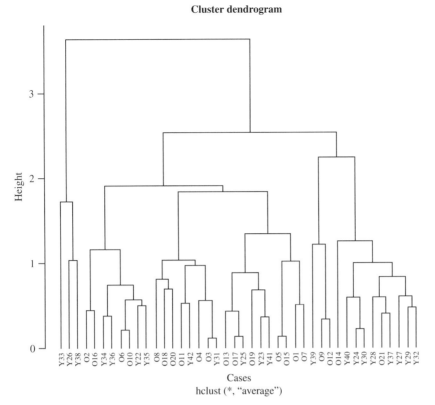

FIGURE 13.1   Hierarchical cluster analysis tree (dendrogram).

The resulting tree is shown in Figure 13.1. The scale on the y-axis shows the distance between cases and clusters of cases. How many clusters or groups of patients are there? There certainly appear to be three clusters of cases comprised of one small cluster of three cases (Y26, Y33, and Y38) and two large groups, each having smaller subclusters. You can, of course, follow all clusters and observe how they split into smaller and smaller subclusters, and it can be subjective as to how many clusters exist. On closer inspection, if we drew a horizontal line at about 1.8 on the y-axis, we would yield six subclusters of cases. We will use more robust methods to determine how many potential groups of cases we have, but there is a method called `cutree()` that you can use to assign cases to a defined number of clusters in your tree (Code 13.3).

```
> membership<-cutree(ntree, k=6)
> membership
 01  02  03  04  05  06  07  08  09 010 011 012 013 014 015 016 017
  1   2   3   3   1   2   1   3   4   2   3   4   1   5   1   2   1
018 019 020 021 Y22 Y23 Y24 Y25 Y26 Y27 Y28 Y29 Y30 Y31 Y32 Y33 Y34
  3   1   3   5   2   1   5   1   6   5   5   5   5   3   5   6   2
Y35 Y36 Y37 Y38 Y39 Y40 Y41 Y42
  2   2   5   6   4   5   1   3
```

Code 13.3.

The `cutree()` function takes the `hclust` tree object as a parameter along with a number specifying the numbers of clusters (*k*) that you want to split the tree into. Here we have asked `cutree()` to split the tree into six clusters, and you can see from the output that cases fall into one of the six clusters, where the cutoff height is roughly 1.9. This tree doesn't give us any clue as to why the cases cluster as they do. We can use a more sophisticated visualization tool, called a heatmap, in combination with our tree to reveal the patterns of object values associated with each cluster.

13.3.3 Combining the Tree with a Heatmap

A heatmap is a color or greyscale chart where colors represent the values of the objects for each case. The order of objects in the heatmap will depend on the dendrogram. There is a function called `heatmap()` in the stats package where the whole HCA procedure for numerical datasets can be accomplished using just one line of code. This is because the `heatmap()` function allows the `dist()` and `hclust()` functions to be parsed as arguments. The color palettes are pretty basic though, and you can improve on these using functions from the RColorBrewer package.

Before we draw the heatmap, we can create a nice greyscale color palette (Code 13.4).

```
> library(RColorBrewer)

# create an object to hold the low-end color of the palette
> low <- col2rgb("black")/255

# create an object to hold the high-end color of the palette
> high <- col2rgb("white")/255

# create vectors for the range of red, green, and blue values in the
# palette - note these ranges will give shades of grey.
> r <- seq(low[1], high[1], len = 50)
> g <- seq(low[2], high[2], len = 50)
> b <- seq(low[3], high[3], len = 50)
> pallette<-rgb(r, g, b)
```

Code 13.4.

The second and third lines in Code 13.4 create two objects that hold the low- and high-end colors of the palette, which in this case are black and white, respectively. The palette used in `heatmap()` follows the RGB color system and can be varying shades of red, green, and blue. We will use greyscale though for our palette, and so we create three vectors for r, g, and b, which actually hold varying shades of grey.

With the heatmap color palette object set up, we are ready to create the heatmap function and trees for both case and variable objects (Code 13.5).

```
> ndata.mat <- as.matrix(ndata.sc)
> ntree.map <- heatmap(ndata.mat, col = pallette, distfun =
function(ndata.mat)dist(ndata.mat), hclustfun =
function(ndata.mat)hclust(ndata.mat, method="average"), labRow = labels,
cexCol = 0.5, cexRow = 0.6, scale= "column", margin=c(6,2))
```

Code 13.5.

The first line creates a data matrix from the `ndata` data frame because `heatmap()` expects a matrix as input. The second line looks a bit complex, but is quite straightforward. First, note the syntax in `heatmap()` for parsing the `dist()` and `hclust()` functions as arguments. Each function name is preceded by the term `function(x)`, where `x` is the data matrix, and the arguments to these functions are also specified. For our analysis we have chosen average linkage as our tree building algorithm. The graphic parameters include `cexCol` and `cexRow`, which allow you to adjust the size of the label characters for both types of objects you have clustered (i.e., patient IDs and variable).

The resulting dendrograms and heatmap are shown in Figure 13.2. The heatmap colors range from white to black and display shades of grey. We know from Code 13.4 that black represents low scores for the numerical values of proteinA, proteinB, and tumor size. Similarly, white represents high scores. There are different shades of grey for each object, where the scores become higher as the grey shade becomes lighter.

We see a horizontal dendrogram on the left of the heatmap showing the clusters of cases, and a vertical dendrogram above showing the clustering of objects. The most meaningful dendrogram is that of the cases, and you can see that the tree is identical by the way cases cluster to that shown in Figure 13.1, even if the order appears at first sight to be different. Glance across to the heatmap alongside the tree, and patterns immediately jump out that correspond to the cluster patterns in the tree.

Working from the top of the cases tree, we can see that clusters exist according to the staining patterns of proteinA, proteinB, and tumor size. The top cluster with cases

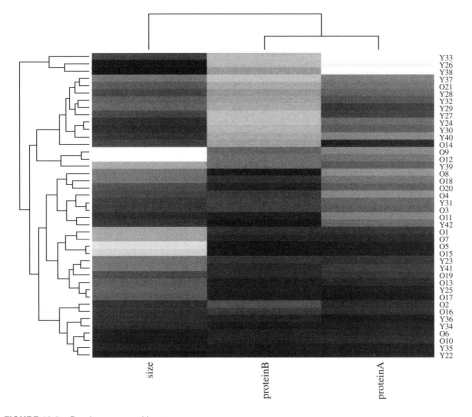

FIGURE 13.2 Dendrograms and heatmap.

Y33, Y26, and Y38 is distinct from other clusters due to high proteinA expression, higher proteinB expression, and small tumor size. The next cluster sees cases group by high proteinB expression. The three cases in the third cluster down differ from the second cluster due to larger tumor size. The fourth cluster has higher proteinA expression but a lower proteinB expression. The last two clusters both have lower proteinA and proteinB expression and a smaller tumor size. It appears as though clusters of cases occur primarily due to proteinA and proteinB staining followed by tumor size.

Let us now look at how these patterns relate to patient age categories. We won't attempt at this stage to draw firm conclusions from the HCA results about whether age-related patterns exist, but certain associations may be apparent. For instance, the second cluster contains ten cases, eight (80%) of which are elderly. Although we may be tempted to look for age associations with the smaller subclusters, this is pointless because we are going to apply more robust multivariate analysis methods to determine the optimal number of groups.

13.4 Exploring the Data Structure Further with Principal Components Analysis

PCA is a statistical method routinely used to analyze interrelationships among large numbers of objects, revealing common underlying factors or components. PCA examines the correlations between the original objects and condenses the information contained within these objects into a smaller group of components with minimal loss of information. Thus, in our study, PCA could reduce a large group of individual objects (the patients in this case) into smaller groups (components) of objects related according to their expression patterns for proteins along with tumor size.

The following is a brief semitechnical overview of PCA, but don't be put off by terminology, as PCA is extremely easy to perform and the results are easy to interpret. The key thing to keep in mind is that when you apply PCA you will end up with groups of unrelated objects called components. The steps involved in PCA are summarized in List 13.5.

List 13.5 Steps involved in PCA.
- Select variables (objects).
- Calculate a correlation or covariance matrix.
- Extract linear combinations (components) from the matrix.
- Retain key components.
- Optionally perform a rotation of components.
- Interpret results with visual aids.

The linear combinations of objects produced by PCA are extracted from either a correlation or covariance matrix. The associations of an object with each component depends on the correlation values (loadings) calculated by PCA. You can think of the loading as a measure of the correlation between the object and the component (the variance shared by the component and the object is equal to the square of the correlation). R will provide the loadings for each object in the form of a matrix. It is also possible to obtain the scores of each independent object (i.e., the variable observations and not the patients) on each component.

Try to picture components as being lines of best fit through multidimensional space that represent the similarities of the objects. The first component represents the best linear combination of objects or, in other words, the particular combination of objects that accounts for more variance in the data as a whole than any other linear combination of objects. So the first component may be described as the single best summary of linear relationships within the data. The second component is derived from the proportion of the variance remaining after the first component has been extracted and is, therefore, the second best linear combination of objects and so on. Returning to our multidimensional-space, lines-of-best-fit analogy, once the line for the first component is drawn, the line for the second component is drawn orthogonally to the first and so on.

Initially, the number of components derived from PCA is equal to the number of original objects, and so criteria are required to help decide which components describe most of the variation within the objects and should be retained. It was mentioned that the square of the loading of an object on a component is equal to the variance. Summation of the squared loadings of each object on each component gives a value called the eigenvalue. An important aspect of PCA is that combinations of objects within each component are derived from shared homologies and there is no intercorrelation between components, which are therefore independent of each other.

If the results of a PCA are not interpretable, you can perform an "orthogonal" or "oblique" rotation of the components. The need to sometimes carry out a rotation reflects the fact that PCA doesn't always give a reliable result. Again, returning to the multidimensional-space, lines-of-best-fit analogy, by applying a rotation, you are rotating components about their origin to try to improve the accuracy of the lines of best fit.

Criteria are required to help decide how many components describe most of the variation within the objects and should be retained, and a number of these criteria will be discussed. The loadings of objects on individual components may be visualized using scatterplots. These plots in our study will reveal the relationships between protein expression scores and patients and thus the underlying structure within the data.

We will use PCA as a second method for estimating patient groups to confirm the clusters we saw using HCA. We will hope that there will be good agreement between the two methods to strengthen our hypothesis about age-related biologic breast cancer groups.

13.4.1 Performing a PCA

There is a whole range of functions to compute a PCA in different packages in R. Two functions are available in the base R stats package for performing a PCA. The `prcomp()` function is considered the most stable and preferred method and computes a singular value decomposition. The second method is `princomp()` that computes an eigen analysis and won't be considered here. The function `prcomp()` creates an object of class "prcomp" and allows you to parse parameters to center and scale the object data as well as rotate the components. Code 13.6 shows how we can perform a PCA

using `prcomp()` on the `ndata` numerical data matrix created in Section 13.2 and return the loadings held within a value called `rotation`.

```
> ndata.pca<- prcomp(t(ndata.sc), center=F)
> ndata.pca$rotation
             PC1          PC2          PC3
01   -0.09440090  -0.163470778  -0.013695790
02   -0.04252273   0.095373275   0.175069338
03    0.02395540   0.050337665  -0.069056523
04    0.06005506   0.006255121  -0.181386318
05   -0.23647251  -0.223821726  -0.053484425
06   -0.12265508   0.225394034   0.049309487
07   -0.15628016  -0.147113025  -0.004959628
08   -0.01833840  -0.063600376  -0.362199177
09    0.05718654  -0.401828707  -0.110275507
010  -0.11015378   0.244258914   0.012886929
011   0.01588458   0.123460969  -0.252619335
012   0.04700425  -0.361289004  -0.061208709
013  -0.15753596   0.017048944  -0.038346028
014   0.03455066   0.004218567   0.345117910
015  -0.23804516  -0.203832869  -0.041142217
016  -0.06913075   0.107809422   0.081497891
017  -0.19529674   0.023010651   0.039415199
018   0.00861495  -0.080483176  -0.190210508
019  -0.09639880   0.039740467  -0.017455699
020  -0.03835764   0.013032376  -0.185648358
021   0.16225014  -0.078458540   0.114180214
Y22  -0.16805503   0.298920729   0.041529429
Y23  -0.08570066  -0.066690571   0.023796865
Y24   0.17899059   0.034596770   0.183783381
Y25  -0.19372409   0.003021794   0.027072991
Y26   0.32440312   0.171210473  -0.072510553
Y27   0.10984597  -0.034546907   0.309541310
Y28   0.12803571  -0.028634455   0.113826253
Y29   0.07090717  -0.102420324   0.264294561
Y30   0.16076810   0.058861368   0.200739668
Y31   0.03579881   0.047918800  -0.083554754
Y32   0.10460575  -0.086353072   0.174329161
Y33   0.35197307  -0.031096962  -0.210430862
Y34  -0.16391077   0.157337392   0.018724968
Y35  -0.18064480   0.221946155   0.031041211
Y36  -0.15478315   0.130116568   0.099626209
Y37   0.16714077  -0.140648437   0.116362202
Y38   0.37589681   0.287014156  -0.222840827
Y39   0.04626643  -0.203797014   0.023030723
Y40   0.19437821   0.016818926   0.063863492
Y41  -0.09085242  -0.060387042  -0.064504509
Y42  -0.04525258   0.100769446  -0.273509664
```

Code 13.6.

First, note that we have transposed the standardized data frame using `t()` because
the `pca()` function carries out PCA on the data columns. We have also switched off
the center parameter because our data is already standardized. There is also a scaling
parameter in `prcomp()`, which is set to FALSE by default. `prcomp()` has returned
three components, which equals the number of data values for the objects. The loadings for each object on the components are given for all three components. The scores
of the "observations" on each component can also be returned using the "x" value of
the `prcomp` object (Code 13.7).

```
> ndata.pca$x
                PC1        PC2        PC3
size      -1.679254 -6.1695086 -0.3424474
proteinA   5.697315 -0.5405114 -2.8720117
proteinB   5.600207 -1.3000783  2.8191279
```

Code 13.7.

Trying to interpret how objects group according to loadings on components at this stage is tricky. Thus, we use scatterplots to visualize the data by plotting the loadings for pairs of components at a time. More on that in Section 13.4.3, but first we need to decide how many components to retain—in other words, how many components reliably describe the groupings in the data.

13.4.2 Choosing the Number of Components to Retain

If you retain all components for interpretation, you will end up assessing meaningless, uninformative data. Exclude too many and you will lose valuable information. There are a couple of ways that can help you decide on an adequate number of components to extract.

Each component has an eigenvalue associated with it. We said previously that each component is like an axis (we actually referred to it as a line of best fit) through the multidimensional data. An eigenvalue (also called a latent root) represents the variance of that axis. A variance of 1 or above is considered important, whereas a component with variance less than 1 is considered less important. Therefore, a quick way of checking which components are important is to calculate the eigenvalues of each one. The prcomp object contains a vector ($sdev) for the standard deviation for each component. Thus, to retrieve the variances, and hence the eigenvalues, for each component, all we need to do is calculate the squares of the elements in the sdev vector (Code 13.8).

```
> eigenvalues<-ndata.pca$sdev^2
> eigenvalues
[1] 33.320802 20.022596  8.156602
```

Code 13.8.

The variances of each component can also be visualized using a scree plot (Figure 13.3):

```
> screeplot(ndata.pca)
```

We would conclude that we should keep the first two components, as both have eigenvalues greater than 1. Consequently, we ignore the third.

You can also sum up the eigenvalues and calculate the percentage of variance each component contributes to the overall variance. This can be obtained using the summary() command on the prcomp object (Code 13.9).

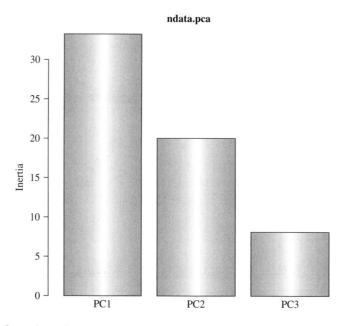

FIGURE 13.3   Scree plot to show component variances.

```
> summary(ndata.pca)
Importance of components:
                         PC1   PC2   PC3
Standard deviation     5.772 4.475 2.856
Proportion of Variance 0.542 0.326 0.133
Cumulative Proportion  0.542 0.867 1.000
```

Code 13.9.

Again, if we look at the "proportion of variance" line of the summary, we see the first two components contribute 87% of the variation between them and should be retained. In more complex analyses, there may be many more components, and a rough rule of thumb is to retain components that contribute at least 5% of the total variation. With this rule then, all three components should be retained.

A more robust method of estimating the number of useful components is to use a method called "broken stick," which is available in the `screeplot.prcomp()` function of the `vegan` package. The broken stick method has been recommended as a stopping rule in PCA, where components should be retained as long as observed eigenvalues are higher than corresponding random broken stick components.[7,8] The broken stick represents the expected variances of components under a random model. The method is visual, where a scree plot is generated with the broken stick superimposed on the bars representing the components (Figure 13.4). Code 13.10 shows how to do this.

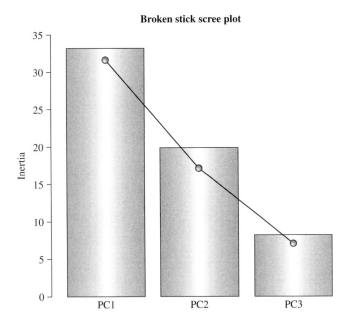

FIGURE 13.4   Scree plot with broken stick.

```
> library(vegan)
> screeplot.prcomp(ndata.pca, bstick = TRUE, bst.col = "black", type =
"barplot", main="Broken stick scree plot")
```

Code 13.10.

In addition to parsing the `prcomp` object, we can set parameters to show the broken stick and its style. All we need to do is see which bars, representing the variance of components, have a height greater than the height of the stick at the same position on the *x*-axis. The plot provides, once again, evidence that all three components should be retained for interpretation.

13.4.3 Interpreting Principal Components

We interpret components according to the loadings of the objects. The pattern of how objects load onto components by value gives us a clue as to how many groups of objects may exist. It is simple to visualize the loadings using scree plots for pairs of components, but there is a useful function called `loadings.pca` in the `labdsv` package that will output loadings for specified components and only print values of higher loadings (Code 13.11).

```
> library(labdsv)
> loadings<-pca(t(ndata))
> loadings.pca(loadings, dim=3)
```

```
Loadings:
      PC1     PC2     PC3
01            0.138   0.306
02    0.133   0.140  -0.732
03
04           -0.131   0.113
05            0.221
06
07            0.176
08           -0.179
09                   -0.180
010
011          -0.193
012   0.101           0.163
013
014   0.268   0.239   0.203
015           0.223
016
017           0.170
018
019
020
021   0.253          -0.187
Y22
Y23           0.123
Y24   0.293
Y25           0.168
Y26   0.271  -0.280  -0.130
Y27   0.306   0.183   0.131
Y28   0.227          -0.141
Y29   0.260   0.200   0.133
Y30   0.288           0.200
Y31
Y32   0.241   0.113
Y33   0.233  -0.318
Y34
Y35
Y36           0.147
Y37   0.259
Y38   0.234  -0.454
Y39   0.134          -0.154
Y40   0.249          -0.126
Y41
Y42          -0.161
```

Code 13.11.

The dim parameter tells the function how many components to print. Have a look
at the pattern of returned loadings across the three components and then compare these
scores to the heatmap in Figure 13.2. The first component (PC1) shows scores for objects that occur in the top two clusters of the HCA tree (with the exception of O2).
Thus, we can assume that PC1 explains high proteinA expression in samples (interestingly, O2 actually has the highest value in the lower clusters of the HCA tree). The
second component (PC2) shows a mixture of positive and negative loadings for objects. Again, comparing to the heatmap, we see that objects with positive loadings
have low proteinB values and larger tumor size, whereas objects with negative loadings have smaller tumor size. Notice how objects that have no loadings printed have
intermediate proteinA and proteinB values. Interpretation of the third component
(PC3), which explains less of the variation, is more difficult to interpret, as the loadings are less than those for other components.

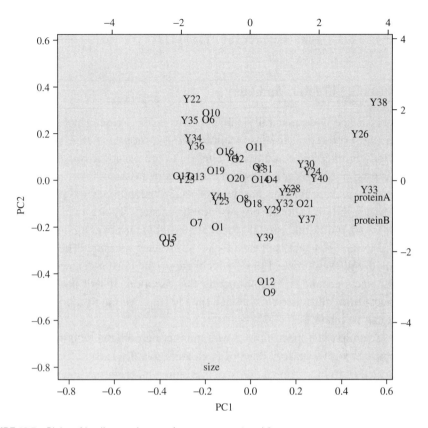

FIGURE 13.5 Biplot of loadings and scores for components 1 and 2.

Loadings of objects, as well as scores for all observations on pairs of components, can be displayed graphically using the `biplot()` function (Figure 13.5). The code to generate a biplot for the loadings on the first two components in the `ndata.pca` object is shown in Code 13.12.

```
> biplot(ndata.pca, choices = c(1,2), var.axes = FALSE, arrow.len = 0, cex=0.6,
col = c(gray(0.1),gray(0)))
```

Code 13.12.

The `choices` parameter allows us to select which components to plot. Other parameters allow graphic manipulation. The parameter called `arrow.len` switches off the drawing of arrows from the center of the plot to the position of each object label. These arrows clutter up the plot when many objects are shown and in many ways are pointless.

We can see a pattern of groups of objects emerging in the biplot. We have already seen in Code 13.11 that objects with higher loadings on PC1 have higher scores for proteinA. We see this also to be the case for loadings on PC2. Note how the objects "pull away from each other" in the plot to form this pattern of groupings. The pattern of distribution of objects shows good concordance with the six clusters observed using HCA and defined using the `cutree()` function.

13.5 Partitioning Cluster Methods

Cluster analysis methods generally fall into two basic classes: hierarchical cluster analysis, as we explored in the last section, and partition cluster analysis. Whereas HCA provides an interconnected tree of clusters to represent the relatedness between objects, partition clustering methods assign objects in a dataset into discrete clusters.

In partition clustering, one has to prespecify the number of clusters and the method will then initially allocate objects into the clusters. Typically in these methods, cluster centers (in multidimensional space) are initially chosen at random, and objects are assigned to a cluster that is closest by some distance measure. The partition clustering method will then proceed to iteratively adjust the cluster centers to best fit the associated objects around it by minimizing the variation. It will also keep moving around objects from cluster to cluster each time to improve the fit until no further improvement can be made.

Because you have to specify the cluster number beforehand, you may be wondering if you have to know, a priori, how many clusters actually exist in the data. You may also be puzzled as to how you could use a partition clustering method to explore data if you do not have a priori knowledge of structure. In an exploratory analysis, you can estimate the number of clusters first using HCA before applying a partition clustering method to more robustly cluster the objects. You can, as we shall see, perform a series of partition clustering analyses specifying different cluster numbers and testing the "strength" of the clusters by how well objects fit in the given number.

As a rule, unmodified partition clustering methods are limited to application on numerical datasets. The most commonly applied partition clustering method is called *k*-means. Partition Around Medoids (PAM) is another popular partition method that is less sensitive to outliers and noise than *k*-means. Both of these methods provide "hard clusters," meaning that an object is assigned to one cluster only. An alternative approach to partition objects is fuzzy clustering, whereby each object is assigned to every cluster but with a degree of membership (hence the fuzziness).

In the following subsections, we will apply *k*-means, PAM, and fuzzy clustering to the age-related breast cancer dataset to obtain the optimal number of clusters of patients. As we do this, we can compare the clustering results to those of HCA and PCA.

13.5.1 *K*-Means Clustering

The implementation of *k*-means clustering in R partitions objects into *k* clusters so that the sum of squares between each object point and the cluster center point (to which the object is assigned) is minimized. The center is called the centroid and is the mean point in the cluster.

The `kmeans()` function allows you to parse a data frame or matrix as input and select how many clusters you want the method to create using the `centers` parameter. Let's turn now to our example, where we suspect after application of HCA that there are six clusters in the breast TMA dataset (Section 13.3). Application of kmeans() to the standardized ndata object by selecting six clusters is shown in Code 13.13.

```
> ndata.km<-kmeans(ndata.sc, centers=6)
> ndata.km
K-means clustering with 6 clusters of sizes 10, 6, 10, 2, 6, 8

Cluster means:
        size    proteinA    proteinB
1   0.6846099   0.2358301   1.0694247
2   0.0684015   0.5836599  -0.4310110
3   0.7470634  -0.7966169  -0.8037367
4  -0.5769518   0.6112654  -0.9696044
5  -0.9933088   1.4854407   1.3372915
6  -0.9516731  -1.0036584  -0.7694193

Clustering vector:
 01  02  03  04  05  06  07  08  09 010 011 012 013 014 015 016 017
  3   6   2   2   3   6   3   2   1   6   4   1   3   1   3   6   3
018 019 020 021 Y22 Y23 Y24 Y25 Y26 Y27 Y28 Y29 Y30 Y31 Y32 Y33 Y34
  2   3   2   1   6   3   5   3   5   1   1   1   5   2   1   5   6
Y35 Y36 Y37 Y38 Y39 Y40 Y41 Y42
  6   6   1   5   1   5   3   4

Within cluster sum of squares by cluster:
[1] 11.7338871  1.9468817  5.9412147  0.1384142  8.7740621  2.9894686

Available components:
[1] "cluster"  "centers"  "withinss"  "size"
```

Code 13.13.

You can use the centers parameter to either specify the number of clusters, as we have done, or to set the actual initial cluster centers for the algorithm. If the number of clusters is parsed, random initial cluster centers are generated. There are a number of different *k*-means clustering algorithms; three are implemented in kmeans(): Hartigan and Wong, Lloyd, and Forgy. The default algorithm is Hartigan and Wong, and this is considered to be the most robust.

If we look at the output in Code 13.13, we see a lot of information. First, the means are given for each cluster by observation. Then we see the all-important clustering vectors, which tell us which clusters the objects fall into. You should compare the clusters to the HCA groupings observed in Figure 13.2 and the cluster assignments produced by cutree() (Code 13.5). There doesn't appear to be very good agreement between the two methods. You should repeat *k*-means a number of times, as you will often get a different result due to the random element of initiating the cluster centers.

Instead, we are going to turn to the more robust partitioning clustering method called Partition Around Medoids, or PAM.

13.5.2 Partition Around Medoids (PAM)

PAM is similar to the *k*-means method of partition clustering, and the concept is the same. Instead of calculating the sum of squares, PAM minimizes the sum of dissimilarities, which gives rise to the term *medoid* instead of *centroid*. PAM is available using the pam() function in the cluster library. The distance metrics available for pam() are "euclidean" and "manhattan," although, instead of parsing a data frame, you can actually parse a dissimilarity matrix by setting the diss parameter to TRUE. Other options allow you to set initial medoids and standardize the data before calculating dissimilarity.

We can apply PAM to the ndata.sc standardized data object selecting six clusters using the k parameter (Code 13.14).

```
> library(cluster)
> ndata.pam<-pam(ndata.sc,k=6)
> pam
Medoids:
      ID      size     proteinA    proteinB
07     7   1.17174754  -0.7966169  -0.6979245
06     6  -1.20148732  -0.9622501  -0.8409140
020   20   0.04758366   0.3076045  -0.7551203
012   12   2.17100432   0.6388710   0.5603823
Y28   28  -0.07732344   0.4180267   1.0751443
Y26   26  -1.57620861   1.9639367   1.3897210

Clustering vector:
 01  02  03  04  05  06  07  08  09 010 011 012 013 014 015 016 017
  1   2   3   3   1   2   1   3   4   2   3   4   1   5   1   2   1
018 019 020 021 Y22 Y23 Y24 Y25 Y26 Y27 Y28 Y29 Y30 Y31 Y32 Y33 Y34
  3   3   3   5   2   1   5   1   6   5   5   5   5   3   5   6   2
Y35 Y36 Y37 Y38 Y39 Y40 Y41 Y42
  2   2   5   6   4   5   3   3

Objective function:
   build     swap
0.6962969 0.6709248

Available components:
 [1] "medoids"   "id.med"    "clustering" "objective" "isolation"
 [6] "clusinfo"  "silinfo"   "diss"       "call"      "data"
```

Code 13.14.

The output is a pam.object with components that are similar in style to that of kmeans(). The key result to look at once again is the clustering vector and the pattern in which objects group. As we did with kmeans(), we can compare the pattern of objects in the six PAM clusters to the patterns observed using HCA. The PAM cluster membership is almost identical to that of the cluster membership of HCA using cutree(), with the only exceptions being objects O19 and Y41.

These results differ from those of the less robust *k*-means approach, but combined with the information obtained using HCA and PCA, we are starting to build a picture of the underlying structure of our dataset. How can we be sure that there are six clusters of objects? After all, we based our assumptions for PAM using a decision, albeit subjective, from HCA that six groupings exist.

One major advantage of using the pam() function is that the "fit" of objects in clusters is automatically calculated when the pam object (called pam.object) is created. Note that one of the "Available components" printed in Code 13.14 is called "silinfo." The pam object is a class derived from another class, known as a partition object, which provides the silinfo list. The silinfo list is created automatically when pam() calls a function called silhouette(). Silhouette objects, and the plots you can create using this information, are one of the most powerful weapons in your data analysis armory. In the next section, we will look at how we can create silhouette plots to tell us the optimal number of clusters in our dataset of interest.

13.6 Calculating the Number of Clusters with Silhouette Plots

The `pam()` algorithm produces a "silhouette value" for each object in a cluster. The silhouette value will be a number between -1 and 1.[9] To see these values for the `ndata.pam` object, you need to call the `silinfo` list, as shown in Code 13.15.

```
> ndata.pam$silinfo
$widths
     cluster neighbor   sil_width
015       1        3   0.52297991
07        1        3   0.51107348
05        1        3   0.50317352
01        1        3   0.30503345
Y25       1        2   0.30370086
013       1        2   0.20207326
017       1        2   0.18043038
Y23       1        3   0.14636457
06        2        3   0.65727986
Y35       2        1   0.62820332
010       2        3   0.60427884
Y22       2        3   0.59211088
Y34       2        1   0.56253501
Y36       2        1   0.50360334
016       2        3   0.32455678
02        2        3   0.25636763
020       3        1   0.50775982
04        3        5   0.45567220
03        3        5   0.44670843
Y31       3        5   0.43670736
018       3        4   0.42389967
011       3        2   0.42294562
08        3        1   0.40091295
Y42       3        2   0.39111854
019       3        2  -0.04757343
Y41       3        1  -0.17480060
012       4        5   0.71717846
09        4        5   0.68874535
Y39       4        5   0.21394214
Y27       5        3   0.62470637
Y24       5        3   0.60775848
021       5        4   0.59046660
Y28       5        3   0.58383476
Y30       5        3   0.57696394
Y32       5        4   0.57613271
Y29       5        4   0.48821940
Y40       5        6   0.47682061
014       5        2   0.40457839
Y37       5        4   0.39742325
Y38       6        5   0.54069397
Y26       6        5   0.47702295
Y33       6        5   0.30075051

$clus.avg.widths
[1] 0.3343537 0.5161170 0.3263351 0.5399553 0.5326905 0.4394891

$avg.width
[1] 0.4364846
```

Code 13.15.

The output displays the cluster number that each object falls into as well as the silhouette value, referred to as the width. A silhouette value of 1 means that the object is well classified in the cluster; 0 means that the object lies between clusters, and −1 means that it is badly classified in a cluster.

If you have hundreds of values, you wouldn't want to check each one, but instead would prefer to have an understanding about the overall fit of objects in each cluster. The output provides such a value, called $clus.avg.widths, displayed for each cluster directly under the list of individual silhouette scores. These scores represent the mean silhouette widths in each cluster. Notice the scores for each of the six clusters, which range approximately between 0.32 and 0.54. Working on a scale of −1 to 1, this gives us a feeling of how good our clustering really is.

The output also goes one step further and calculates the mean of the silhouette scores across all clusters. This quality index is called the silhouette coefficient (SC). An SC score greater than 0.51 suggests that a reasonable structure has been found, whereas a score greater than 0.71 suggests a strong structure. The SC for the six-cluster analysis is 0.44, suggesting that we may be close to a reasonable structure but fall slightly short of the cutoff of 0.51.

All this information can be displayed graphically using a silhouette plot (Figure 13.6). The plot shows the silhouette values as horizontal bars ordered according to the fit in each cluster (i.e., the bar width). You can see that the overall SC value is reduced because two objects in the third cluster have negative values. This suggests that these two cases (O19 and Y41) do not really fit in this cluster. Thus, you may be tempted to assess seven clusters to see if you have a better cluster fit. In fact, we can produce SC

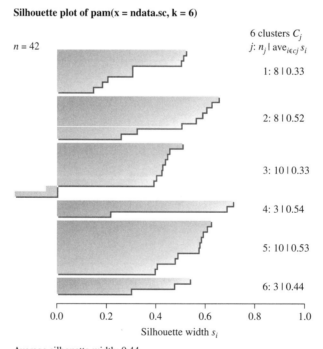

FIGURE 13.6   Silhouette plot to show cluster structures.

values for a range of cluster numbers and see which result gives the highest value (Code 13.16).

```
> sc<-as.vector(rep(0,9))
> sc[1]<-pam(ndata.sc,k=2)$silinfo$avg.width
> sc[2]<-pam(ndata.sc,k=3)$silinfo$avg.width
> sc[3]<-pam(ndata.sc,k=4)$silinfo$avg.width
> sc[4]<-pam(ndata.sc,k=5)$silinfo$avg.width
> sc[5]<-pam(ndata.sc,k=6)$silinfo$avg.width
> sc[6]<-pam(ndata.sc,k=7)$silinfo$avg.width
> sc[7]<-pam(ndata.sc,k=8)$silinfo$avg.width
> sc[8]<-pam(ndata.sc,k=9)$silinfo$avg.width
> sc[9]<-pam(ndata.sc,k=10)$silinfo$avg.width
> names(sc)<-c(2:10)
> sc
        2         3         4         5         6         7         8
0.3649119 0.2835467 0.3079100 0.3825019 0.4364846 0.4198778 0.4109958
        9        10
0.3710640 0.3198754
```

Code 13.16.

We have created a range of SC values for between 2 and 10 clusters. We first created a vector to hold each of the nine numbers. Then we put each SC value in the vector, which we have called `sc`. To do this, we used the `pam()` function each time and in the same line of code return the `avg.width` value, which, as we saw, is a value of the `silinfo` list. By printing the `sc` vector, we can discover the cluster number that yields the highest SC score. The reason we created the `sc` vector is that we can then use this to graphically present the data. A good plot style is the bar plot (Figure 13.7):

```
> barplot(sc, main="Silhouette Coefficients for cluster sizes: 2
to 10", xlab="Cluster Number", ylab="SC")
```

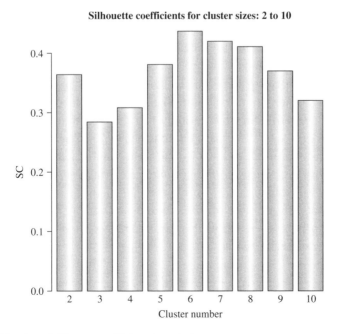

FIGURE 13.7   Bar chart of silhouette coefficients in each cluster.

The SC values for low-number clusters are relatively small, rising to a maximum at six with a score of 0.4364846. The SC values then tail off as the cluster number increases. Using a combination of PAM, silhouette coefficients, and plotting, we now have even more confidence about the underlying structure of the data in that there are six subgroups of patients. We should be cautious about how "tight" these clusters are because we did not achieve a high SC value for the most optimal cluster number. We are left wondering whether, despite six clusters of objects existing, some objects may actually fall between clusters to some degree.

13.7 Fuzzy Clustering

Although PAM provided us with "hard clusters," the silhouette approach allowed us to see that a few objects did not fit well into certain clusters. Partition clustering methods assign objects in a dataset to only one cluster, thus producing a hard clustering.

Fuzzy clustering, on the other hand, allows for some ambiguity in the data, which often occurs in practice.

In fuzzy clustering, each object is "spread over" various clusters, and the degree to which an object belongs to different clusters is quantified by means of membership coefficients, which range from 0 to 1, with the stipulation that the sum of their values is 1. Most importantly, objects are spread over all clusters, where membership coefficients reveal how each object "fits" into each group (where the sum of membership coefficients for an object across all clusters is 1). For instance, an object may be assigned membership coefficients of 0.72 to cluster A, 0.21 to cluster B, and 0.07 to cluster C, meaning that the object has a 72% membership in cluster A and can be assigned to that group with some confidence. Thus, although the degree of membership of an object to each cluster is given, hard clustering may still be achieved by assigning the object to the group with the highest membership.

There are also a number of ways you can evaluate the fit of objects to fuzzy clusters, including the silhouette approach described for PAM. The "hardness" of the fuzzy clustering may be measured using Dunn's partition coefficient, which is the sum of all squared membership coefficients divided by the number of observations (k), its value always being between $1/k$ and 1. Thus, 1 is a completely hard clustering.

The fuzzy clustering method implemented in the R cluster package called FANNY is a generalization of k-medoids partitioning (similar to PAM), and the methodology is detailed in Kaufman and Rousseeuw.[9] The function is called fanny() and takes a number of parameters, as we can see by applying the method to the breast cancer TMA dataset. We parse the standardized data object, ndata.sc, and tell fanny() that we want to use six clusters (k) and the membership exponent (memb.exp) that influences the fuzziness of the clustering (Code 13.17). The default membership exponent is set to 2, but this has been reduced to obtain a less fuzzy membership of objects in each cluster.

```
> fanny(ndata.sc,k=6, memb.exp=1.5)
Fuzzy Clustering object of class 'fanny' :
m.ship.expon.      1.5
```

```
objective       15.31429
tolerance       1e-15
iterations      59
converged       1
maxit           500
n               42
Membership coefficients (in %, rounded):
    [,1] [,2] [,3] [,4] [,5] [,6]
01    66   18    8    3    4    1
02     7   33   17   27   13    3
03     2    7   82    3    5    1
04     2    4   84    2    5    2
05    90    5    2    2    1    0
06     0    2    1   96    0    0
07    82   12    3    2    1    0
08    12   14   59    5    6    3
09    32   15   19    6   20    7
010    1    3    2   92    1    0
011    4   10   70    8    5    4
012   33   15   19    6   21    6
013    4   86    3    5    1    0
014    8   14   12    9   53    4
015   90    5    2    1    1    0
016    5   38   15   35    6    2
017    8   77    4    9    2    1
018    8   11   70    3    6    2
019    4   77    9    7    2    1
020    4   13   76    4    3    1
021    1    2    3    1   91    2
Y22    2    5    3   87    2    1
Y23   18   62   11    4    4    1
Y24    2    3    5    2   83    5
Y25    8   79    4    7    1    0
Y26    0    1    1    1    2   95
Y27    3    4    5    2   84    2
Y28    1    2    4    1   91    1
Y29    5    6    7    3   76    2
Y30    2    4    7    3   79    6
Y31    2    5   84    3    5    1
Y32    2    2    4    1   91    1
Y33    3    4    7    3   11   72
Y34    2    9    2   85    1    0
Y35    1    3    1   93    1    0
Y36    3   21    4   69    2    1
Y37    3    4    7    2   80    4
Y38    1    1    2    1    3   91
Y39   21   15   23    5   33    4
Y40    3    4   10    3   71   10
Y41   17   60   15    4    3    1
Y42    6   16   62   10    4    3
Fuzzyness coefficients:
dunn_coeff normalized
 0.6081122  0.5297347
Closest hard clustering:
 01  02  03  04  05  06  07  08  09 010 011 012 013 014 015 016 017
  1   2   3   3   1   4   1   3   1   4   3   1   2   5   1   2   2
018 019 020 021 Y22 Y23 Y24 Y25 Y26 Y27 Y28 Y29 Y30 Y31 Y32 Y33 Y34
  3   2   3   5   4   2   5   2   6   5   5   5   3   5   6   4
Y35 Y36 Y37 Y38 Y39 Y40 Y41 Y42
  4   4   5   6   5   5   2   3

Available components:
 [1] "membership"  "coeff"    "me mb.exp" "clustering" "k.crisp"
 [6] "objective"   "convergence" "diss"  "call"  "silinfo" "data"
```

Code 13.17.

The output is quite comprehensive and lists first a number of parameter values (for a full description of each, type "?fanny"). Then the table of membership coefficients is printed for each cluster. Each number shows the degree of membership of each object in the cluster represented as a percentage. For example, the object O1 has 66% membership with cluster 1, 18% membership with cluster 2, and so on. Just below is Dunn's coefficient with a value of 0.6081122 showing that we have a reasonably hard clustering. This is followed by a table showing the hard cluster assignment for each object, where each cluster number reflects the highest cluster membership.

The membership coefficients for an object indicate how well it fits in a cluster. Most objects have a high membership for one of the clusters, with the exception of five objects (O2, O9, O12, O16, and Y39). These objects are more fuzzy, never having a membership coefficient greater than 38%.

We have now applied four different clustering approaches to the dataset and can think about comparing the results of each method. In the next section, we will assess the observed clusters and attempt to obtain a consensus result as to which patient subgroupings are real.

13.8 Comparing the Clustering Results to Obtain a Consensus

Compare the hard clusters from FANNY to the PAM clusters and the HCA clusters produced by `cutree()`. You will see much similarity in how objects group, but also some differences. This is a reminder that different clustering methods often produce different groupings. In fact, if you go one step further using the silhouette method, you will see that using `fanny()`, the most optimal number of clusters is actually seven. It is important to realize the importance that, when applying different clustering methods, a consensus result is often the main aim.

To obtain a consensus cluster result, we just need to see how cluster patterns for objects vary between methods. Figure 13.8 shows the same HCA plot from Figure 13.2 but with objects labeled if they cluster differently between methods and if they were classed as fuzzy by FANNY (i.e., did not have a high membership coefficient for at least one cluster). The six clusters predicted by `cutree()` are also separated by lines and labeled from A to F. This is a simple but effective way to obtain a measure of concordance between methods. Seven objects group differently by method, which gives an agreement of 83% (35/42). Of the seven objects that cluster differently by method, five of these are also the fuzzy objects. The two nonfuzzy objects in group E cluster differently between `pam()` and `cutree()`; otherwise, these two methods give identical clustering patterns overall. The objects in group C, which have a large tumor size, should probably be classed as a robust cluster despite the degree of fuzziness. Group F also has objects that cluster identically using `pam()` and `cutree()` but two fuzzy objects (O2 and O16). The fuzziness in O2 and O16 likely arises due to a higher proteinB score relative to the other objects in the cluster.

Let's now compare the consensus cluster patterns and ambiguities to the way objects grouped using PCA. The scatterplot of Figure 13.5 showed the loadings of objects on principal components one and two. In Figure 13.9 we have redrawn the biplot and superimposed on top of this the consensus clusters A to F. When discussing the

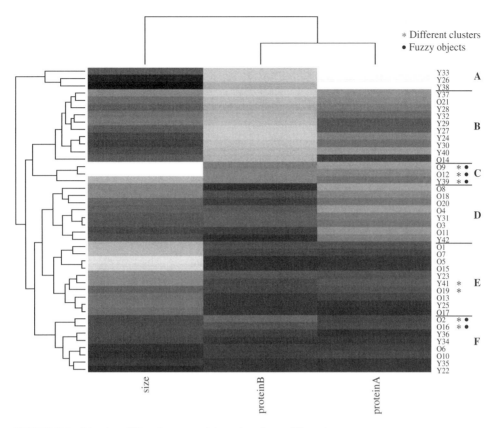

FIGURE 13.8   Mapping of "fuzzy" cases and those that cluster differently.

PCA clustering, it was suggested to think of PCA as "pulling away" objects from each other according to how different they were. Figure 13.9 allows us to visualize this process by looking at the way the consensus clusters "separate." Just like the HCA heatmap and dendrograms, PCA biplots can explain why clusters occur.

We stated previously that PC1 associates with proteinA and proteinB levels. Consensus group A occurs due to high values of both. Groups E and F both have lower values for proteinA and proteinB; hence, their negative correlation with PC1. PC2 associates with tumor size, and we see group C, with negative correlation, pulling away due to larger tumor size in the objects. Conversely, group F has small tumor sizes, and objects pull in the opposite direction.

The biplot does not give us a good feel for any difference between groups B and D, however. This is simply because the difference between these two groups resides in the variation explained by the third component, PC3. If we create a biplot for PC1 and PC3 by changing the `choices` parameter in `biplot()`, we create the plot shown in Figure 13.10. Consensus groups B and D have been drawn on the biplot. The distinction between the two groups now becomes clear as PC3 separates objects as to whether they have high proteinA values or high proteinB values. Group B has a positive correlation on PC3 due to high proteinB values, whereas group D has a negative correlation with PC3 due to high proteinA values.

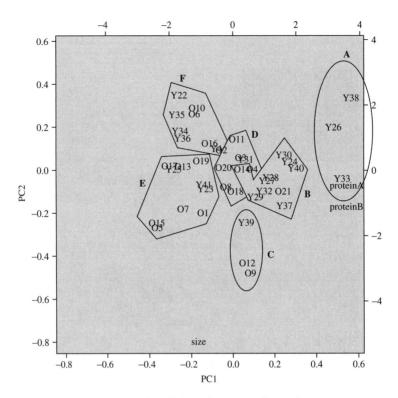

FIGURE 13.9 Biplot of components 1 and 2 with fuzzy clusters superimposed.

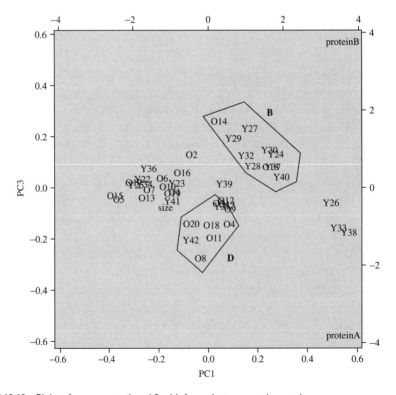

FIGURE 13.10 Biplot of components 1 and 3 with fuzzy clusters superimposed.

By applying different clustering and data reduction methods, we have been able to generate a consensus result describing the way objects group. By using HCA and PCA, we were able to visualize potential groupings and estimate the number of clusters. By applying partitioning methods, we were able to robustly assign objects to clusters and assess any ambiguities by looking at the fuzziness of objects. Finally, we were able to return to the plots generated by HCA and PCA to visualize the consensus groupings of objects. Hopefully, this analysis has demonstrated how one should attempt to combine clustering methods to explore data to reveal the underlying structure of objects.

The one remaining thing to establish is whether the consensus groups associate with any age pattern. Group A contained all young patients. Young patients also made up 80% (8/10) of objects in Group B. Groups D and E contained 75% (6/8) and 70% (7/10) old patients, respectively. Objects in group F contained 50% of each age group, and group C only contained three "fuzzy" patients. Only the foolish would draw immediate conclusions from the results because the dataset is so small. This result does, however, provide patterns that should encourage a much larger study with perhaps many hundreds of cases.

13.9 Clustering of Categorical Data

The focus of the chapter so far has been the application of clustering and data reduction methods to numerical data. As medical data can often be ordinal or nominal, we will sometimes need to analyze datasets that contain just these data types. We may even want to explore mixed numerical and categorical datasets. To do this, we can use a number of the approaches described. All we have to do is change the way in which we calculate distance measures and use metrics that allow for the different data types.

In this section we will create a dataset from the original TMA data that contains mixed numerical and factor data. We won't attempt to interpret the results as we did in the preceding sections, but will just work through the process of applying clustering methods to mixed datasets.

13.9.1 Creating a Mixed Dataset with Numerical and Nominal Data

To create the dataset, we follow the same procedure as detailed in Section 13.2 but assign five variables to a data frame object called cdata (Code 13.18).

```
# mixed numerical & categorical objects - proteinA, proteinB
# & her2
> data<-data[order(data$label),]
> cdata<-data[,c(2,3,7,8,9)]
> labels<-c(paste("O", c(1:21), sep=""), paste("Y", c(22:42), sep=""))
> row.names(cdata)<-labels
```

Code 13.18.

We ensure first that the data object is ordered by label. After creating the `cdata` object, we then create labels and assign these to the data frame row names. Because the dataset contains factor variables, we don't apply a standardization procedure this time.

The variables that we have entered into the dataset are er, pgr, proteinA, proteinB, and her2. The er, pgr, and her2 variables are nominal in that the values increase by one unit but the units are not proportional. It is critical that the type of each column in the data frame is recognized correctly before calculating distance metrics. You can, of course, check the type using the `summary()` function, and you may find that each is of numerical type. To ensure that the factor variables are indeed factors, you can apply the code shown in Code 13.19.

```
> cdata [,c(3)]<-as.factor(cdata[,3])
> cdata<-data[,c(2,3,7,8,9)]
> cdata [,c(1)]<-as.factor(cdata[,1])
> cdata [,c(2)]<-as.factor(cdata[,2])
> cdata [,c(5)]<-as.factor(cdata[,5])
> summary(cdata)
 er      pgr        proteinA         proteinB       her2
1:12   1: 6    Min.   : 0.00   Min.   : 0.00    0:27
2:27   2:34    1st Qu.: 8.00   1st Qu.:11.50    1: 4
3: 3   3: 2    Median :21.50   Median :31.50    2: 4
               Mean   :22.43   Mean   :42.40    3: 7
               3rd Qu.:34.00   3rd Qu.:82.50
               Max.   :70.00   Max.   :96.00
```

Code 13.19.

13.9.2 Creating a Distance Matrix with `daisy()` for Clustering

The distance metrics supplied with the `dist()` function are not suited for use on datasets containing nominal data. You can, however, use a function called `daisy()` available in the cluster package. `daisy()` also allows you to calculate euclidean and manhattan metrics, but if one or more objects are nominal (factor), the function will calculate Gower's distance (type "?daisy" for more details on these metrics). The `daisy()` function will accept either a matrix or a data frame object. The syntax for `daisy()` is similar to `dist()`:

```
> cdist<-daisy(cdata)
```

13.9.3 Performing Cluster Analysis on a `daisy()` Distance Matrix

To apply HCA, we can create the distance matrix, call the clustering algorithm, and plot the dendrogram using one line (Code 13.20)

```
> plot(hclust(daisy(cdata)),hang=-1, cex=0.5, xlab="Cases", main="HCA using
mixed numerical and nominal data")
```

Code 13.20.

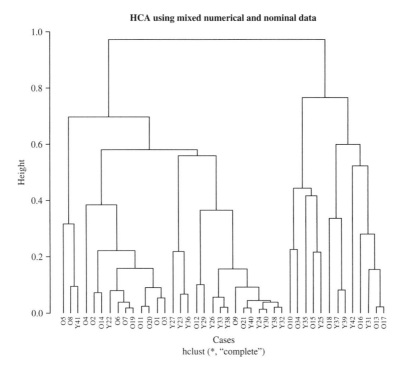

FIGURE 13.11 Hierarchical cluster analysis of mixed numerical and categorical variables.

We have left the clustering method as complete, which is the default in `hclust()`. The dendrogram is shown in Figure 13.11. The `heatmap()` function, however, cannot be used because it expects a matrix of numerical data.

PCA can only be performed on numerical variables in `prcomp()`, but we can apply PAM and FANNY to a distance matrix provided by `daisy()`. Both `pam()` and `fanny()` can be applied in the same way as for numerical data (Code 13.21).

```
# Perform PAM
> pam(daisy(cdata), k=3)
Medoids:
      ID
[1,] "19" "019"
[2,] "21" "021"
[3,] "13" "013"
Clustering vector:
 O1  O2  O3  O4  O5  O6  O7  O8  O9 O10 O11 O12 O13 O14 O15 O16 O17
  1   1   1   1   1   1   1   1   2   3   1   2   3   2   3   3   3
O18 O19 O20 O21 Y22 Y23 Y24 Y25 Y26 Y27 Y28 Y29 Y30 Y31 Y32 Y33 Y34
  1   1   1   2   1   1   2   3   2   2   2   2   3   2   2   3
Y35 Y36 Y37 Y38 Y39 Y40 Y41 Y42
  1   1   2   2   2   2   1   3
Objective function:
    build      swap
0.1778005 0.1751842

Available components:
[1] "medoids"    "id.med"     "clustering" "objective"  "isolation"
[6] "clusinfo"   "silinfo"    "diss"       "call"
```

```
# Perform FANNY
> fanny(daisy(cdata), k=3)
Fuzzy Clustering object of class 'fanny' :
m.ship.expon.             2
objective        2.755753
tolerance           1e-15
iterations             33
converged               1
maxit                 500
n                      42
Membership coefficients (in %, rounded):
    [,1] [,2] [,3]
01    63   11   27
02    53   12   35
03    47   10   42
04    36   29   35
05    42   33   25
06    66   13   21
07    68   11   21
08    36   37   26
09    26    9   65
010   25   56   18
011   54   13   34
012   36   22   42
013   24   60   16
014   30   10   60
015   29   52   19
016   24   58   18
017   24   59   17
018   29   49   23
019   68   11   22
020   59   12   30
021   13    5   82
Y22   60   17   23
Y23   48   25   27
Y24   15    6   79
Y25   29   54   17
Y26   22   10   68
Y27   36   22   42
Y28   15    5   80
Y29   35   22   43
Y30   15    5   80
Y31   26   52   23
Y32   16    6   78
Y33   25   13   63
Y34   22   63   15
Y35   44   31   25
Y36   48   28   24
Y37   28   29   42
Y38   26   12   62
Y39   31   34   35
Y40   15    6   80
Y41   38   37   25
Y42   27   53   20
Fuzzyness coefficients:
dunn_coeff normalized
 0.4486067  0.1729100
Closest hard clustering:
 01  02  03  04  05  06  07  08  09 010 011 012 013 014 015 016 017
  1   1   1   1   1   1   1   2   3   2   1   3   2   3   2   2   2
018 019 020 021 Y22 Y23 Y24 Y25 Y26 Y27 Y28 Y29 Y30 Y31 Y32 Y33 Y34
  2   1   1   3   1   1   3   2   3   3   3   3   2   3   3   2
Y35 Y36 Y37 Y38 Y39 Y40 Y41 Y42
  1   1   3   3   3   3   1   2
Available components:
 [1] "membership"  "coeff"       "memb.exp"    "clustering"  "k.crisp"
 [6] "objective"   "convergence" "diss"        "call"        "silinfo"
```

Code 13.21.

13.10 Other Clustering Approaches in R

There are a variety of clustering and data reduction methods implemented in R, and in this chapter we have only looked at some of the mainstream functions. A good starting point to explore other clustering methods is the CRAN Task View for cluster analysis and finite mixture models (http://cran.r-project.org/web/views/Cluster.html).

Listed among alternative approaches for HCA are `agnes()` and `Diana()` from the cluster package, which provide, respectively, an alternative method to `hclust()` and a method for divisive hierarchical clustering. The cluster package also provides a function called `clara()` for clustering large datasets. A number of packages developed specifically for analyses of certain datasets, such as microarray data, utilize these functions. Another function that performs HCA using `hclust()` and provides P Values for each cluster is `pvclust()`, available in the package of the same name. The P Value gives an indication of the strength of the cluster and is computed using bootstrapping.

We have covered a number of partition clustering methods in this chapter, but there are several variations. The package `flexclust` provides a function called `kcca()` that performs *k*-centroids cluster analysis. Other functions in the package allow for neural gas and QT clustering. The `som` package provides a function called `som()` for performing self-organizing maps, a method that has become popular for the unsupervised analysis of microarray data. Another fuzzy clustering approach is implemented in the `e1071` package called `cmeans()` and is based on fuzzy *c*-means clustering.

There are also a number of model-based clustering methods available in the `mclust`, `bayesm`, and `MFDA`. These approaches require a sound knowledge of the statistical methodology that lies beneath. Alternative data reduction functions may be found in the `stats` package as well as `amap`. Multidimensional scaling and sammon mapping may be performed using the `cmdscale()` and `sammon()` functions. Functions for binary and multiple correspondence analysis may be found in the `ca` package. The `stats` package has a function called `factanal()` to perform factor analysis, as does the `FactoMineR` package, which is worth exploring for variations on a number of data reduction techniques.

Finally, in this chapter we have applied data analysis methods to tissue microarray data, which often has peculiar properties, including an often-skewed distribution. Random forest clustering, a nonparametric method, has been successfully applied to TMA data and is proposed as a robust method for the clustering of immunohistochemical data. Importantly, the method allows for a dataset comprised of both categorical and numerical data, and would be well suited for application to the datasets studied in this chapter.

13.11 Summary

In summary, it is possible to apply different clustering methods to biomedical data in R. You can apply diverse clustering methods in combination to the same dataset, such as HCA and PCA, as well as partition clustering methods, such as PAM and fuzzy clustering. Using this strategy, you can generate a consensus result by closely scrutinizing the grouping of objects generated by PCA by each method. HCA and PCA provide you with good visualization tools.

13.12 Questions

1. What do we mean by "exploring" data?

2. How can we order data in a data frame by index using a single variable?

3. Why is it useful to standardize data prior to analysis using the `scale()` function?

4. What are the four main steps of a hierarchical cluster analysis? What is the minimum number of R functions required to perform these steps?

5. What do the distance metrics specify in the `dist()` function?

6. What question does the `cutree()` function attempt to answer?

7. What does the `t()` function achieve as an argument to `prcomp()`?

8. What is the purpose of the `screeplot()` function?

9. How do partition methods differ from hierarchical cluster analysis?

10. In terms of usage, what are the key differences between k-means, PAM, and fuzzy clustering?

11. What do we mean by the concept "consensus result" in relation to multivariate analysis?

12. What is the method available as an argument to the `daisy()` function that allows you to generate a distance matrix from categorical data?

14 | Survival Analysis

14.1 Background

For many studies in medicine, we are interested in assessing the time it takes for an event to happen. Very often, the event is an outcome, such as diagnosis of death, but the outcome may also be other measurable parameters, such as onset of disease or relapse of disease. There is a blanket term that describes the period leading to the event, called "survival time." Furthermore, survival analysis is the term used to describe the investigation into the patterns of these events that occur within one or more cohorts in a study.

In dealing with the analysis of survival data, we are interested in the length of time it takes a patient to reach an event rather than simply the fact that the event has or hasn't occurred. Survival times are calculated from a start point, such as the beginning of treatment for a patient or the actual date of a study. However, not all patients may be monitored throughout the time-course of a study. For example, in a drug treatment clinical trial, a patient may elect to drop out of the study halfway through or their treatment may have to be changed. Such a patient is termed "censored." Thus, when analyzing our data, we need to account for censored patients.

Many survival studies set out to measure time periods in more than one group of patients. Analysis of survival data requires specialized statistical methods if we are to compare two or more groups of patients. There are a number of reasons for this. Survival data is rarely normally distributed, but is skewed, often with many early events and few late ones. Primarily, though, we need methods that deal with the fact that the event does not occur for all patients who, at the end of the study, have a time to event that is unknown.

Survival analysis is carried out to calculate two types of probabilities from the data. The first of these is the probability of survival from a starting point (e.g., diagnosis of disease) to a specified endpoint (e.g., 10 years from diagnosis). The second is the hazard or probability that a patient at a given time-point will have the event. In publications of survival analyses, you will often see survival data presented as survival curves, where the probabilities of survival along the curve are calculated by the Kaplan-Meier method. The nonparametric log-rank test can be used to test for differences between survival curves of different groups. The hazard, of course, is the endpoint, such as death. Generally, two groups in a study will differ by at least one factor, such as new drug treatment versus placebo. You will want to know if the hazard is different between groups across all time-points when taking into account this factor and possibly other differing factors, such as patient age.

This chapter provides an overview of how you can carry out a basic survival analysis on patient data using R. Although we will keep our analysis basic, R has some powerful packages for survival analysis. Our example data will come from the breast cancer dataset that contains variables for alive or dead status (event) as well as survival time in months (time to event). We'll set the scene by considering 10-year patient survival patterns in breast cancer and how these relate to tumor attributes. We can then examine survival and hazard between subgroups of patients defined by these attributes when accounting for other factors, such as age.

Ensure first that the survival library is loaded:

```
> library("survival")
```

14.2 The Breast Cancer Survival Dataset

In Section 8.7, we created a dataset object called data.all that was a data frame containing 10 objects, as shown in List 14.1.

List 14.1 Variables in the 10-year survival dataset.

- ageatdiagnosis
- alivestatus
- survivaltime
- yearofdiagnosis
- size
- grade
- nodesexam
- nodespos
- surgery
- radiotherapy

The dataset can be loaded from the tab-delimited text file, "data.all.txt," that you created in Chapter 8; a copy also accompanies the book:

```
> data.all<-read.table("data.all.txt", header=T)
```

Each variable is described in Section 5.2. As discussed in Section 8.7, make sure that each object in the data.all data frame is of the correct type, particularly the alivestatus and grade objects, which must be factors. There are 62,618 rows representing individual patients contained within the dataset.

The tumor is described for each patient by size, grade, and the number of nodes examined (nodespos). As we shall see, these attributes are useful for predicting survival in breast cancer. Treatment data is given for both surgery and radiotherapy, which have, of course, a major influence on a patient's survival. Survival time is measured in months for the survivaltime attribute, and the patient's mortality status is binary, representing "alive" (0) or "dead" (1).

14.3 Breast Cancer and Survival—The Nottingham Prognostic Index

To predict patient survival in cancer, we need reliable prognostic factors. One of the most widely used methods for predicting survival in breast cancer involves calculation of the Nottingham Prognostic Index (NPI). Originally derived in 1978,[10] the NPI is basically an equation that yields an index based on prognostic factors that are independently significant. These factors are the maximum diameter of the tumor, the grade (1 to 3) of the tumor, and a score representing the number of nodes positive for metastasis.[11]

NPI = 0.2 × size (cm) + grade (1-3) + lymph node score (1-3)

The lymph node scores are 1 point for no positive nodes, 2 points for one to three positive nodes, and 3 points for four or more positive nodes. The NPI scores will range between 2.08 and up. The original NPI system divided patients into three groups that differed (according to NPI score) by the likelihood of survival.[11] The current system recommends six groups, as shown in Code 14.1.[12]

```
-Excellent prognostic group (EPG)      2.08-2.4
-Good (GPG)                            2.42≤3.4
-Moderate I (MPG I)                    3.42≤4.4
-Moderate II (MPG II)                  4.42≤5.4
-Poor (PPG)                            5.42≤6.4
-Very poor (VPG)                       ≥6.5
```

Code 14.1.

One of our goals will be to compare survival data between these groups, so we need to calculate the NPI for each patient. The breast cancer survival dataset will need a few simple alterations first so as to assign the correct scores for tumor grade, nodes positive, and size. First, you should copy the `data.all` dataset into a new data frame called `data.npi` to which we will create a new variable for the NPI scores (Code 14.2).

```
> data.npi<-data.all
> npi<-c(rep(0,nrow(data.npi)))
> data.npi<-cbind(data.npi, npi)
```

Code 14.2.

We then create a vector called `npi`, which, using the `rep()` function, we fill with zeros, and this has a length the same as `data.npi`. We finally bind the vector to `data.npi`.

Next, we change the `grade` variable to hold scores needed for the NPI scoring system. The grading system for breast cancer described in Chapter 5, ranging between I and IV, is based on the scoring system in the United States. The NPI was developed using data from the United Kingdom, which has a slightly different grading system: I (low-grade), II (moderate-grade), and III (high-grade). Grades III and IV in our U.S.-derived dataset represent poorly differentiated tumors, which are all high-grade. We can, therefore, reclassify the Grade IV cases in our dataset to Grade III to suit the NPI scoring system:

```
> data.npi$grade<-ifelse(data.npi$grade==4,3,data.npi$grade)
```

We can use the `ifelse` statement to loop through the `data.npi$grade` variable (which is the third variable in the data frame), changing the value of any case with a "4" to a "3."

To change the `nodespos` values to scores of 1, 2, or 3, we can again use `ifelse` to loop through each case, checking the value each time and replacing it with the corresponding score:

```
> data.npi$nodespos<-ifelse(data.npi$nodespos<1, 1,
ifelse(data.npi$nodespos<4, 2, 3))
```

This time, however, we use nested `ifelse` statements to check and replace values. Once this is complete, you should ensure that the `nodespos` and `grade` variables are actually assigned as factors if you want to try modeling your data later.

The NPI scores for each case can now be calculated and placed in the `npi` variable:

```
> data.npi$npi<-as.numeric(data.npi$nodespos) +
as.numeric(data.npi$grade) + (data.npi$size/10)*0.2
```

Note that when calculating the NPI in R we need to carry out mathematical operations on the `grade` and `nodespos` variables, so we ensure each type is numeric to avoid any error. The values for tumor size in the dataset are in mm, but the NPI scores are calculated using cm, so we also need to divide the size values by 10.

With the NPI scores generated, we can now classify each case according to the six prognostic groups (Code 14.3).

```
> data.npi$npi<-ifelse(data.npi$npi<=2.4, 1,ifelse(data.npi$npi <=3.4, 2,
ifelse(data.npi$npi<=4.4, 3, ifelse(data.npi$npi<=5.4, 4,
ifelse(data.npi$npi<=6.4, 5, 6)))))
> data.npi$npi<-as.factor(data.npi$npi)
```

Code 14.3.

These six groups provide a prognosis but do not tell us the proportion of patients who survive within each group. Calculating these figures from such a large dataset would provide a likelihood of survival if a patient fell into a particular group. We can calculate and visualize this information by creating a table. First, create a vector called `groups` and fill with zeros:

```
> groups<-c(rep(0,6))
```

Then create a table to hold the counts of patients who survive 10 years and those who don't (`alivestatus`) according to their NPI status (`npi`) (Code 14.4).

```
> npi.table<-table(data.npi$npi, data.npi$alivestatus)
> npi.table

      0     1
1  6548   149
2 15573   792
3 15395  1792
4  9266  2431
5  4220  2512
6  1820  2120
```

Code 14.4.

Remember that the "0" column denotes surviving patients. We then use this table to calculate the percentage of patients who survive in each NPI group and place this in a vector called `percent.surviving`:

```
> percent.surviving<-100*(npi.table [,1]/( npi.table [,1]+
npi.table [,2]))
```

Last, we convert the vector into a data frame and assign the NPI group labels as names for each row (Code 14.5).

```
> groups<-as.data.frame(percent.surviving, row.names=c("EPG", "GPG", "MPG1",
"MPG2", "PPG", "VPG"))
> groups
      percent.surviving
EPG             97.77512
GPG             95.16040
MPG1            89.57351
MPG2            79.21689
PPG             62.68568
VPG             46.19289
```

Code 14.5.

We can compare these survival trends for American patients with similar survival data for patients in the United Kingdom. Survival figures for Nottingham City Hospital in the United Kingdom are shown in Code 14.6 and can be added to the prognostic group table for a direct comparison.[12]

```
> uk.data<-c(96,93,81,74,50,38)
> groups2<-cbind(groups2, uk.data)
> groups2
      percent.surviving uk.data
EPG             97.77512      96
GPG             95.16040      93
MPG1            89.57351      81
MPG2            79.21689      74
PPG             62.68568      50
VPG             46.19289      38
```

Code 14.6.

The survival pattern is quite similar between the two datasets, with a slight increase in survival in the American data (although this may reflect a more recent diagnostic period for the American data and a far stricter inclusion criteria for the UK study).

14.4 Generating Survival Curves

A survival curve allows us to visualize the percentage of survival in a group of patients as a function of time. The Kaplan-Meier method is the one most commonly used to generate a survival curve. Over a specified time period, the Kaplan Meier method will recalculate the fraction of patients who are still alive at every time-point. In the case of the breast cancer dataset, we can apply the Kaplan-Meier method to calculate the fraction of patients still alive every month over a period of years we have data for. The fraction of patients alive at each monthly time-point will be the number still alive at the end of that month divided by the number alive just before that month.

The follow-up time for patients in our breast cancer dataset spans 13 years, from 1990 to 2002 inclusive. The date of diagnosis for each patient will fall within and between these years. Therefore, some patients will have a 13-year follow-up and some will have died within this time. The remaining patients were alive at the end of 2002 but have a follow-up less than 13 years, as they were diagnosed later than 1990. This last group is known as "right censored."

We can familiarize ourselves with the concepts of survival curves by plotting survival times for the patients in the six different NPI groups. The first step is creating an object using the Surv() function to hold the survival time (data.npi$survivaltime) and survival status (data.npi$alivestatus) data (Code 14.7).

```
#create a survival object
> surv.obj<-Surv(data.npi$survivaltime, data.npi$alivestatus)
> surv[1:10]
 [1] 110+ 100+  70+  31   47  100+  87+  86+ 138+  86+
```

Code 14.7.

When we output the first 10 elements of surv.obj, we see the survival time for each patient followed by a cross in some cases. The cross indicates that the patient is alive at the end of the specified time period, whereas those that lack the cross did not survive. In order to calculate the nonparametric Kaplan-Meier estimate of a survival curve, we need to parse the survival object to a function called survfit() (Code 14.8).

```
> surv.fit <- survfit(surv.obj~ npi, data=data.npi)
> names(surv.fit)
 [1] "n"           "time"        "n.risk"        "n.event"
 [5] "surv"        "type"        "ntimes.strata" "strata"
 [9] "strata.all"  "std.err"     "upper"         "lower"
[13] "conf.type"   "conf.int"    "call"
```

Code 14.8.

We can calculate Kaplan-Meier estimates for survival curves for each of the six NPI groups. To do this, we actually parse a formula that includes the terms for surv.obj and npi (remember that response and predictor variable terms are separated by a "~"). You must also specify the dataset to which the terms belong. This will create a survfit object (that we call surv.fit), which will contain the cumulative survival for each of the NPI group survival curves. This data is held in one of the object's components, called `surv`. You can see that the `survfit` object contains a number of other components too (type "?survfit.object" for a description of each).

If you want to know the fraction of patients surviving after 13 years for each NPI group, you need to access the `surv` component. However, this component is just one long continuous list of survival fractions for all the NPI groups, and we don't know which elements represent the beginning or end of a group. To retrieve final survival fractions, you can use one of the other components, called `strata` (Code 14.9).

```
> surv.fit$strata
npi=1 npi=2 npi=3 npi=4 npi=5 npi=6
  151   155   156   156   156   156
last.element <-c(1:6)
x<-0
for(i in 1:6){
x<-x + surv.fit $strata[i]
last.element[i]<-x
}
> npi.frac<- surv.fit$surv[as.integer(last.element)]
[1] 0.9576113 0.9122728 0.8407750 0.7338455 0.5463734 0.3776011
```

Code 14.9.

By printing `surv.fit$strata`, we see the number of elements for the time-points in each of the NPI groups. Now we can use this information to determine the elements in the `surv` component that contain the final survival fractions in each group. We just need to create a vector, called `last.element`, to hold the fraction data and then use a loop to place the number of each group's final element into `last.element`. We can then use `last.element` as an index to retrieve the fraction data, placing it into a vector called `npi.frac`.

We are now ready to create survival curves for each NPI group (Code 14.10).

```
> labels<-c("EPG", "GPG", "MPG1", "MPG2", "PPG", "VPG")
> plot(surv.fit, main="Survival Curve for NPI Groups", xlab="Survival Time
(months)", ylab="Proportion Surviving")
> axis(side=4,at=fit$surv[as.integer(npi.frac)],labels=labels, tick=F, las=1,
cex.axis=0.7, hadj=0.5)
```

Code 14.10.

Before we plot the curves, we can create a vector of labels for the NPI groups. Then, using the `plot()` and `axis()` functions, we parse the `surv.fit` object along with other parameters for group label, plotting along the right axis. The NPI group labels are positioned correctly on the *y* axis by the `at` parameter, using each group's final survival fractions (`npi.frac`). The resulting survival plot is shown in Figure 14.1. Notice that cases that are censored are represented as horizontal lines on each curve. The differential survival patterns over time in each group are easily visualized in this way.

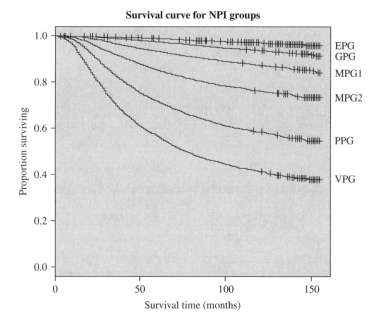

FIGURE 14.1   Survival plot for the NPI groups.

14.5 Comparing Survival Curves

In a survival analysis, you often have a situation where there are two survival curves and you want to know whether there is any significant difference between the survival distributions of each. A common example from the literature would be a comparison of two curves generated from clinical trial data, where one represents the survival distribution of patients who have been treated with a new drug and the other for patients receiving the old drug. Returning to our NPI dataset, we may, after viewing the Kaplan-Meier plot in Figure 14.1, question whether there is any significant difference between the survival distributions of the excellent (EPG) and good (GPG) prognosis groups. The statistical test most commonly used for this purpose is the log-rank test (also known as the Mantel-Haenszel), which can be applied to two or more survival curves using the `survdiff()` function in the survival package.

Calculating a log-rank test using `survdiff()` is straightforward. We need to parse a formula to `survdiff()` that includes a survival object created using `Surv()` and a predictor variable to tell the function what the groups are. Code 14.11 shows how we do this for the EPG and GPG survival curves.

```
> survdiff(Surv(survivaltime, alivestatus) ~ npi, subset=npi==1 | npi==2,
data=data.npi)
Call:
survdiff(formula = Surv(survivaltime, alivestatus) ~ npi, data = data.npi,
    subset = npi == 1 | npi == 2)
```

```
          N Observed Expected (O-E)^2/E (O-E)^2/V
npi=1  6697      149      268      52.9        74
npi=2 16365      792      673      21.1        74

 Chisq= 74.1  on 1 degrees of freedom, p= 0
```

Code 14.11.

For the formula, both `survivaltime` and `alivestatus` variables are parsed to `Surv()` and the predictor variable is `npi`. So as only to perform a log-rank test on the EPG and GPG curves, we can specify a parameter in the function called `subset`. By setting `subset` to either "npi==1" or (|) "npi==2," only the cases in EPG and GPG are used. Similarly, if we set `npi` to 3 or 4 using `subset`, only the two moderate survival groups would be compared. You could, of course, select more than two values of `npi` to compare more groups at once.

We could have set a parameter called `rho` to select either the log-rank test (rho=0, which is the default) or another test, which is a modification of the Gehan-Wilcoxon test (rho=1). The log-rank test is nonparametric and provides us with a chi-square value (74.1) and corresponding P Value. The P Value of zero is a clear indication that the EPG and GPG survival curves are significantly different.

14.6 Survival Analysis with Risk Factors

The log-rank test was useful to directly compare survival curves where we weren't accounting for any other variables. We can also analyze the effects of one or more risk factors on survival curves in one or more groups using Cox's proportional hazards regression model. Risk factors are also referred to as predictor variables, explanatory variables, or covariates in references and text books. These predictor variables can be categorical or numerical. The term proportional means that at any given time-point, the predictor variable reduces or increases the hazard by the same amount as at any other time-point.

To demonstrate building a Cox regression model, we will try to determine whether differential survival between EPG patients can be explained by treatment factors, including surgery and radiotherapy, as well as age at diagnosis. The surgery variable (List 5.6) is categorical, as each value codes a specific type of surgery. The vast majority of cases have one of two codes, which are "20" (partial mastectomy) or "50" (modified radical mastectomy), so we will include these types in our analysis. The radiotherapy variable is binary, where a "1" codes for beam and a "3" for no radiotherapy. The age at diagnosis is a continuous variable. Essentially, we're asking Cox's regression to tell us whether surgery, radiotherapy, and age at diagnosis are independently associated with survival after adjusting for the other covariates.

We can build a Cox's regression model using the `coxph()` function from the survival package (Code 14.12). Once more, we need to create a survival object using `Surv()` as part of the formula needed by `coxph()`. The covariates on the right side of the formula include `npi` as well as `radiotherapy` and `surgery`. These are separated by a " | " to symbolize the "addition" of each term to the model. Because we need to be selective in which values we include for `npi` (EPG = 1) and `surgery` (20, 50) we specify these values in the `subset` parameter.

```
> cox<-coxph(Surv(survivaltime, alivestatus) ~
surgery+radiotherapy+ageatdiagnosis, subset=((npi==1) & (surgery==50 |
surgery==20)), data=data.npi)
> summary(cox)
Call:
coxph(formula = Surv(survivaltime, alivestatus) ~ surgery + radiotherapy +
    ageatdiagnosis, data = data.npi, subset = ((npi == 1) & (surgery ==
    50 | surgery == 20)))

  n= 6261
                   coef exp(coef) se(coef)      z        p
surgery20       -0.4343     0.648   0.3401  -1.28  2.0e-01
radiotherapy3    0.1726     1.188   0.3452   0.50  6.2e-01
ageatdiagnosis   0.0376     1.038   0.0081   4.64  3.5e-06

               exp(coef) exp(-coef) lower .95 upper .95
surgery20          0.648      1.544     0.333      1.26
radiotherapy3      1.188      0.841     0.604      2.34
ageatdiagnosis     1.038      0.963     1.022      1.05

Rsquare= 0.007   (max possible= 0.309 )
Likelihood ratio test= 43.1  on 3 df,    p=2.38e-09
Wald test            = 41.3  on 3 df,    p=5.71e-09
Score (logrank) test = 43.1  on 3 df,    p=2.35e-09
```

Code 14.12.

The output provides a wealth of information for each covariate (information for unused factor values has been removed for clarity). We can begin to interpret the results by focusing on the `surgery` variable. There is an entry called "surgery20," which means the information that applies to this row is based on a comparison between survival in patients who have had a partial mastectomy (surgery20) and a radical mastectomy (surgery50). The key column is "exp(coef)," which provides the estimated hazard ratio and is the exponent of the coefficient (coef). For surgery, we see that excellent-prognosis patients who have had a radical mastectomy have approximately one and a half (1.554) times the risk for death than those same-category patients who have had a partial mastectomy. This figure is taken from the exp(-coef) because it makes more sense than saying "the hazard ratio of partial mastectomy to radical mastectomy is 0.648." Also provided are the 95% lower and upper confidence intervals, and when looking at these we see that they fall to either side of 1. This result, coupled with an insignificant coefficient (P = 2.0E-01), suggests that differing surgery is not associated with survival in EPG patients.

The hazard ratio (1.188) for radiotherapy is also not significant (P = 6.2E-01). The factor value shown is "radiotherapy3," which suggests that there is no increased hazard of death for EPG patients who have not had radiotherapy compared to those who have. The hazard ratio for age at diagnosis is extremely interesting. Despite being extremely close to 1 (1.038), the coefficient is actually significant (P = 3.5E-06) with quite a tight confidence interval too. The three tests at the end of the output test whether the variable coefficients are all different from zero (which indeed they are in this case).

Thus, it appears that the age of a patient is significantly associated with survival in an EPG group. Because the `ageatdiagnosis` variable is continuous and the coefficient (coef) is positive, we can deduce that there is a small increased risk of death as an EPG patient's age increases. If the other covariates are kept constant then for each increasing year of age, the monthly hazard of death for an EPG patient is increased, on average, by a factor of 1.038 (or 3.8%).

Finally, we can assess how well our model fits the data. One way of doing this is to test the assumption of proportional hazard for each covariate. This can be done by parsing the `coxph` object to a function called `cox.zph()` (Code 14.13).

```
> cox.zph(cox)
                  rho   chisq    p
ageatdiagnosis -0.00265 0.00123 0.972
surgery20      -0.06126 0.71270 0.399
radiotherapy3  -0.06600 0.79759 0.372
GLOBAL             NA   0.82792 1.000
```

Code 14.13.

In our model, the assumption of proportional hazard for each covariate is satisfied because no significant P Value is observed.

14.7 Summary

The analyses described in this chapter were a brief introduction to survival analysis using R. We were able to derive survival curves and test for differences between survival curves for different cohorts of patients. We were also able to construct a model to investigate the effects of different covariates on survival. Although covering the most widely used techniques, we have only scratched the surface of survival analysis. Indeed, many other functions are available in the survival package, as well as other packages, such as Design.

14.8 Questions

1. What is a prognostic factor? Can it be of any data type (numerical, categorical, etc.)?

2. What type of object must we supply to the `survfit()` function?

3. In an object created by `survfit()`, what does the strata component tell us?

4. What type of statistical test does `survfit()` perform on two survival curves?

5. What type of model does the `coxph()` function create, and under what circumstances would we use this function?

6. What R function can be used to test the fit of a hazards model?

15 Data Mining and Predictive Modeling with R and Weka

15.1 Background

In the last decade, we have been introduced to new concepts that have had a direct impact on the marriage between biology and medicine. Evidence-based medicine is the term now applied to utilizing research findings in response to clinical questions for improved decision making. Another concept is translational medicine, which can mean a number of things, including the translation of molecular research into new patient therapies. Key to translational research is the discovery and evaluation of new biomarkers.

The rise of such new domains in modern medicine mirrors the almost exponential improvement in computing power over the last two decades. The advancement of medicine now relies upon the collection, storage, and analysis of massive datasets. Data may come directly from the clinic or from high-throughput technologies in the laboratory. Indeed, we have looked at the benefits of building heterogeneous biomedical datasets in previous chapters. The biomedical world is, however, already in a position where the amount of data available far exceeds our abilities to analyze it. Huge efforts are being made to develop algorithms and procedures for better storage, processing, and analysis of biomedical data. Progress is being made due to the contribution of computer science, which extends the multidisciplinary aspects of medicine.

As we talk of relationships between medicine and other disciplines, we must not forget that the field of statistics has had the longest relationship with medicine for utilizing data. Statistics has, for a long time, provided models for new means of diagnosis, prognosis, and management of disease. If medicine and biology provide the data, statistics and computer science provide the tools to gain knowledge. In biomedicine, however, the relationship between medics and biologists is a much happier one than that between computer scientists and statisticians. Whereas medicine and bioinformatics have adopted the concept of "data mining" massive datasets, statistics has not. Data mining is not really a term used by statisticians.

One will often see in the literature a distinction between the use of statistical approaches and machine learning as two separate "types" of methods for analyzing data. Machine learning is a term often used to describe methods that have been developed within computer science. This may be surprising to many nonspecialists who consider data mining to be the application of both statistical and machine learning approaches.

Ironically, both statistical and machine learning methods are often applied in an analysis to achieve the same objective—to yield a predictive model. In bioinformatics, one would apply statistical and machine learning methods to the same dataset as part of the same analysis. If we are neither a statistician nor a computer scientist, should we care about any differences of opinion about what data mining actually is? Not really, but be aware that although the fundamentals are often the same, the use of terminology between statistics and machine learning can be quite different. For example, a statistician may talk about an independent variable, whereas a computer scientist may refer to this as an attribute.

Data mining, as a term, can mean two different things: (1) knowledge discovery or descriptive modeling or (2) predictive modeling. Data mining methodologies are often applied to large biomedical datasets to yield new models that improve diagnosis, prognosis, or management of disease. We will focus on predictive modeling methods in this chapter. For what follows, the term "model" can be used to describe any equation, mathematical object, or set of rules that allows us to make a prediction. We can apply both multivariate statistical and machine learning procedures to generate predictive models in biomedicine. The literature is ever-expanding, with papers revealing how data mining approaches have yielded new gene signatures (i.e., models) that can improve diagnosis or prognosis of disease.[13,14,15] But it is not just molecular data that is subject to data mining in medicine. Pathoclinical datasets have also been mined successfully,[16,17,18] and there is frequently debate as to whether new gene signature markers will be more powerful than clinical markers.[19]

You should already know by now from your experience with R that using open-source software is more flexible and comprehensive than any commercial platform when it comes to data analysis. Data mining is no exception. Dedicated open-source software for data mining includes Weka (http://www.cs.waikato.ac.nz/~ml/weka/) and RapidMiner (http://rapid-i.com/content/blogcategory/10/69). R, as you would expect, has some excellent functions for data mining in a variety of packages. It is worth visiting the Task View list for machine learning and statistical learning on the CRAN website, where there are more than 30 packages listed.

The advantage that dedicated tools like Weka (written in Java) have over R is that they are faster. If you are analyzing datasets with hundreds of thousands of rows, R is not for you. However, as we have learned in previous chapters, R is not just a set of statistical tools but a complete data processing and analysis environment with great connectivity to other open-source resources. Just by using a package called RWeka, you can combine the power of R and Weka to create a powerful, open-source, and free data mining platform for biomedical data. Although we will install Weka, we will only need to interact with R to perform our data mining. Aside from the advantage of speed, Weka also provides us with a huge number of data mining methods in one package with a standard form of output.

The analytical focus in this chapter will be prognostic modeling in breast cancer. Whereas the last chapter was devoted to survival analysis, the methods used in this chapter are used to generate new survival models. The methods that we apply can be used in a range of predictive problem domains, including diagnosis and response to treatment. These methods were selected because they are applicable to datasets that contain binary response variables as well as numerical and/or categorical predictor variables. Also, at the end of the chapter is an example of using multiple regression to analyze datasets with continuous response variables.

We will begin by familiarizing ourselves with the concepts of predictive modeling by applying a commonly used statistical method, logistic regression, to the breast cancer dataset. At the same time, we will look at some of the issues with the use of response classes for predictive modeling and how we can attempt to overcome these problems. We will then use the optimized data as the subject of analysis by machine learning methods to generate different models that predict 10-year survival in breast cancer patients. To generate machine learning models, we will process our data in R, send it to Weka for analysis, and then visualize the results in R. When a model is generated, it needs to be validated and this, too, can be performed in Weka. The validated models will tell us which is most accurate as a prognostic tool for predicting outcome of breast cancer.

Before you begin, you should load the breast cancer 10-year survival dataset as described in Section 14.2.

15.2 R and Weka

15.2.1 An Introduction to Weka

Weka is decribed as a suite of machine learning tools written in Java and developed by the University of Waikato in New Zealand.[20] The Weka project began in 1993 and was converted into a Java platform in 1997. Between 2000 and 2008, there were around 1.2 million downloads of the software. Weka contains a huge array of data mining methods for classification, regression, clustering, and association rules (Table 15.1). Algorithms for classification, regression, etc. are called "classifiers" in Weka. There are also tools for data pre-processing and visualization of data, models, and evaluation.

As a software package, Weka is user-friendly; an excellent introduction to both the tool and data mining is provided by Witten and Frank.[20]

Table 15.1   Data Mining Methods Available in Weka.

Group	Classifier
Bayes	AODE, BayesNet, ComplementNaiveBayes, NaiveBayes, NaiveBayesMultinomial, NaiveBayesSimple, NaiveBayesUpdateable
Functions	LeastMedSq, LinearRegression, Logistic, MultilayerPerceptron, PaceRegression, RBFNetwork, SimpleLinearRegression, SimpleLogistic, SMO, SMOreg, VotedPerceptron, Winnow
Lazy	IB1, IBk, KStar, LBR, LWL
Trees	ADTree, DecisionStump, Id3, J48, LMT, M5P, NBTree, RandomForest, RandomTree, REPTree, UserClassifier
Rules	ConjunctiveRule, DecisionTable, JRip, M5Rules, Nnge, OneR, Part, Prism, Ridor, ZeroR
Misc	Hyperpipes ,VFI
Meta	AdaBoostM1, AdditiveRegression, Bagging, Decorate, AttributeSelectedClassifier, Filtered Classifier, ClassificationViaRegression, CVParameterSelection, CostSensitiveClassifier, Grading, LogitBoost, MetaCost, MultiBoostAB, MultiClassClassifier, MultiScheme, OrdinalClassClassifier, RacedIncrementalLogitBoost, RandomCommittee, RegressionByDiscretization, Stacking, StackingC, ThresholdSelector, Vote

15.2.2 Installing Weka

Weka requires Java 1.4 or higher and is, therefore, cross-platform. Weka is issued under the GNU General Public License and available from http://www.cs.waikato.ac.nz/ml/weka/. At the time of writing, the current release of Weka is version 3.4. To download Weka for Windows, navigate to the download page from the home page. You can either download a self-extracting executable file that includes Java 1.4 (Java VM 1.4) or one that does not include Java in case you have it already installed. Once redirected to the download page, you can download the file of choice and follow the instructions for installation. Both Weka and Java should install without problems, but if one is encountered, you can check for the solution on the Weka Wiki site accessible from the home page.

15.2.3 Talking to Weka from R

Once Weka is installed, you can download the RWeka package:

```
> install.packages("RWeka")
```

You can think of the RWeka package as an interface between R and Weka in the same way that RMySQL allows R to talk to MySQL.

Once the library is loaded:

```
> library(RWeka)
```

you are faced with a few simple choices as to how you use Weka.

Each available Weka classifier can be accessed by R in a standard way by typing the function name and parsing a formula—the data frame that contains the variables, or attributes, as Weka calls them—and an optional list of control parameters.

The general function call is:

```
> model<-Function_Name(formula, data, control)
```

There will be many examples of calls to Weka functions in the sections that follow. The value returned is a list of components (Table 15.2).

Again, the output and RWeka functions for evaluating models will be discussed in later sections.

Not all Weka classifiers have predefined functions in the package. Those that do are shown in Table 15.3.

These R classifier function objects inherit from the Weka_classifier class. To call a classifier that isn't in Table 15.3, you can create your own function object using the make_Weka_classifier() function. All you have to do is parse the Weka classifier name as a path, which will create an interface function object just like those in

Table 15.2   Components Returned by Calling Weka Classifiers.

Classifier	Reference to a Java object listing the model
Predictions	Numeric vector/factor containing the predictions of the model for the training instances
Call	The function call made by R to Weka

Table 15.3 Classifiers with Dedicated Functions in RWeka.

Group	Classifier
Functions	LinearRegression, Logistic, SMO
Lazy	IBk, LBR
Rules	JRip, M5Rules, OneR, PART
Trees	J48, LMT, M5P, DecisionStump
Meta	AdaBoostM1, Bagging, LogitBoost, MultiBoostAB, Stacking

the list. As an example, to create a function object called "rf" that will interface the `RandomForest` classifier, you would type:

```
> rf <- make_Weka_classifier("weka/classifiers/
trees/RandomForest")
```

Each classifier in Weka has a number of parameters that can be set. You can list the available parameters using the `WOW()` function whereby you parse the name of the classifier. Table 15.4 shows the parameters available for the J48 classifier determined by `WOW()`.

To set individual parameters for a classifier, you can specify the values using `Weka_control()`:

```
Model<-J48(response ~ ., data = data.10yr, control =
Weka_control(R = TRUE, M = 5)).
```

Weka has its own text-based file format for storing data called Weka Attribute-Relation File Format, with the file extension .arff. If you intend to use Weka as a stand-alone package, you may want to save your datasets from R in ARFF format using the `write.arff()` function:

```
> write.arff(data.10yr, file = "data.10yr.arff")
```

Table 15.4 Retrieving Available Classifier Parameters Using `WOW(48)`.

U	Use unpruned tree.
C	Set confidence threshold for pruning. (default 0.25) Number of arguments: 1.
M	Set minimum number of instances per leaf. (default 2) Number of arguments: 1.
R	Use reduced error pruning.
N	Set number of folds for reduced error pruning. One fold is used as pruning set. (default 3) Number of arguments: 1.
B	Use binary splits only.
S	Do not perform subtree raising.
L	Do not clean up after the tree has been built.
A	Laplace smoothing for predicted probabilities.
Q	Seed for random data shuffling. (default 1) Number of arguments: 1.

ARFF files can also be read into R using the `read.arff()` function:

```
> test<-read.arff("data.10yr.arff")
```

15.3 Generating Predictive Models in R

Whether we use a statistical or machine learning approach to model building, there is a general approach that one follows (List 15.1).

List 15.1 General approach for predictive model building.

- Create training and testing datasets.
- Apply chosen method to training set and generate model.
- Evaluate model on test set.
- Repeat with other methods.
- Compare performance between methods.

Providing a technical detail for each method is beyond the scope of a single chapter. The point of this chapter is to introduce each method in a nontechnical way. Each method is presented to simply demonstrate its use and application to a biomedical dataset for predictive model building in R and Weka. Readers are advised to obtain a sound understanding of each method before applying it to their data. Fortunately, typing the method name in an Internet search engine will provide much freely available information.

Before we begin to consider machine learning methods, we will go through the process of model building with a statistical approach using logistic regression. Both the steps involved and the derived model from logistic regression are relatively simple to understand and serve as a useful introduction to modeling.

In the examples that follow, although the algorithmic processes of model building will differ widely, the input and output of each model will be the same. To build a model, we will need to specify a formula that includes the response variable and the predictor variables. The formula can then be parsed to the function of choice. Because we want to build models to predict outcome in breast cancer patients, we need to consider variables from the patient dataset that are prognostic. As we saw in the previous chapter, tumor size, tumor grade, and number of nodes positive are useful prognostic factors for breast cancer, so we will build models based on these predictor variables. Remember that the methods we apply have been selected because they can deal with mixed sets of continuous and categorical predictor variables.

To specify a model formula in R, we use a special syntax. If we were to parse the prognostic predictors and the survival status variables from the 10-year survival dataset, the formula would be:

alivestatus ~ size + grade + nodespos, data = data.10yr

The tilde (~) symbol separates the response variable, which always comes first, from the predictor variables. The predictor variables, or terms, are added to the formula using the addition (+) symbol. If we were to specify interactions between predictor variables, we would use a colon (:) (e.g., size:grade + nodespos). If we want to specify the terms and interactions between them, we use the multiplication symbol (*) (e.g., size * grade + nodespos), which is the equivalent of size + grade + nodespos + size:grade. A term can also be ignored using the minus symbol (−).

The output from each model in the case of binary response variables will be the same. If the response variable in the dataset is binary and encoded as 0 or 1, the predicted value from the model for new cases will also be 0 or 1.

15.3.1 Logistic Regression

We have already seen in Chapter 11 how we can apply linear regression to measure the degree of relationship between a dependent response variable and an independent predictor variable. Linear regression can be extended for two or more predictor variables, as we will see when we consider multiple regression later in the chapter. There are certain assumptions for linear regression that fail when the response variable is not continuous. If the response variable in our dataset is binary (dichotomous), we need to consider using generalized linear models.

The generalized linear family of models are an extension of linear modeling that allow for the response variable to follow a non-normal probability distribution. They also do not require equal variance within each class. One of the most commonly used generalized linear models is logistic regression. Logistic regression may be applied to a dataset where there is a nonlinear relationship between the response variable and one or more predictor variables.

Logistic regression is easy to perform in both R and Weka. Because the method can be called using a function in the preinstalled stats package, there is really no need to perform the analysis in Weka.

15.3.2 Performing Logistic Regression in R Using `glm()`

We will begin by performing logistic regression on the complete 10-year survival dataset held within the `data.10yr` data frame (ensuring that factor variables have been set). In R syntax, we will use the following formula for our model:

> alivestatus ~ size + nodespos + grade

We can use `summary()` and `length` on the `alivestatus` factor to determine the number of rows for each outcome (i.e., patients alive or dead after 10 years), as well as the total number of rows (Code 15.1).

```
> summary(data.10yr$alivestatus)
    0     1
11718  3476
> length(data.10yr$alivestatus)
[1] 15194
```

Code 15.1.

Notice that the number of patients alive after 10 years (0) is more than three times the number of patients that have died (1).

To create a logistic regression model, we call the glm() function (Code 15.2). Aside from the formula, there are many parameters you can parse to glm(), which can be viewed by typing "?glm." The glm() function allows you to call different models within the generalized linear family, so we need to specify logistic regression by setting the family parameter to binomial. We also ensure that the data parameter is set to data.10yr. glm() creates an object of class glm, which contains a list of many components. Using summary(), we can retrieve key components related to the model, including the coefficients and relative statistics.

```
> lr.glm <- glm(alivestatus ~ size + nodespos + grade, data = data.10yr,
family="binomial")
> summary(lr.glm)

Call:
glm(formula = alivestatus ~ size + nodespos + grade, family = "binomial",
    data = data.10yr)

Deviance Residuals:
    Min       1Q    Median        3Q       Max
-5.0801   -0.6829   -0.5286   -0.3118    2.5217

Coefficients:
             Estimate Std. Error z value Pr(>|z|)
(Intercept) -3.219505   0.104260 -30.879  < 2e-16 ***
size         0.027535   0.001376  20.004  < 2e-16 ***
nodespos     0.135413   0.005643  23.997  < 2e-16 ***
grade2       0.853755   0.107022   7.977 1.49e-15 ***
grade3       1.336117   0.105627  12.649  < 2e-16 ***
grade4       1.395016   0.129778  10.749  < 2e-16 ***
---
Signif. codes:  0 '***' 0.001 '**' 0.01 '*' 0.05 '.' 0.1 ' ' 1
(Dispersion parameter for binomial family taken to be 1)

    Null deviance: 16342  on 15193  degrees of freedom
Residual deviance: 14051  on 15188  degrees of freedom
AIC: 14063
Number of Fisher Scoring Iterations: 4
```

Code 15.2.

glm() provides a model that is an equation to predict whether a patient will survive 10 years or not given the values of tumor size, nodes positive, and grade. The parameters for the equation and corresponding statistics are given in a matrix provided by the "Coefficients" list. The first column lists the predictor variables, where factors (in this case, grade) are split according to level. The second column provides an estimate of the coefficients for the equation. The remaining columns provide the standard error for the coefficients, z statistic values (estimate divided by the standard error and referred to as Wald Test in other software packages), and P Values dictating whether the association between survival status and predictor variable is significant. All predictor variables are reliable predictors of outcome, each being highly significant.

The "Deviance Residuals" list summarizes the range of residual differences between predicted and actual outcome values.

15.3.3 Evaluating the Model

To evaluate the predictive ability of the model, we can use the `predict()` function to predict the probability of outcome for all cases in the `data.10yr` dataset (Code 15.3).

```
> pr<-predict(lr.glm, newdata=data.10yr, type="response")
> for(i in 1:length(pr))ifelse(pr[i]>0.5, pr[i]<-1, pr[i]<-0)
> confmat<-table(data.10yr$alivestatus, pr, dnn=c("actual","predicted"))
> confmat
       predicted
actual      0     1
      0 11301   417
      1  2697   779
```

Code 15.3.

The predicted values are put into a vector called `pr`. These values are probabilities and will be between 0 and 1. If a case has a value greater than 0.5, it is reasonable to predict that this patient will not survive 10 years (i.e., patient has a less than 50% chance of surviving). The next line loops the length of `pr` to convert each predicted value to 0, if less than or equal to 0.5, or 1, if equal or greater than 0.5. The third line creates a contingency table or confusion matrix, called `confmat`, of four counts. The columns represent predicted values, while the rows represent actual values. Each column and row has a header of 0 (alive) or 1 (dead).

The table summarizes the error rate of the model. The first cell of the top row tells us that 11,580 cases have been correctly predicted as being alive after 10 years (TP = true positives). The second cell of the top row shows 138 cases incorrectly classified as not surviving 10 years even though they did (FN = false negatives). The first cell of the bottom row shows 3,150 cases incorrectly classified as surviving 10 years but who had actually died (FP = false positives). Finally, the last cell shows 326 cases correctly classified as dying within 10 years (TN = true negatives).

From here, we can estimate the accuracy of the model where:

Accuracy = (TP + TN) / (TP + TN + FP + FN)

which in R gives a value of 0.795 (Code 15.4).

```
> accuracy<-(11301+779)/nrow(data.10yr); accuracy
[1] 0.7950507
#recall
> recall<-11301/(11301+417); recall
[1] 0.9644137
#precision
> precision<-(11301)/(11301+2697); precision
[1] 0.8073296
> TNR<-779/(2697+779); TNR
[1] 0.2241082
```

Code 15.4.

Medics are often more familiar with the terms sensitivity (TP / (TP + FN)) and specificity (TN / (FP + TN)). The accuracy of the model is 79.5%, which suggests the model is rather good at predictions. The recall, or true positive rate, tells us that 96.4% of cases that survive 10 years were correctly predicted. The precision, or proportion of cases correctly predicted as surviving, is also high at 80.7%. These figures so far give an ever greater confidence in the predictive power of our model. There are two problems, however, that we have not accounted for.

If we examine the true negative rate (TNR), which is the proportion of cases correctly predicted as dying within 10 years, we observe a value of 22.4%, which is a modeling disaster. There is an extremely high cost of predicting that a patient will survive 10 years when in all likelihood she will not. When applying your model, you will wrongly predict that 77% of cases with a poorer prognosis actually have a good prognosis. The reason for the bias toward accurate prediction of good-prognosis patients is that the dataset was heavily biased in number for these patients. Only about 23% of patients in the entire dataset died within 10 years, and this is referred to as the class imbalance problem.[21]

A second problem is that we have not evaluated our model on a different dataset. It is one thing to test the accuracy of a model on the dataset on which it was built, but if our model has been "overfitted" to that data, then it may perform poorly on new cases. We used our entire dataset to build the model, whereas it would have been wiser to split the data into a training dataset and a testing dataset. The model can then be built or trained on one dataset and tested on another. This approach is called the holdout method. If a dataset is small and splitting is not practical, we can use an alternative approach called *k*-fold cross-evaluation. The next few sections look at how we can address the class imbalance to redevelop our model and create datasets for model evaluation.

15.4 Dealing with Class Imbalance

Class imbalance is clearly a problem if the minority class has equal or more importance than the majority class. It would be fair to say that the class representing patients who died within 10 years has more importance than the class stating survival. The reason for this is that the cost of wrongly predicting survival for a patient who will die is much higher than wrongly predicting a patient who will die.

One way of accounting for this cost is to use a cost-sensitive classification method that tries to reduce the rate of false positives or false negatives. Another widely used approach is to resample from the original data so as to obtain a new dataset containing cases with a similar distribution for each class. In the case of the 10-year survival dataset, we could create a new dataset with all the deceased cases and then undersample the remaining cases.

We know that there are 3476 cases that make up the minority (dead) class and 11,718 cases in the majority class (alive). The first thing we need to do is order the dataset according to the `alivestatus` variable:

```
> data.10yr<-data.10yr[order(data.10yr$alivestatus),]
```

To under-sample this majority class, we can choose 3476 random cases using the sample function (Code 15.5). The vector of random numbers, called rand.0, can then be used to sample the data.10yr dataset and create a new data frame we call alive. Furthermore, we can create a second data frame called dead to hold the 3476 cases that are classified as dead.

```
> rand.0<-sample(1:11718, 3476)
> alive <- data.10yr[rand.0,]
> dead <- data.10yr[11718:15194,]
```

Code 15.5.

We now have two data frames of equal size, each containing 3476 rows.

15.5 Splitting Datasets for Training and Testing

The holdout method is frequently used to create training and testing datasets for model evaluation when sufficient data is available. Typically, the dataset is split randomly into a training set of two-thirds of the data with the remaining third used for the test set.

To create a split dataset for the holdout method, we first need to combine into a data frame called training.split the alive and dead datasets created in the last section (Code 15.6). We can then use sample() once more to randomly select 70% of row numbers in training.split and put these values into a vector called split. These random values can then be used as the array elements of training.split to select a random sample equating to 70% of this data frame, which is put into a new data frame called split.train. The remaining 30% of the training.split dataset is put into a data frame called split.test.

```
> training.split <-rbind(alive, dead)
> split <- sample(nrow(training.split), floor(nrow(training.split) * 0.7))
> split.train <-training.split[split, ]
> split.test <-training.split[-split, ]
```

Code 15.6.

We now have two data frames, one for model training (split.train) and one for model testing (split.test).

15.6 *K*-Fold Cross-Validation

If a dataset has insufficient cases for splitting, then one can apply *k*-fold cross-evaluation. This method is actually a powerful and common model validation procedure in its own right, regardless of the size of the dataset. It is an extension of the holdout method in that the dataset is split for model training and testing. Ten-fold cross-evaluation is often used, where the dataset is divided into 10 parts and the model trained on nine-tenths of the data before calculating the error rate. Once this is complete, the next tenth of the data is held back while the model is continued to be trained on the remaining nine-tenths. Again the error rate is calculated. This process is repeated until all 10 parts have been held out and the resultant model will have been trained and evaluated on 10 different individual datasets. Each case will have been included nine times in the overall training process and once in the evaluation process. The mean of the 10 error rates is then calculated to provide an overall estimate of the true error.

Whether you use the holdout or k-fold cross-evaluation procedure, it is always wise to retain a third fraction of the complete dataset for testing. In other words, hold some data back so that you can apply the model to data that is not part of the training or testing process. We will be using 10-fold cross-evaluation extensively in subsequent sections, so we need to create a balanced dataset but hold back a small random sample on which to apply the models we generate. The small dataset will hold just 200 cases. In Section 15.4, we created two data frames called `alive` and `dead` that contained equal numbers of cases. We can randomly sample 100 cases from each of these data frames to create our mini dataset (Code 15.7).

```
> split.alive<-sample(nrow(alive), nrow(alive)-100)
> split.dead<-sample(nrow(dead), nrow(dead)-100)
> train.full <-rbind(alive[split.alive,], dead[split.dead,])
> test.200 <-rbind(alive[-split.alive,], dead[-split.dead,])
```

Code 15.7.

As you can see in the first and second lines, we actually sample random numbers for the complete `alive` or `dead` data frames minus 100 cases in each. This data is temporarily put into the `split.alive` and `split.dead` vectors before being used to retrieve the necessary rows from `alive` and `dead`, which is then combined into the `train.full` dataset. The remaining rows in the `alive` and `dead` data frames are combined into the `test.200` dataset. We will use these datasets throughout the rest of this chapter.

15.7 Repeating Logistic Regression with Split Balanced Dataset

We can now repeat the logistic regression approach we applied in Section 15.3.1, but this time using the split balanced training set called `split.train` (Code 15.8).

```
> lr.split<-glm(formula = alivestatus ~ size + nodespos + grade, family =
"binomial", data = split.train)
> summary(lr.split)

Coefficients:
             Estimate Std. Error z value Pr(>|z|)
(Intercept) -1.902583   0.133830 -14.216  < 2e-16 ***
```

```
size        0.032049  0.002469  12.982  < 2e-16 ***
nodespos    0.143501  0.010209  14.056  < 2e-16 ***
grade2      0.632924  0.137478   4.604 4.15e-06 ***
grade3      1.081525  0.136741   7.909 2.59e-15 ***
grade4      1.053683  0.180353   5.842 5.15e-09 ***
---
Signif. codes:  0 '***' 0.001 '**' 0.01 '*' 0.05 '.' 0.1 ' ' 1
Number of Fisher Scoring Iterations: 5
```

Code 15.8.

The output just shows the coefficients, and once more we see that all predictor variables are highly significant. Using the `split.test` for evaluation, we can immediately see that we have improved the model (Code 15.9).

```
> pr.split<-predict(lr.split, newdata=split.test, type="response")
> for(i in 1:length(pr.split))ifelse(pr.split[i]>0.5, pr.split[i]<-1,
pr.split[i]<-0)
> confmat<-table(split.test$alivestatus, pr.split, dnn=c("actual","predicted"))
> confmat
       predicted
actual   0   1
     0 816 212
     1 392 666
> accuracy<-(816+666)/nrow(split.test); accuracy
[1] 0.7104506
> precision<-(816)/(816+392); precision
[1] 0.6754967
> recall<-816/(816+212); recall
[1] 0.7937743
> TNR<-666/(666+392); TNR
[1] 0.6294896
```

Code 15.9.

Whereas the accuracy, recall, and precision are decreased to 71%, 79.4%, and 67.5%, respectively, the true negative rate is increased to 63.0%.

Next, we will look at whether we can improve the model further using 10-fold cross-evaluation.

15.8 Repeating Logistic Regression Using 10-Fold Cross-Validation

To evaluate a logistic regression model using cross validation, we first need to create the model using the `train.full` dataset:

```
lr.fold<-glm(formula = alivestatus ~ size + nodespos + grade,
family = "binomial",  data = train.full)
```

To calculate the cross-validation estimate of accuracy, we can use a function called `CVbinary()` from a package called `DAAG`. Firstly, make sure `DAAG` is installed:

```
> install.packages("DAAG")
```

Then we can parse the `lr.fold` object to `CVbinary()` and examine the output (Code 15.10).

```
> library(DAAG)
> CVbinary(lr.fold)

Fold:  6 2 10 4 9 1 7 8 5 3
Internal estimate of accuracy = 0.693
Cross-validation estimate of accuracy = 0.694
```

Code 15.10.

With an accuracy of 69.4%, this is slightly lower than that for the model gener-ated using a split dataset. However, this model has undergone a rigorous validation process, and we can now use the model to predict survival in the small dataset of 200 cases (Code 15.11).

```
> pr.fold<-predict(lr.fold, newdata=test.200, type="response")
> for(i in 1:length(pr.fold))ifelse(pr.fold[i]>0.5, pr.fold[i]<-1,
pr.fold[i]<-0)
> confmat<-table(test.200$alivestatus, pr.fold, dnn=c("actual","predicted"))
> confmat
      predicted
actual  0  1
     0 82 18
     1 37 63
```

Code 15.11.

The predicted accuracy is 72.5%, and we can conclude that the model has per-formed quite well on this independent dataset. But we can now ask ourselves how this model might compare to those produced using machine-learning methods.

15.9 Decision Trees

Decision trees are often used in tutorials and introductory texts to machine learning because they are relatively easy to understand and visualize. A decision tree is like a flowchart comprised of a series of questions. Decision trees are popular because they often provide models with good accuracy, but also allow the mechanics of the model to be visualized. They are popular in medicine because the concept of a decision tree reflects the procedure used by a doctor when asking a patient a series of questions un-til arriving at a diagnosis or prognosis. There are a wide range of decision tree algo-rithms that can predict both binary and multiple outcomes.

You start at the bottom of the tree and work your way through the tree by an-swering yes or no to these questions, branching your way through until you arrive at a leaf with an answer. The starting point and points where a tree will branch are called the root node and internal nodes, respectively. The endpoints of each branch are called leaf, or terminal, nodes.

When a tree model is built, you can then apply it to new cases to predict an out-come. If a tree is small, you could do this by hand, but in reality you will more likely make predictions algorithmically. Like any model, a decision tree will predict an outcome based on the values of the input attributes. So let's consider in more detail what happens every time you arrive at an internal node. The example that follows builds a tree model using the same three predictor attributes used for logistic regression. Thus, our tree will have a series of internal nodes where questions are asked sequentially about the tumor size, tumor grade, and number of nodes positive for that patient. There may be a number of leaves in the tree, but each one will either predict 0 or 1, i.e., survival or death within 10 years.

Although we won't look at technical aspects of decision trees, you should be aware that tree building has a number of parameters that can affect the overall performance of the model. First, there are a number of ways a tree can be grown and these algorithms have advantages and disadvantages. Methods include ID3, CART, and C4.5, which is known as J48 in Weka. Splitting at branches can be either binary (i.e., two branches) or multiple. Trees can be pruned either during the building process or after the initial tree is built. Pruning has the effect of removing subtrees or preventing them from growing and reducing the effectiveness of the model. Most often, pruning is carried out after the tree is built.

15.9.1 Build the Model

To build a decision tree model for the 10-year survival dataset, we will use the J48 algorithm in Weka. The function is called J48() and is already implemented in RWeka. Parameters that could be parsed using Weka_control() within the J48() function were shown in Table 15.4.

If we just accept the default parameters and not use Weka_control(), we only need to parse the formula to J48() and create an object called tree.full (Code 15.12). It is vital that the alivestatus variable be declared as a factor or Java will throw an error. The tree can then be displayed in the console by typing the object name. It is also possible to use the party or RGraphViz (interfacing GraphViz) packages to plot the tree.

```
> train.full$alivestatus<-as.factor(train.full$alivestatus)
> train.full$grade<-as.factor(train.full$grade)
> tree.full<-J48(alivestatus ~ size + nodespos + grade, data = train.full)
> tree.full
J48 pruned tree
------------------

nodespos <= 0
|   size <= 20: 0 (2485.0/631.0)
|   size > 20
|   |   grade = 1: 0 (43.0/13.0)
|   |   grade = 2
|   |   |   size <= 25: 0 (162.0/69.0)
|   |   |   size > 25: 1 (218.0/97.0)
|   |   grade = 3: 1 (636.0/314.0)
|   |   grade = 4
|   |   |   size <= 38: 0 (63.0/27.0)
|   |   |   size > 38: 1 (20.0/5.0)
nodespos > 0
|   nodespos <= 5
|   |   size <= 23
|   |   |   grade = 1: 0 (70.0/18.0)
```

```
|   |   |   grade = 2
|   |   |   |   size <= 19: 0 (255.0/101.0)
|   |   |   |   size > 19: 1 (131.0/56.0)
|   |   |   grade = 3: 1 (430.0/181.0)
|   |   |   grade = 4: 0 (61.0/29.0)
|   |   size > 23: 1 (1073.0/298.0)
|   nodespos > 5: 1 (1106.0/175.0)

Number of Leaves  :     14
Size of the tree  :     23
```

Code 15.12.

If we are predicting survival for a new patient, the decision tree may be read by starting at the bottom left of the tree. We follow the line to the first node, asking if the number of nodes positive is equal to or greater than 0. If not, we follow the line up the tree (through nodespos <=0, which is true) to the next node, which asks if the tumor size is less than or equal to 20. If true, we reach a leaf where the outcome is 0, which means the patient is predicted to survive 10 years. The numbers that follow the outcome score in parentheses show the correct and incorrect classifications in the model for this leaf. The precision for this node is 79.9% (2485 / (2485 + 631)).

For a second example of reading the tree, we will return to the root node. Let us say another patient has a nodes positive value greater than 0. Now we reach an internal node asking if the number of positive nodes is less than or equal to 5. If true, we go to the next node that queries size. If false, we follow the line to a leaf that predicts a 1 and poor prognosis. In this case, the precision is 86.3% (1106 / (1106 + 175)), which is encouraging given that there is a high cost of wrongly predicting good survival in a patient with poor outlook.

15.9.2 Evaluate the Model

We can evaluate the model using 10-fold cross-validation. Again we will let Weka do this using the evaluate_Weka_classifier() function and parsing the tree.full model and the original train.full dataset used to build the model (Code 15.13).

```
###############################################
# 10-fold cross validation on full dataset #
###############################################
> tree.full.eval<-evaluate_Weka_classifier(tree.full, train.full, class=T)
> tree.full.eval

=== Summary ===

Correctly Classified Instances     4739               70.1762 %
Incorrectly Classified Instances   2014               29.8238 %
Kappa statistic                       0.4035
Mean absolute error                   0.397
Root mean squared error               0.4455
Relative absolute error              79.3987 %
Root relative squared error          89.106  %
Total Number of Instances          6753

=== Detailed Accuracy By Class ===
```

```
TP Rate   FP Rate   Precision   Recall   F-Measure   ROC Area   Class
  0.667     0.263      0.717     0.667      0.691       0.753     0
  0.737     0.333      0.688     0.737      0.712       0.753     1

=== Confusion Matrix ===

     a     b    <-- classified as
  2251  1126 |    a = 0
   888  2488 |    b = 1


################################################
# Testing the model on an independent dataset #
################################################

> tree.200.eval<-evaluate_Weka_classifier(tree.full, test.200, class=T)
> tree.200.eval

=== Summary ===

Correctly Classified Instances        143              71.5    %
Incorrectly Classified Instances       57              28.5    %
Kappa statistic                         0.43
Mean absolute error                     0.3897
Root mean squared error                 0.4402
Relative absolute error                77.9459 %
Root relative squared error            88.039  %
Total Number of Instances             200

=== Detailed Accuracy By Class ===

TP Rate   FP Rate   Precision   Recall   F-Measure   ROC Area   Class
  0.66      0.23       0.742     0.66       0.698       0.763     0
  0.77      0.34       0.694     0.77       0.73        0.763     1

=== Confusion Matrix ===

   a   b    <-- classified as
  66  34 |    a = 0
  23  77 |    b = 1
```

Code 15.13.

The object we create, called `tree.full.eval`, is a list with a number of components. First, we see a summary for the evaluation that tells us that the overall accuracy is 70.2%. What follows in this list are a number of statistics and error values. The Kappa statistic measures the agreement between values predicted by the model and those observed, but also accounts for the proportion of agreement due to chance. Four values that estimate the errors of prediction are also given, which would be useful in a situation where the prediction was a numeric value as opposed to nominal, as in this case.

You may have noticed in the function call `evaluate_Weka_classifier()` that we set a parameter called class to True. By doing this, we allow the "Detailed Accuracy By Class" table to be displayed, which allows us to assess the prediction error rates, much the same as we did for logistic regression. The first row shows values for the good prognosis class 0 and the second row values for class 1. The True positive rate (recall) for class 1 is 73.7%, although the false positive rate at 33.3% is slightly higher than the class 0. The *F*-measure is a figure equal to twice the product of recall and precision divided by the sum of recall and precision.

The ROC area is a widely used evaluation measure produced using ROC curves. ROC (receiver operating characteristic) curves are plots of the percentage of true positives for a class against the percentage of false positives for the same class. The plot will be a curved line whereby the accuracy of predicting a class will be equal to the area under the curve. The ROC area for both classes for the decision tree is 75.3%. A confusion matrix is also provided after the summary statistics.

With the decision tree model evaluated, we can now go one step further and test it on the small but independent dataset of 200 cases. We use the `evaluate_Weka_classifier()` once more for 10-fold cross-evaluation, parsing the `tree.full` model object and the `test.200` data frame. As you will probably have realized, the advantage of using RWeka is that with just a couple of lines of code you can generate comprehensive output for both your model and its evaluation. The output shows that there is excellent agreement with the predictive ability demonstrated when evaluating on the full dataset.

15.10 Support Vector Machine (SVM)

The next modeling approach we will apply to the survival dataset is a method called support vector machine (SVM). SVM has its roots in statistical learning theory and has become extremely popular for predictive modeling tasks, particularly as it deals very well with high-dimensionality data.[22] SVM often outperforms other robust modeling methods on the same dataset.

To begin a brief introduction to SVM, picture all the instances in a multivariate dataset as points plotted in multidimensional space (at best, you could probably picture a 3-D scatterplot!). Then imagine that to separate instances according to the class in which they fall, we need to draw a line between them. A linear model is one that draws a linear boundary through instances of different classes in such a multidimensional space. Linear models, however, must be built using numerical attributes, whereas modeling many biomedical datasets requires class separation by nonlinear boundaries.

SVM is clever in that it transforms a dataset of training instances from a nonlinear space into linear space so we can apply linear models. Linear boundaries can then be drawn between instances of different classes. Whereas a boundary is linear in the transformed space, it is actually nonlinear in the original space. In multidimensional space, a boundary between instances is a hyperplane. You could draw a number of hyperplanes between instances, and it is the job of SVM during the training process to find the most optimal hyperplane. These training patterns are called support vectors.

SVM is considered a powerful approach to modeling, but has drawbacks. Execution time can be extremely long compared to other techniques. Another drawback is that many coefficients can be generated in the model during the transformation, a problem that can lead to overfitting.

15.10.1 Build the Model

The SVM algorithm implemented in Weka is called SMO (sequential minimal optimization). A significant factor in the SVM model-building process is parameter adjust-

Table 15-5   Some Control Parameters for SMO().

C	The complexity constant C (default 1)
	Number of arguments: 1
N	Whether to 0=normalize/1=standardize/2=neither (default 0=normalize)
	Number of arguments: 1
L	The tolerance parameter (default 1.0e-3)
	Number of arguments: 1
P	The epsilon for round-off error (default 1.0e-12)
	Number of arguments: 1

ment. A number of important parameter options are shown in Table 15.5. An SVM model can be generated using RWeka's inbuilt SMO() function (Code 15.14). Aside from the formula, we set a training parameter to use RBF (radial basis function) kernels via the Weka_control() function and accept all other parameters with their default values.

```
> smo.full<-SMO(alivestatus ~ size + nodespos + grade,data = train.full ,
control = Weka_control(K =
"weka.classifiers.functions.supportVector.RBFKernel"))
> smo.full$classifier
  0.1791 *  <0.045226 0 0 1 0 0 > * X]
-        1      *  <0.090452 0.010309 0 1 0 0 > * X]
+        1      *  <0.095477 0.051546 0 0 0 1 > * X]
+        1      *  <0.105528 0.010309 0 0 1 0 > * X]
-        1      *  <0.060302 0 0 0 1 0 > * X]
+        1      *  <0.221106 0.030928 0 1 0 0 > * X]
-        1      *  <0.246231 0.010309 0 0 1 0 > * X]
-        1      *  <0.070352 0 0 1 0 0 > * X]
+        1      *  <0.110553 0 0 0 0 1 > * X]
```

Code 15.14.

Code 15.14 shows a partial output of the classifier. In reality, you would generate a number of SVM models for the dataset for optimization by adjusting key parameters, including the complexity constant.

15.10.2 Evaluate the Model

Ten-fold cross-validation of the SVM model is performed using the evaluate_Weka_classifier() function (Code 15.15). We also test the model on the small 200-instance test set. In this case, the evaluation on the test set shows that the model is less accurate than when applied on the full dataset. It must be stressed again, though, that when you attempt to model with SVM, much consideration should be given to parameter adjustment but being mindful of the possibility of overfitting.

```
############################################
# 10-fold cross validation on full dataset #
############################################

> smo.full.eval<-evaluate_Weka_classifier(smo.full, numFolds=10,train.full,
class=T)
```

```
> smo.full.eval
=== 10 Fold Cross Validation ===

=== Summary ===

Correctly Classified Instances          4620              68.414 %
Incorrectly Classified Instances        2133              31.586 %
Kappa statistic                            0.3683
Mean absolute error                        0.3159
Root mean squared error                    0.562
Relative absolute error                   63.1719 %
Root relative squared error              112.4028 %
Total Number of Instances               6753

=== Detailed Accuracy By Class ===

TP Rate   FP Rate   Precision   Recall  F-Measure   ROC Area  Class
  0.697     0.329       0.679    0.697      0.688      0.684    0
  0.671     0.303       0.689    0.671      0.68       0.684    1

=== Confusion Matrix ===

    a     b    <-- classified as
 2355  1022  |    a = 0
 1111  2265  |    b = 1


###############################################
# Testing the model on an independent dataset #
###############################################

> smo.200.eval<-evaluate_Weka_classifier(smo.full, numFolds=10,test.200,
class=T)
> smo.200.eval
=== 10 Fold Cross Validation ===

=== Summary ===

Correctly Classified Instances           133               66.5   %
Incorrectly Classified Instances          67               33.5   %
Kappa statistic                            0.33
Mean absolute error                        0.335
Root mean squared error                    0.5788
Relative absolute error                   67       %
Root relative squared error              115.7584 %
Total Number of Instances                200

=== Detailed Accuracy By Class ===

TP Rate   FP Rate   Precision   Recall  F-Measure   ROC Area  Class
  0.59      0.26        0.694    0.59       0.638      0.665    0
  0.74      0.41        0.643    0.74       0.688      0.665    1

=== Confusion Matrix ===

  a   b   <-- classified as
 59  41  |   a = 0
 26  74  |   b = 1
```

Code 15.15.

15.11 Ensemble Methods: Improving Models Using Meta Learners

So far, we have generated predictive models using different methods and predicted classes for new cases using a single classifier each time. It is possible to improve the accuracy of prediction by combining the predictions of multiple models. Methods that combine models in this way are called ensemble, or meta learners.

Meta learners work by generating multiple base classifiers from a set of training data. These base classifiers each predict the class for training instances before the meta learner takes a vote as to which prediction is best. In this section, we will investigate two meta learner approaches: boosting and bagging.

15.11.1 Boosting

Boosting builds a series of models all of the same type (such as decision trees), where new models are influenced by how well previous models have performed. It is an iterative process where new models focus on remedying the inaccuracies of those models built in earlier iterations. In other words, instances in the training set that were previously wrongly classified will be given priority by the boosting algorithm in subsequent model building.

Weka has a number of boosting methods implemented, including a version of the commonly used AdaBoost called AdaBoost.M1. The output is a series of models, and the voting process sees models as having different weights for their contribution to prediction according to the model accuracy.

Build the Model

As an example, we will apply boosting to the survival dataset using the J48 decision tree as our model-building algorithm. To implement AdaBoost.M1, we call the `AdaBoostM1()` function and set the classifier algorithm parameter (W) to "J48" using `Weka_control()` (Code 15.16).

```
> boost.full<-AdaBoostM1(alivestatus ~ size + nodespos + grade,data =
train.full, control = Weka_control(W = "J48"))
> boost.full
AdaBoostM1: Base classifiers and their weights:

J48 pruned tree
------------------

nodespos <= 0
|   size <= 20: 0 (2485.0/631.0)
|   size > 20
|   |   grade = 1: 0 (43.0/13.0)
|   |   grade = 2
|   |   |   size <= 25: 0 (162.0/69.0)
|   |   |   size > 25: 1 (218.0/97.0)
|   |   grade = 3: 1 (636.0/314.0)
|   |   grade = 4
|   |   |   size <= 38: 0 (63.0/27.0)
|   |   |   size > 38: 1 (20.0/5.0)
nodespos > 0
|   nodespos <= 5
|   |   size <= 23
|   |   |   grade = 1: 0 (70.0/18.0)
|   |   |   grade = 2
|   |   |   |   size <= 19: 0 (255.0/101.0)
|   |   |   |   size > 19: 1 (131.0/56.0)
|   |   |   grade = 3: 1 (430.0/181.0)
|   |   |   grade = 4: 0 (61.0/29.0)
```

```
|   |     size > 23: 1 (1073.0/298.0)
|   nodespos > 5: 1 (1106.0/175.0)

Number of Leaves  :      14
Size of the tree :      23
Weight: 0.86


J48 pruned tree
------------------

nodespos <= 3
|   size <= 13
|   |   grade = 1
|   |   |   nodespos <= 0: 0 (203.15/38.56)
|   |   |   nodespos > 0
|   |   |   |   nodespos <= 1: 0 (15.93/1.68)
|   |   |   |   nodespos > 1
|   |   |   |   |   nodespos <= 2: 1 (8.13/1.42)
|   |   |   |   |   nodespos > 2: 0 (2.85)
|   |   grade = 2
|   |   |   nodespos <= 0: 0 (519.71/149.21)
|   |   |   nodespos > 0
|   |   |   |   size <= 8: 0 (18.57/5.03)
|   |   |   |   size > 8
|   |   |   |   |   nodespos <= 2
|   |   |   |   |   |   nodespos <= 1
|   |   |   |   |   |   |   size <= 11
|   |   |   |   |   |   |   |   size <= 9: 0 (2.14)
|   |   |   |   |   |   |   |   size > 9: 1 (27.96/7.84)
|   |   |   |   |   |   |   size > 11: 0 (15.0/5.03)
|   |   |   |   |   |   nodespos > 1: 1 (30.34/8.55)
|   |   |   |   |   nodespos > 2: 0 (9.77/3.35)
|   |   grade = 3: 0 (466.73/195.61)
|   |   grade = 4: 1 (60.14/24.94)
|   size > 13
|   |   size <= 21
|   |   |   grade = 1
|   |   |   |   nodespos <= 1: 0 (130.3/46.94)
|   |   |   |   nodespos > 1
|   |   |   |   |   nodespos <= 2
|   |   |   |   |   |   size <= 19: 0 (2.14)
|   |   |   |   |   |   size > 19: 1 (6.45/1.42)
|   |   |   |   |   nodespos > 2: 1 (2.39/0.71)
|   |   |   grade = 2: 1 (764.02/357.84)
|   |   |   grade = 3
|   |   |   |   nodespos <= 0
|   |   |   |   |   size <= 20: 1 (543.1/209.47)
|   |   |   |   |   size > 20: 0 (26.74/9.97)
|   |   |   |   nodespos > 0
|   |   |   |   |   nodespos <= 1: 0 (136.68/37.76)
|   |   |   |   |   nodespos > 1
|   |   |   |   |   |   nodespos <= 2
|   |   |   |   |   |   |   size <= 18: 0 (52.56/15.67)
|   |   |   |   |   |   |   size > 18: 1 (16.43/5.03)
|   |   |   |   |   |   nodespos > 2
|   |   |   |   |   |   |   size <= 17: 1 (10.23/1.68)
|   |   |   |   |   |   |   size > 17: 0 (24.1/10.69)
|   |   |   grade = 4: 1 (112.2/43.46)
|   |   size > 21
|   |   |   nodespos <= 0
|   |   |   |   grade = 1
|   |   |   |   |   size <= 52: 1 (39.61/17.81)
|   |   |   |   |   size > 52: 0 (3.56)
|   |   |   |   grade = 2
|   |   |   |   |   size <= 25: 1 (159.48/60.56)
|   |   |   |   |   size > 25: 0 (248.83/86.21)
|   |   |   |   grade = 3: 0 (729.11/219.45)
|   |   |   |   grade = 4: 1 (85.21/32.61)
```

```
|  |  |    nodespos > 0
|  |  |  |   grade = 1
|  |  |  |  |   nodespos <= 2
|  |  |  |  |  |   nodespos <= 1
|  |  |  |  |  |  |   size <= 22: 1 (3.35)
|  |  |  |  |  |  |   size > 22: 0 (11.69/4.27)
|  |  |  |  |  |   nodespos > 1: 1 (10.48/3.81)
|  |  |  |  |   nodespos > 2: 0 (6.45/0.71)
|  |  |  |   grade = 2
|  |  |  |  |   size <= 29: 0 (143.59/41.32)
|  |  |  |  |   size > 29
|  |  |  |  |  |   nodespos <= 2
|  |  |  |  |  |  |   nodespos <= 1: 0 (68.99/27.07)
|  |  |  |  |  |  |   nodespos > 1
|  |  |  |  |  |  |  |   size <= 30: 1 (4.27)
|  |  |  |  |  |  |  |   size > 30: 0 (50.84/20.66)
|  |  |  |  |  |   nodespos > 2: 1 (35.5/13.41)
|  |  |  |   grade = 3: 1 (509.73/234.71)
|  |  |  |   grade = 4: 1 (64.63/28.25)
nodespos > 3: 1 (1369.93/467.83)

Number of Leaves  :      44
Size of the tree :      79
Weight: 0.57
Number of performed Iterations: 10
```

Code 15.16.

The model is held in the `boost.full` object, and when this is called, a series of decision trees is printed in the Console window. Only the first two are shown here.

Evaluate the Model

We can perform 10-fold cross-validation in the usual way (Code 15.17). The boosted model is then evaluated on the small test set. We can conclude that the boosting procedure has performed relatively well on the independent test set, with improved accuracy and values for error relative to the decision tree model generated without boosting.

```
###############################################
# 10-fold cross validation on full dataset #
###############################################

> boost.full.eval<-evaluate_Weka_classifier(boost.full, numFolds=10,train.full,
class=T)
> boost.full.eval
=== 10 Fold Cross Validation ===

=== Summary ===

Correctly Classified Instances        4694                69.5098 %
Incorrectly Classified Instances      2059                30.4902 %
Kappa statistic                          0.3902
Mean absolute error                      0.3901
Root mean squared error                  0.4476
Relative absolute error                 78.025  %
Root relative squared error             89.5146 %
Total Number of Instances             6753

=== Detailed Accuracy By Class ===
```

```
 TP Rate   FP Rate   Precision   Recall   F-Measure   ROC Area   Class
  0.678     0.287      0.702      0.678     0.69        0.759      0
  0.713     0.322      0.688      0.713     0.7         0.759      1

=== Confusion Matrix ===

     a    b    <-- classified as
  2288 1089 |   a = 0
   970 2406 |   b = 1


##############################################
# Testing the model on an independent dataset #
##############################################

> boost.200.eval<-evaluate_Weka_classifier(boost.full, numFolds=10,test.200,
class=T)
> boost.200.eval
=== 10 Fold Cross Validation ===

=== Summary ===

Correctly Classified Instances          146              73      %
Incorrectly Classified Instances         54              27      %
Kappa statistic                          0.46
Mean absolute error                      0.3299
Root mean squared error                  0.4353
Relative absolute error                 65.9841 %
Root relative squared error             87.0528 %
Total Number of Instances               200

=== Detailed Accuracy By Class ===

 TP Rate   FP Rate   Precision   Recall   F-Measure   ROC Area   Class
  0.76      0.3        0.717      0.76      0.738       0.796      0
  0.7       0.24       0.745      0.7       0.722       0.796      1

=== Confusion Matrix ===

   a  b    <-- classified as
  76 24 |   a = 0
  30 70 |   b = 1
```

Code 15.17.

15.11.2 Bagging

The second meta learner we will consider is called bagging. Bagging differs from boosting mainly by building models independently of each other and offering no weights to models in the voting procedure. Bagging is a process of bootstrapping whereby random datasets of the same size are produced from a training dataset by sampling with replacement. Each of the models generated votes on the prediction from each training or test instance. Bagging is less prone to overfitting the data than is boosting.

Build the Model

In similar fashion to the boosting example, we will apply bagging using the J48 decision tree. The function required to call the bagging algorithm in Weka is called Bagging(), and again we parse the necessary formula and set the the classifier algorithm parameter (W) to "J48" (Code 15.18).

```
> bag.full<-Bagging(alivestatus ~ size + nodespos + grade,data = train.full,
control = Weka_control(W = "J48"))
> bag.full
All the base classifiers:

J48 pruned tree
------------------

nodespos <= 0
|   size <= 17: 0 (1966.0/431.0)
|   size > 17
|   |   grade = 1: 0 (86.0/21.0)
|   |   grade = 2
|   |   |   size <= 27: 0 (407.0/144.0)
|   |   |   size > 27: 1 (190.0/80.0)
|   |   grade = 3: 1 (850.0/408.0)
|   |   grade = 4
|   |   |   size <= 29
|   |   |   |   size <= 18: 1 (12.0/4.0)
|   |   |   |   size > 18: 0 (52.0/17.0)
|   |   |   size > 29: 1 (34.0/9.0)
nodespos > 0
|   nodespos <= 5
|   |   size <= 20
|   |   |   grade = 1: 0 (52.0/12.0)
|   |   |   grade = 2
|   |   |   |   size <= 15: 0 (187.0/58.0)
|   |   |   |   size > 15
|   |   |   |   |   size <= 17
|   |   |   |   |   |   nodespos <= 2: 1 (23.0/4.0)
|   |   |   |   |   |   nodespos > 2
|   |   |   |   |   |   |   size <= 16: 0 (5.0/1.0)
|   |   |   |   |   |   |   size > 16: 1 (9.0/3.0)
|   |   |   |   |   size > 17: 0 (122.0/56.0)
|   |   |   grade = 3: 1 (338.0/149.0)
|   |   |   grade = 4: 0 (47.0/23.0)
|   |   size > 20: 1 (1257.0/386.0)
|   nodespos > 5
|   |   size <= 24
|   |   |   size <= 20: 1 (296.0/55.0)
|   |   |   size > 20
|   |   |   |   size <= 21: 0 (14.0/2.0)
|   |   |   |   size > 21
|   |   |   |   |   nodespos <= 9
|   |   |   |   |   |   grade = 1: 1 (2.0)
|   |   |   |   |   |   grade = 2: 0 (8.0/3.0)
|   |   |   |   |   |   grade = 3: 0 (15.0/4.0)
|   |   |   |   |   |   grade = 4: 1 (1.0)
|   |   |   |   |   nodespos > 9: 1 (24.0/2.0)
|   |   size > 24: 1 (756.0/97.0)

Number of Leaves  :     25

Size of the tree :     43



J48 pruned tree
------------------

nodespos <= 0
|   size <= 15: 0 (1809.0/377.0)
|   size > 15
|   |   grade = 1: 0 (106.0/29.0)
|   |   grade = 2
|   |   |   size <= 25: 0 (471.0/178.0)
|   |   |   size > 25: 1 (213.0/101.0)
```

```
|   |   grade = 3
|   |   |   size <= 40: 0 (810.0/386.0)
|   |   |   size > 40: 1 (99.0/29.0)
|   |   grade = 4
|   |   |   size <= 24: 0 (47.0/20.0)
|   |   |   size > 24: 1 (44.0/14.0)
nodespos > 0
|   nodespos <= 4
|   |   size <= 21
|   |   |   grade = 1: 0 (60.0/16.0)
|   |   |   grade = 2: 0 (343.0/131.0)
|   |   |   grade = 3: 1 (346.0/133.0)
|   |   |   grade = 4: 0 (43.0/17.0)
|   |   size > 21: 1 (1085.0/332.0)
|   nodespos > 4: 1 (1277.0/215.0)

Number of Leaves  :       14
Size of the tree :       23
```

Code 15.18.

Evaluate the Model

Ten-fold cross-validation is performed on the bagged decision tree model (Code 15.19). Finally, the bagged model is evaluated on the small test set.

```
###############################################
# 10-fold cross validation on full dataset #
###############################################

bag.full.eval<-evaluate_Weka_classifier(bag.full, numFolds=10,train.full,
class=T)
> bag.full.eval
=== 10 Fold Cross Validation ===

=== Summary ===

Correctly Classified Instances        4649              68.8435 %
Incorrectly Classified Instances      2104              31.1565 %
Kappa statistic                          0.3769
Mean absolute error                      0.4022
Root mean squared error                  0.4479
Relative absolute error                 80.4367 %
Root relative squared error             89.5732 %
Total Number of Instances             6753

=== Detailed Accuracy By Class ===

TP Rate    FP Rate    Precision    Recall    F-Measure    ROC Area    Class
  0.735      0.358       0.673       0.735       0.702        0.758       0
  0.642      0.265       0.708       0.642       0.673        0.758       1

=== Confusion Matrix ===

    a     b    <-- classified as
 2481   896 |    a = 0
 1208  2168 |    b = 1


###############################################
# Testing the model on an independent dataset #
###############################################
```

```
> bag.200.eval<-evaluate_Weka_classifier(bag.full, numFolds=10,test.200,
class=T)
> bag.200.eval
=== 10 Fold Cross Validation ===

=== Summary ===

Correctly Classified Instances          144            72     %
Incorrectly Classified Instances         56            28     %
Kappa statistic                           0.44
Mean absolute error                       0.383
Root mean squared error                   0.4492
Relative absolute error                  76.5934 %
Root relative squared error              89.8306 %
Total Number of Instances               200

=== Detailed Accuracy By Class ===

TP Rate   FP Rate   Precision   Recall  F-Measure   ROC Area   Class
  0.73      0.29       0.716     0.73      0.723       0.736     0
  0.71      0.27       0.724     0.71      0.717       0.736     1

=== Confusion Matrix ===

  a   b    <-- classified as
 73  27  |   a = 0
 29  71  |   b = 1
```

Code 15.19.

Once more, we see an improved performance of a classifier on the small test set.

15.12 Accessing Other Classifiers: RandomForest

Up until this point, we have created classifiers using function names already available in the RWeka package. As we saw in Table 15.1, the dedicated function names represent only a proportion of classifiers available in Weka. However, you can use RWeka's make_Weka_classifier() function to create an interface to any Weka classifier. All you need to do is parse the "path" of the classifier as a string to make_Weka_classifier(). You can name the resulting object anything you want. For example, to create a classifier interface for the RandomForest function, we would use the following code:

```
> rf <- make_Weka_classifier("weka/classifiers/
trees/RandomForest")
```

Before using our new classifer function, called rf, we will have a brief look at Weka's RandomForest() function. RandomForest() is based on the same concept as the original Random Forest™ algorithm developed by the late Leo Breiman, who was a statistician at Berkley, and Adele Cutler.[23]

Like boosting and bagging, RandomForest() is a function for an ensemble method, but specifically designed for decision trees. RandomForest() generates many decision trees by bootstrapping. The user can define how many trees to construct, and the collection of trees is called the forest. The term "random" reflects the fact that each tree is built using a training set of random instances, unlike boosting, which creates new trees with a focus on hard-to-classify instances. RandomForest()

classifies a new instance by running it through each tree to make a prediction and then votes by taking the majority class predicted by all trees. There is some confusion on FAQ sites online about how similar RandomForest() actually is to Breiman and Cutler's Random Forest™ algorithm, and it is worth noting that the approach is the same except it uses a variation of REPTree() for tree growing.

We can create a new model using the RandomForest() classifier in the usual way (Code 15.20).

```
> randf.full <- rf(alivestatus ~ size + nodespos + grade, data = train.full,
control = Weka_control(I = 1000))
> randf.full
Random forest of 1000 trees, each constructed while considering 3 random
features.
Out of bag error: 0.3431

> summary(randf.full)

=== Summary ===

Correctly Classified Instances         5064               74.9889 %
Incorrectly Classified Instances       1689               25.0111 %
Kappa statistic                           0.4998
Mean absolute error                       0.3275
Root mean squared error                   0.4003
Relative absolute error                  65.5018 %
Root relative squared error              80.0651 %
Total Number of Instances              6753

=== Confusion Matrix ===

    a    b    <-- classified as
 2701  676 |    a = 0
 1013 2363 |    b = 1
```

Code 15.20.

Using Weka_control(), we tell the RandomForest() function to create 1,000 trees by setting the I parameter to 1000. If we output the randf.full object, we see that the model has an "out-of-bag error" of 0.3431. During the training process, RandomForest() will split the dataset into two-thirds training and one-third testing datasets. The test dataset contains the "out-of-bag" instances. The out-of-bag error (oob) is the error rate calculated by RandomForest() for the test dataset. This error is useful because it means we do not necessarily have to perform cross-validation. It is also suggested that the random forest approach is not susceptible to overfitting. Note that the important number of patients who are predicted to survive but actually die is 0.30 (1013/2363).

15.13 Predicting the Probability of Outcome

If you are an oncologist and have read up to this point, you may be questioning the usefulness of predictive models that only tell you whether a patient will survive 10 years or die within that time. A binary predictor, "live" or "die," may not be informative enough. You may prefer to make a prognosis and be able to predict the likelihood of survival or death within a chosen time span.

Most machine-learning methods provide a binary outcome. We can, however, combine machine-learning methods with Bayes theorem, which is a simple means of calculating conditional probabilities. Bayes theorem is itself used in predictive modeling, particularly the naïve Bayes classifier. Weka implements a function called NBTree() that combines the naïve Bayes classifier with a decision tree classifier. What this means is that you can build a hybrid model that has the node structure of a decision tree but the leaves are naïve Bayes classifiers that provide the probabilities of the binary outcomes.

As always, the method is best described by an example. We first need to create an interface function for NBTree(), which we will call nbtree:

```
> nbtree <- make_Weka_classifier("weka/classifiers/trees/NBTree")
```

If you apply the NBTree() function to the tumor size, grade, and nodes positive attributes from the train.full dataset, as we have done previously, you will find that the tree model built will only have nodes for grade. This method will work better with factored variables. Before we apply NBTree() we can factor the size and nodespos attributes so that size ranges from 1 to 5 and nodespos ranges from 0 to 5 (Code 15.21).

```
> train.full.factor<-train.full
> for(i in 1:length(train.full.factor$nodespos))
if(train.full.factor$nodespos[i]>5){train.full.factor$nodespos[i]<-5}
> for(i in 1:length(train.full.factor$size))
if(train.full.factor$size[i]>5){train.full.factor$size[i]<-5}
```

Code 15.21.

Then we can parse the formula to the nbtree() interface to create an object called nbtree.full before displaying the output consisting of a decision tree, probability tables for each leaf, and the summary statistics (Code 15.22).

```
> nbtree.full <- nbtree(alivestatus ~ size + nodespos + grade, data =
train.full.factor)
> nbtree.full
NBTree
------------------

size <= 4.5
|    nodespos <= 1.5: NB 2
|    nodespos > 1.5: NB 3
size > 4.5
|    nodespos <= 1.5
|    |    nodespos <= 0.5
|    |    |    grade = 1: NB 7
|    |    |    grade = 2: NB 8
|    |    |    grade = 3: NB 9
|    |    |    grade = 4: NB 10
|    |    nodespos > 0.5: NB 11
|    nodespos > 1.5: NB 12

Leaf number: 2 Naive Bayes Classifier

Class 0: Prior probability = 0.73
size:  Discrete Estimator. Counts =  67   (Total = 67)
```

```
nodespos:  Discrete Estimator. Counts =  67  (Total = 67)
grade:  Discrete Estimator. Counts =  14 29 22 5  (Total = 70)


Class 1: Prior probability = 0.27
size:  Discrete Estimator. Counts =  25  (Total = 25)
nodespos:  Discrete Estimator. Counts =  25  (Total = 25)
grade:  Discrete Estimator. Counts =  5 9 12 2  (Total = 28)


Leaf number: 3 Naive Bayes Classifier
Class 0: Prior probability = 0.11: Class 1: Prior probability = 0.89
Leaf number: 7 Naive Bayes Classifier
Class 0: Prior probability = 0.86: Class 1: Prior probability = 0.14
Leaf number: 8 Naive Bayes Classifier
Class 0: Prior probability = 0.71: Class 1: Prior probability = 0.29
Leaf number: 9 Naive Bayes Classifier
Class 0: Prior probability = 0.59: Class 1: Prior probability = 0.41
Leaf number: 10 Naive Bayes Classifier
Class 0: Prior probability = 0.57: Class 1: Prior probability = 0.43
Leaf number: 11 Naive Bayes Classifier
Class 0: Prior probability = 0.46: Class 1: Prior probability = 0.54
Leaf number: 12 Naive Bayes Classifier
Class 0: Prior probability = 0.25: Class 1: Prior probability = 0.75


Number of Leaves  :      8
Size of the tree :      13
```

Code 15.22.

Full details are provided for the first two naïve Bayes classifiers only. The decision tree has nodes akin to what we saw for the J48 trees earlier. The leaves, however, provide a number (NB x) corresponding to the naïve Bayes classifiers that follow the tree. For example, if we use the model to classify a new instance where the prediction is made by a leaf labeled "NB 2," we can look at the tables to find that the probability of survival is 0.73.

The model can be evaluated in the usual way (Code 15.23).

```
############################################
# 10-fold cross validation on full dataset #
############################################

> nbtree.full.eval<-evaluate_Weka_classifier(nbtree.full,
numFolds=10,train.full, class=T)
> nbtree.full.eval
=== 10 Fold Cross Validation ===

=== Summary ===

Correctly Classified Instances       4717            69.8504 %
Incorrectly Classified Instances     2036            30.1496 %
Kappa statistic                      0.397
Mean absolute error                  0.3618
Root mean squared error              0.4462
Relative absolute error              72.3503 %
Root relative squared error          89.2364 %
Total Number of Instances            6753

=== Detailed Accuracy By Class ===

TP Rate   FP Rate   Precision   Recall  F-Measure  ROC Area  Class
 0.736     0.339     0.685      0.736     0.71      0.769      0
 0.661     0.264     0.715      0.661     0.687     0.769      1
```

```
=== Confusion Matrix ===

    a    b    <-- classified as
 2487  890 |   a = 0
 1146 2230 |   b = 1


#################################################
# Testing the model on an independent dataset #
#################################################

> nbtree.200.eval<-evaluate_Weka_classifier(nbtree.full, numFolds=10,test.200,
class=T)
> nbtree.200.eval
=== 10 Fold Cross Validation ===

=== Summary ===

Correctly Classified Instances          140              70      %
Incorrectly Classified Instances         60              30      %
Kappa statistic                           0.4
Mean absolute error                       0.3828
Root mean squared error                   0.4538
Relative absolute error                  76.5526 %
Root relative squared error              90.7628 %
Total Number of Instances               200

=== Detailed Accuracy By Class ===

TP Rate   FP Rate   Precision   Recall   F-Measure   ROC Area   Class
  0.66      0.26       0.717      0.66      0.688        0.745     0
  0.74      0.34       0.685      0.74      0.712        0.745     1

=== Confusion Matrix ===

  a  b    <-- classified as
 66 34 |   a = 0
 26 74 |   b = 1
```

Code 15.23.

Finally, we can create a function that can be used to quickly make predictions for new cases using the NBTree model (Code 15.24).

```
NBTree.model<-function(data){
#factorize size and nodespos

for(i in 1:length(data$nodespos)) if(data$nodespos[i]>5){data$nodespos[i]<-5}
for(i in 1:length(data$size)) if(data$size[i]>5){data$size[i]<-5}

prob<-NULL
for(i in 1:nrow(data)){
    pred<-0
    if(data$size[i] <= 4.5){
        ifelse(data$nodespos[i]<=1.5, pred<-0.73, pred<-0.11)
    }
    if(data$size[i] > 4.5){
        if(data$nodespos[i]<=1.5){
            if(data$nodespos[i]>0.5) pred<-0.46
            if(data$nodespos[i]<=0.5){
                if(data$grade[i]==1) pred<-0.86
                if(data$grade[i]==2) pred<-0.71
                if(data$grade[i]==3) pred<-0.59
                if(data$grade[i]==4) pred<-0.5/
```

```
                    }
                }
            if(data$nodespos[i]>1.5) pred<-0.25
            }
        prob<-c(prob, pred)
        }
    }
    return(prob)
    }
```

Code 15.24.

The function can then be used to make predictions, such as with the independent dataset (Code 15.25).

```
> out<-NBTree.model(test.200)
> table(out)
out
0.11 0.25 0.46 0.57 0.59 0.71 0.73 0.86
   1   69   24    8   41   46    3    8
```

Code 15.25.

15.14 Predicting a Numerical Response

All the models generated so far have involved "classification" problems where a binary response or class is predicted. What if we needed to build a predictive model where the response variable was numeric and continuous? Analyses requiring this type of predictive modeling approach are often termed regression problems. One way to do this is to use the well-established statistical method of multiple regression. Multiple regression is easy to perform in R, as well as in Weka through RWeka. There are also machine-learning methods that create numerical response models, most notably regression trees, and these, too, can be accessed using RWeka.

Let us consider a scenario where we want to determine from patient data whether the number of lymph nodes examined for a breast cancer patient is influenced by the size of the tumor and the patient's age. In this particular model, all attributes are continuous variables, but binary variables could also be used. The corresponding variable names in the `train.full` data frame are `nodesexam` (nodes examined), `size` (tumor size), and `ageatdiagnosis` (patient's age).

In this last section, we will use both regression trees and multiple regression to build models to predict nodes examined. Both approaches will be implemented using functions available in Weka, but we will also perform multiple regression using the `lm()` function in R.

15.14.1 Regression Trees Using Weka

You can think of regression trees as the equivalent of classification trees, such as J48, but able to predict a continuous response variable. With an introduction to decision trees already given in Section 15.9, we can launch straight into the problem of building a regression tree to predict nodes examined from our dataset.

A regression tree implementation in Weka suitable for the task is called M5P. Because there is no readymade function in RWeka for M5P, we need to create an interface, called `m5p`, using `make_Weka_classifier()`:

```
> m5p <- make_Weka_classifier("weka/classifiers/trees/M5P")
```

Build the Model

To the `m5p` object we parse the formula that contains `nodesexam` as the response variable, with `size` and `ageatdiagnosis` as predictor variables (Code 15.26).

```
> m5p.full<-m5p(nodesexam ~ size + ageatdiagnosis, data=train.full)
> m5p.full
M5 pruned model tree:
(using smoothed linear models)

ageatdiagnosis <= 54.5 : LM1 (2951/100.367%)
ageatdiagnosis >  54.5 :
|   ageatdiagnosis <= 73.5 : LM2 (2918/100.115%)
|   ageatdiagnosis >  73.5 : LM3 (884/88.299%)

LM num: 1
nodesexam =
        0.029 * size
        - 0.0003 * ageatdiagnosis
        + 16.0199

LM num: 2
nodesexam =
        0.0113 * size
        - 0.0007 * ageatdiagnosis
        + 15.5648

LM num: 3
nodesexam =
        0.0001 * size
        - 0.2189 * ageatdiagnosis
        + 31.0797

Number of Rules : 3
```

Code 15.26.

When we view the output, we see that the final tree utilizes only the `ageatdiagnosis` variable. Essentially, the model separates patients into three groups, depending on their age. If a patient is aged below 54.5 years, the number of nodes examined may be predicted by the equation:

$$0.029 \times size - 0.0003 \times ageatdiagnosis + 16.0199$$

Likewise, two more predictive equations exist if the patient is aged between 54.5 and 73.5 years or older than 73.5 years.

Evaluate the Model

We can evaluate the regression tree model in the usual way (Code 15.27).

```
###########################################
# 10-fold cross validation on full dataset #
###########################################

> m5p.full.eval<-evaluate_Weka_classifier(m5p.full, numFolds=10,train.full,
class=T)
> m5p.full.eval
=== 10 Fold Cross Validation ===

=== Summary ===

Correlation coefficient                  0.1437
Mean absolute error                      5.2778
Root mean squared error                  6.9205
Relative absolute error                 98.8641 %
Root relative squared error             98.9554 %
Total Number of Instances               6753
```

Code 15.27.

There is no binary outcome this time, however, so we cannot evaluate the model for accuracy based on class.

15.14.2 Multiple Regression Using lm()

In Chapter 11, we performed a simple regression of a single predictor variable on a response variable (see Section 11.10). Linear regression can be extended so that more than one predictor variable can be used to predict response. In this way, multiple regression is just another tool in the predictive modeling armory. The goal of multiple regression is to generate a regression equation of the form:

$$y = \beta 0 + \beta 1 x 1 + \beta 2 x 2 + \beta 3 x 3 + \dots$$

where y is the predicted response value, such as nodes examined; $\beta 0$ is a constant term, which is the value of y when all x's are zero (the intercept); $x1$, $x2$, $x3$, etc. are the values of the predictor variables; and $\beta 1$, $\beta 2$, etc. are the partial regression coefficients. Multiple regression, despite its relative simplicity as a statistical procedure, allows predictor variables to be continuous and nonindependent. So far in this chapter we have always made the assumption that there is no interaction between variables and, thus, have not accounted for this in our models. We have assumed that the effects of the predictors have been additive. To determine whether interactions exist between predictor variables, you can add interaction terms to the formula. We have not done this in the previous models for no other reason than to keep things simple as we explored the different functions. We can explore the existence of interactions using multiple regression and the lm() function that comes in the standard stats package (Code 15.28).

```
> model.lm<-lm(nodesexam~ size + ageatdiagnosis + size:ageatdiagnosis,
data=train.full)
> summary(model.lm)
```

```
Call:
lm(formula = nodesexam ~ size + ageatdiagnosis + size:ageatdiagnosis,
    data = train.full)

Residuals:
     Min       1Q   Median       3Q      Max
-18.6148  -4.6848  -0.8826   3.7787  60.9656

Coefficients:
                      Estimate Std. Error t value Pr(>|t|)
(Intercept)          17.6797540  0.6295582  28.083  < 2e-16 ***
size                  0.0702099  0.0188980   3.715 0.000205 ***
ageatdiagnosis       -0.0364730  0.0107289  -3.400 0.000679 ***
size:ageatdiagnosis  -0.0009838  0.0003284  -2.995 0.002752 **
---
Signif. codes:  0 '***' 0.001 '**' 0.01 '*' 0.05 '.' 0.1 ' ' 1

Residual standard error: 6.928 on 6749 degrees of freedom
Multiple R-Squared: 0.01896,    Adjusted R-squared: 0.01852
F-statistic: 43.47 on 3 and 6749 DF, p-value: < 2.2e-16
```

Code 15.28.

The formula supplied to lm() contains the term "size:ageatdiagnosis" where ":" notation signifies that this term is bilinear. We could just as easily have written "size*ageatdiagnosis" to represent the interaction. The interaction terms must be included after the main effect terms, which are the individual variables. We see from the summary that the bilinear interaction between size and ageatdiagnosis is significant (as are the other predictor variables). The predictor variables individually (main effects) have an even higher level of significance. Next we ask what happens if we remove the interaction term from our model to leave just the main effects—will it improve in terms of significance of the two predictor variables? Code 15.29 shows this function call and the resulting output.

```
> model.lm<-lm(nodesexam~ size + ageatdiagnosis, data=train.full)
> summary(model.lm)

Call:
lm(formula = nodesexam ~ size + ageatdiagnosis, data = train.full)

Residuals:
    Min      1Q  Median      3Q     Max
-17.372  -4.679  -0.886   3.784  60.880

Coefficients:
                 Estimate Std. Error t value Pr(>|t|)
(Intercept)     19.170510   0.385757  49.696  < 2e-16 ***
size             0.015353   0.004662   3.293 0.000996 ***
ageatdiagnosis  -0.062872   0.006122 -10.270  < 2e-16 ***
---
Signif. codes:  0 '***' 0.001 '**' 0.01 '*' 0.05 '.' 0.1 ' ' 1

Residual standard error: 6.932 on 6750 degrees of freedom
Multiple R-Squared: 0.01765,    Adjusted R-squared: 0.01736
F-statistic: 60.65 on 2 and 6750 DF, p-value: < 2.2e-16
```

Code 15.29.

We now see that the two variables are highly significant and that this is a more robust model. The "Estimate column" in the summary provides the coefficients for the regression equation, which is

nodesexam = (0.0154 * size) – (0.0629 * ageatdiagnosis) + 19.1705

15.14.3 Multiple Regression Using LinearRegression in Weka

We can also perform multiple regression using the LinearRegression method in Weka via RWeka. We first need to create an interface object to the method using make_Weka_classifier:

```
> linreg<-make_Weka_classifier("weka/classifiers/functions/
LinearRegression")
```

We can then use the linreg object to parse the formula and build the model (Code 15.30).

```
> linreg.full<-linreg(nodesexam ~ size + ageatdiagnosis, data = train.full)
> linreg.full

Linear Regression Model

nodesexam =

      0.0154 * size +
     -0.0629 * ageatdiagnosis +
     19.1705

> summary(linreg.full)

=== Summary ===

Correlation coefficient                   0.1329
Mean absolute error                       5.2824
Root mean squared error                   6.9306
Relative absolute error                  98.969  %
Root relative squared error              99.1135 %
Total Number of Instances              6753

> linreg.full.eval<-evaluate_Weka_classifier(linreg.full,
numFolds=10,train.full, class=T)
> linreg.full.eval
=== 10 Fold Cross Validation ===

=== Summary ===

Correlation coefficient                   0.129
Mean absolute error                       5.285
Root mean squared error                   6.9342
Relative absolute error                  99.0051 %
Root relative squared error              99.1491 %
Total Number of Instances              6753
```

Code 15.30.

The output provides the regression equation, which, of course, is identical to that produced using lm().

15.15 Summary

In this chapter, we have explored a number of different statistical and machine-learning methods for generating predictive models for datasets with either binary or continuous response variables. A single chapter covering many methods does not permit any more than an introduction to each approach and a "how to" guide to model building in R and RWeka. It is critical that one does not apply classification or regression methods to datasets without having confidence that the methods are indeed suitable for the data.

For the binary outcome survival status dataset, we generated a number of models from diverse methods. This was useful because it gave us a choice of models and indicates which model is superior by assessing the accuracy (as well as other measures). Do we choose the model with the highest overall accuracy? If we just go by accuracy, then the best model is `randomforest` (75.0%). We did, however, express concern about cost of predicting patients to survive 10 years but who actually die. If this is more important than overall accuracy, our best model is produced by bagging (26.5% error) and the worst is the decision tree (33.3% error). The second best error rate for false-positive rate is `randomforest` (30% error), but the cost of wrongly predicting survival for a poor-prognosis patient is still 3.5%, which could equal many patients.

Clearly there is much to think about even after you have generated your models. The most optimal model may be a complex equation, which would never make the desk of an oncologist, unlike the simple Nottingham Prognostic Index seen in the last chapter. How useful is the prediction of any of these models in the clinic compared to the likelihood approach given by NBTree? The take-home message is that you will often read in the literature about a new diagnostic or prognostic model that outperforms a currently used one, but overall accuracy is not always the optimal measure of the usefulness of a model.

15.16 Questions

1. What do you consider the main advantages of combining Weka with R as opposed to data mining solely using R?

2. Why is predictive modeling important in biomedicine?

3. Why do we parse a "data" argument to certain functions in R? How might we avoid having to do this?

4. What is a confusion matrix?

5. What do the four values in a confusion matrix represent?

6. In building a predictive prognostic model, what do we mean by "cost"?

7. Can you describe two ways to address the class imbalance problem?

8. What does the term "fold" mean in relation to cross-validation?

9. What does the `J48()` function generate from the RWeka package?

10. How might you attempt to improve the accuracy of an SVM model using `SMO()`?

11. How do ensemble methods attempt to improve construction of a classifier?

12. Why do we need to use the `make_Weka_classifier()` function to apply the Random Forest method in Rweka?

13. What is the fundamental difference between the output of the `NBTree()` function and the `J48()` function?

14. What is the key difference in the initial argument parsed to the `lm()` function when performing multiple regression as opposed to simple linear regression?

16 | Surveillance of Infectious Disease

16.1 Background

In this chapter we turn our attention to infectious disease and disease surveillance. Once more we encounter a domain in biomedicine that involves researchers and skills from multiple disciplines. Disease surveillance is a high priority for most governments, as we are all threatened to some extent by infectious disease every day of our lives. We will begin with some definitions. An infectious disease is one involving the entry into the body of pathogenic microbial organisms such as bacteria and viruses. A communicable disease is an infectious disease that can be transmitted via contact from person to person. Globally, infectious disease accounts for over one-fifth of all human deaths and about 25% of morbidity.[24]

Management of communicable disease spread is becoming ever more difficult. There is a continuous emergence of infectious disease due to increased population movements, behavioral changes, food production, and transport, as well as other causes associated with globalization and economic development.[25] In addition, curative treatments for bacterial, viral, and parasitic infections have become less effective due to rapid adaptation of microorganisms and the evolution of antimicrobial resistance leading to the return of old communicable diseases and the emergence of new ones.

If a disease is reported within a population at a rate considerably higher than expected given recent levels, it is classed as an epidemic. If an epidemic is restricted to a relatively small area, it is termed an outbreak. If the epidemic is more widely distributed, it is usually referred to simply as an epidemic, and if global, it is a pandemic.

Disease surveillance is the process of monitoring the spread of disease in order to predict the occurrence of outbreak, epidemic, or pandemic circumstances. Surveillance involves the continuous collection, storage, and analysis of data (specific for an outcome) that can be used for planning and deployment of national and international policies and public health programs. Surveillance does not necessarily involve the monitoring of disease, but may also target chronic disease, pollution, occupational exposure, and human behavior such as smoking or drug use. The lead player in monitoring and responding to disease spread globally is the World Health Organization (WHO). In the United States, the Centers for Disease Control and Prevention (CDC) has a primary focus for controlling disease, whereas in Europe the agency with the same responsibility is the European Centre for Disease Prevention and Control (ECDC).

Surveillance programs in individual countries often collect data for infectious diseases on a daily or weekly basis. This ever-growing data is vital for the development of planning services and prevention strategies within each country or locale. But how can we utilize such data for realizing current epidemics or predicting the future? The idea that disease spread is classified as an epidemic if levels are higher than expected is vital to analysis in disease surveillance. The ability to detect an outbreak, epidemic, or pandemic is based on using a number of algorithms that have been developed specifically for this purpose. This chapter is devoted to using a selection of these algorithms that have been implemented in R in a package called surveillance. We will apply these algorithms to an example time series disease dataset in order to detect an outbreak of disease. With little effort, you can apply these methods to your own surveillance data.

First, ensure that you have the surveillance library installed via the CRAN website:

```
> install.packages("surveillance")
```

Ensure that the working directory is set to the surveillance library folder. There is also a useful vignette in the surveillance library's "doc" folder with a walk-through demonstration of testing and applying surveillance algorithms. References for each of the algorithms used in this chapter are provided in the package help files as well as the vignette.

16.2 A Simulated Surveillance Dataset

For the purposes of demonstration, there is a useful function in the surveillance package called sim.pointSource() that allows you to generate a simulation of a point-source outbreak of disease. Point-source outbreaks are a type of "common source outbreak" where the source of exposure is common to all victims. If exposure is simultaneous and exposure time is brief, consisting of a single incubation period, this is classed as a point-source outbreak.

An example of the problems faced in dealing with point-source outbreaks is that of severe acute respiratory syndrome (SARS), which is an untreatable respiratory illness. In 2003, SARS materialized in a Hong Kong hotel, where infected guests went on to seed outbreaks in many other countries.[26]

The sim.pointSource() function generates a point-source outbreak model using a hidden Markov model (HMM) approach. Hidden Markov models are statistical models commonly used in pattern discovery of time series or sequence data. We are not really interested in how we produce our simulated point-source outbreak data, however, as we just require a dataset. To obtain outbreak data, we can parse a number of parameters to sim.pointSource(), as shown in Code 16.1.

```
> epi.data<-sim.pointSource(p = 0.99, r = 0.5, length = 400,  A = 1,alpha = 1,
beta = 0, phi = 0, frequency = 1, state = NULL, K = 1.7)
> names(epi.data)
 [1] "observed" "state"    "A"       "alpha"    "beta"     "K"
 [7] "p"        "r"        "freq"    "start"
```

Code 16.1.

Each of the parameter values (for more information, type "?sim.pointSource") shown in the function call are actually default and are shown just to stress the point that you can tweak each one to change the likelihood of an outbreak(s) in a model. The `length` parameter allows you to set the number of weeks in the model; for our model, we have selected 400 weeks. Outbreaks are included in the model and can occur throughout this period. The function creates a `disProg` object, which is a list containing details of the HMM model as well as the observed values. It is these observed values that show the reported cases for each week within the time period and are the subject of analysis by surveillance algorithms for detection of outbreaks.

16.3 Surveillance Algorithms for Outbreak Detection

We can plot the distribution of reported cases and highlight the defined outbreaks for our simulated data (Figure 16.1) by parsing the `epi.data` object to `plot()`:

```
> plot(epi.data, startyear=2001)
```

In this plot, for display purposes we have defined the start period (`startyear`) for the 400 weeks to be the first week of 2001. Notice that in 2008 there is an extremely high relative frequency of infection in one week and that this is classed in our model to be an outbreak. Also note, however, that the distribution is cyclical (seasonal) and that there are actually two other defined outbreaks around the year 2008. We only know that there are three outbreaks, as our model gives this information, but what if this were real data where we have no prior knowledge of any outbreak? It would be difficult to identify outbreaks from the background seasonal pattern of infection. To do this, we need to apply surveillance algorithms, a number of which are included in the surveillance package.

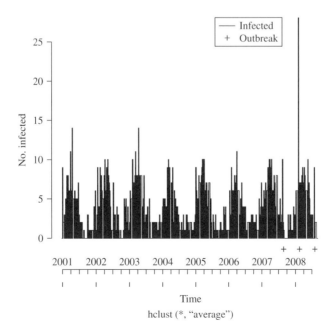

FIGURE 16.1 Distribution of reported cases and outbreaks.

Surveillance algorithms generally work by comparing an observed pattern of infections over time against an expected pattern. The expected number of infections in a given week (or month, etc.) is calculated using observed patterns in a defined window of previous weeks. The weekly expected infection counts are termed reference values. If a particular week has a greater number of observed infections relative to the expected number, the algorithm may define this week as an alarm or an outbreak, depending on the difference.

The four algorithms implemented in the surveillance package are called CDC, RKI, bayes, and farrington. Algorithms differ in how they calculate expected weekly counts. For instance, the bayes algorithm implements a Bayesian approach, whereas the farrington method applies a generalized linear model (GLM). Variations also exist within and between algorithms due to the window size of preceding weeks used to calculate the expected infection count.

We can compare each algorithm in its ability to predict alarms and outbreaks from our simulated data. We saw in Figure 16.1 that three outbreaks occur within the years 2007 and 2008, so we will focus on comparing algorithms for a 100-week period within this time. The first function call we will make is for the CDC surveillance algorithm:

```
> res.cdc<-algo.cdc(epi.data, control = list(range = 300:400,alpha
= 0.025))
```

We parse the epi.data object and a list for control parameters. The object we create, called res.cdc, is of the class survRes (as are the objects produced by the other algorithms). The range parameter specifies the actual weeks included in the analysis, which lies between weeks 300 and 400. The alpha parameter dictates the confidence interval that is used to calculate the upper threshold of the estimated counts, and is set at 97.5%. We can then generate a plot to visualize both the simulated model data over these weeks and the expected counts and calculated upper boundary for outbreak classification (Figure 16.2):

```
> plot(res.cdc, firstweek = 1, startyear = 2007, main="cdc")
```

The startyear parameter is added for x-axis labeling, and the firstweek parameter indicates the number of the first week to plot. The broken line represents the upper boundary and circles the expected infection counts for each week. If an observed value is higher than the corresponding expected value, an outbreak is declared. We see three instances of predicted outbreaks, with the second outbreak-related count clearly elevated above the expected. There are two series of alerts, which correspond to the first two outbreaks.

The second algorithm we will apply is RKI:

```
> res.rki<-algo.rki(epi.data, control = list(range = 300:400))
```

The plot is displayed in Figure 16.3:

```
> plot(res.rki, firstweek = 1, startyear = 2007, main="rki")
```

With this method, we do not see the expected values plotted but look to see if the observed values cross the upper boundary. Again, we see the three predicted outbreaks, and in addition to the alerts associated with each outbreak, there is a fourth alert predicted in an earlier week.

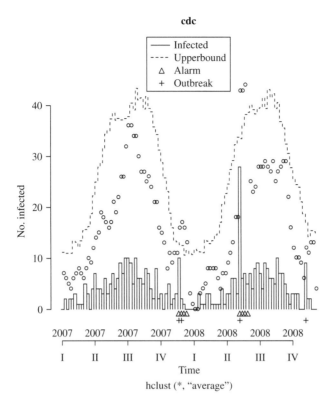

FIGURE 16.2   Simulated CDC model data and expected counts over time.

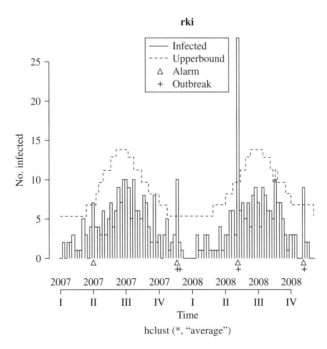

FIGURE 16.3   Simulated RKI model data and expected counts over time.

The third algorithm is bayes:

```
> res.bayes <- algo.bayes(epi.data, control = list(range =
300:400, w = 6, alpha = 0.01))
```

We have parsed a wider range of parameters this time and asked for a window size of six weeks before and after the current week. The alpha parameter is set at 99.0%. The plot display (Figure 16.4) is similar to that for RKI but with a different shape for the upper boundary, particularly between the first and second alerts:

```
> plot(res.bayes, firstweek = 1, startyear = 2007, main="bayes")
```

There is a fifth alert predicted by the bayes method correlating with the lower expected counts at this time.

Finally, the fourth algorithm is farrington:

```
> res.farrington <- algo.farrington(epi.data, control = list(range
= 300:400,alpha = 0.025))
```

Again, when we plot the model (Figure 16.5), we see the same outbreaks predicted:

```
> plot(res.farrington, firstweek = 1, startyear = 2007, main="
farrington")
```

We also see another change in the position of predicted alerts. Over the four algorithms, we have seen correct prediction of outbreaks but variation in how alerts are predicted related to the curve of each upper boundary.

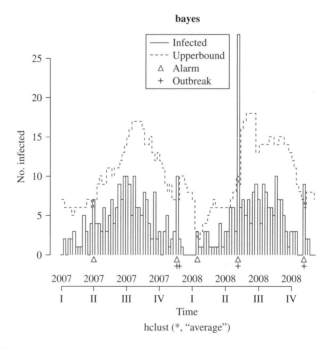

FIGURE 16.4   Simulated bayes model data and expected counts over time.

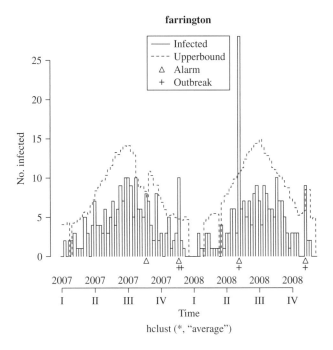

FIGURE 16.5   Simulated farrington model data and expected counts over time.

16.4 Evaluating Surveillance Algorithm Performance

In applying four different surveillance algorithms to a 100-week period, we saw that each generated a different pattern of expected values across the time period. Each also predicted a different pattern of alerts. How can we assess the performance of each model generated and decide which is most accurate? In the surveillance package, there is a function called `algo.compare()` that allows us to do just that. We just need to parse the `epi.data` model object and the same list of control parameters for each algorithm as we had set previously (Code 16.2).

```
> alg<-algo.compare(algo.call(epi.data, control = list(list(funcName = "cdc",
range = 300:400,alpha = 0.025), list(funcName = "rki", range = 300:400,alpha =
0.025), list(funcName = "bayes", range = 300:400,alpha = 0.001), list(funcName
= "farrington", range = 300:400,alpha = 0.025)) ) )
> alg
                TP FP TN FN sens spec      dist       mlag
cdc(4*,0,5)      3  5 92  1  0.75 0.9484536 0.2552588 6.666667
rki(4,0,2)       3  1 96  1  0.75 0.9896907 0.2502125 0
bayes(6,6,0)     3  1 96  1  0.75 0.9896907 0.2502125 0
farrington(3,0,5) 3 1 96 1  0.75 0.9896907 0.2502125 0
```

Code 16.2.

Essentially, we are just repeating the application of each algorithm in one go using another function called `algo.call()` to generate a `survResList` object and evaluating the accuracy of each algorithm at the same time. The `funcName` parameter must be set to select each algorithm. The `algo.compare()` function assesses the accuracy by calling another function called `algo.quality()`, which provides "quality values" for an algorithm. These quality values include true (TP) and false positive (FP) calls and true (TN) and false negative (FN) calls, as well as sensitivity and specificity (see Section 15.3.3). There is no difference in performance between the RKI, bayes, and farrington methods, which all perform better than the CDC method. In reality, all methods predicted the outbreaks correctly and have performed well, and the errors occur due to incorrect prediction of alerts, with the CDC method predicting five alerts.

16.5 Summary

The surveillance package is useful for generating simulated surveillance datasets as well as applying current surveillance algorithms and evaluation of new ones. The package provides good visualization tools to help assess infection distribution over weekly time periods. The package provides further functions for multivariate surveillance data. Using this approach, it is possible to model infection over time in more than one geographical location.

16.6 Questions

1. What is the difference between an epidemic and an outbreak?

2. Aside from generating simulated surveillance datasets, what other uses might you find for the surveillance package?

3. What R function from the surveillance package allows you to generate a simulation of a point-source outbreak of disease?

4. How do the CDC, RKI, bayes, and farrington algorithms implemented in the surveillance package differ?

5. What algorithm is used to compare surveillance algorithm performance?

17 Medical Imaging and R

17.1 Background

The quality of medical care has improved significantly in recent years with the introduction and improvements of imaging technologies. Medical imaging is not new, of course, and finds its origin way back in 1895 when the German physicist Wilhelm Roentgen took the very first x-ray. Within a year, x-ray departments had been established in many hospitals.[27] The next century saw further development of x-ray technology and the introduction of other ionizing radiation-based technologies, including positron emission tomography (PET) and computed tomography (CT). Non-ionizing radiation technologies include ultrasound and magnetic resonance (MR). Medical imaging technologies can be divided into methods that investigate anatomical structures and methods that map physical function. Areas of application include surgery, radiation therapy, radiology, and nuclear medicine.

The real output of these technologies is a single image or series of images allowing the physician to make decisions about accurate diagnosis and mode of treatment. Computerized images produced by such technologies may be 2-D, 3-D, or even 4-D. Accurate processing, analysis, and interpretation of these images within the clinic are critical. Consequently, a whole branch of medical informatics (medical imaging informatics) is now devoted to studying and improving image/data acquisition, processing, manipulation, storage, transmission, security, management, distribution, visualization, and knowledge discovery in medical images.[28]

The multidimensional complexity of medical images has also led to huge efforts to produce acceptable standards and formats for describing and storing image data. Image data formats need to account for metadata such as patient and demographic data, as well as technique information. A major problem is that imaging instruments are produced by different companies and can have their own data format for storing images. Hence, in 1985, a partnership between the American College of Radiology and the National Electrical Manufacturers Association saw the introduction of the Digital Imaging and Communications in Medicine Standard (DICOM, http://medical.nema.org/) format for storing and transmission of medical images.[29] DICOM facilitates the universal sharing of medical image data between hardware, regardless of manufacturer. A key advantage of this standard is that different types of imaging hardware and computers may be integrated within the clinic.

Many commercial and freeware software tools have been developed to allow users to view medical images stored in DICOM format. Many of these tools also allow processing and analysis of image data. Furthermore, DICOM files may easily be converted to other image formats prior to further analysis or publication. Typically, other file formats exist for medical images, including the widely used Analyze 7.5 format (liked for its simplicity) from the Mayo Clinic (http://www.mayo.edu/bir/PDF/ANALYZE75.pdf), and conversion between DICOM and Analyze is often necessary. The Analyze format is particularly used for storage and sharing of magnetic resonance imaging (MRI) data.

With a glut of medical imaging tools available, both commercially and open-source, you may be struggling to find a reason why you would need to use R with your 2-D, 3-D, or 4-D image collection. Indeed, medical images can have heavy memory requirements and are not ideally suited for processing and visualization using R on a standard PC. So what does R offer that gives it an advantage over commercial tools? First, R has great graphics capabilities and, as we shall see, handles 3-D imaging rather well. Second, the statistical analysis of medical images is a huge area of research, and R provides the platform to both develop and apply novel methods of analysis to image data. Third, R is well suited to the analysis of time series data, such as the type associated with functional MRI (fMRI) data.

In this chapter we will investigate the usage of some R packages that have been developed for medical image data. Dedicated packages are few in this field, no doubt reflecting the wealth of existing software. However, R has begun to find a niche in the processing and statistical analysis of fMRI data, which is what this chapter will focus on. We will also take a look at how R can import data using the DICOM standard, extract DICOM metadata, and visualize fMRI images in both 2-D and 3-D.

You should already have downloaded the example fMRI datasets from the accompanying website to a folder called "fMRI Datasets" (see Section 4.4.). Make sure that the R working directory is set to this folder before commencing.

17.2 A Brief Overview of fMRI

fMRI is a relatively new procedure that, using magnetic resonance imaging, can measure small metabolic changes that occur in an active part of the brain. fMRI is now becoming a mainstream diagnostic method for establishing how normal, diseased, or injured regions of the brain function in addition to allowing assessment of risks for brain surgery. Detailed uses of fMRI are shown in List 17.1.

List 17.1 Uses of fMRI.

- Brain mapping: elucidating areas of the brain responsible for critical functions
- Studying the anatomy of the brain
- Assisting in planning of surgery or radiotherapy
- Monitoring tumor progression
- Assessing the effects of neurological disorders (e.g., epilepsy) and stroke

fMRI works by detecting the changes in blood flow and oxygenation that occur due to neuronal activity. Neurons that have an increased activity have an increased demand for oxygen, which causes an increase of blood flow in the active region. Oxygen is delivered to neurons by hemoglobin, which is paramagnetic when deoxygenated but becomes diamagnetic when oxygenated. What this means is that the change in magnetic properties of hemoglobin leads to subtle changes in the magnetic resonance signal of blood, which can be detected by an MRI scanner. This type of imaging is referred to as blood oxygenation level dependent (BOLD) imaging. The change in blood flow is not immediate after the increase in neuronal activity, and there is a delay of approximately five seconds between these two events. As we will see, this delay in BOLD response must be accounted for when generating statistical models for our data.

The fMRI procedure involves the patient performing a task (responding to stimuli), such as tapping a finger at intervals or responding to questions while in the magnet of an MRI scanner. The technique produces a stack of 2-D slices of the brain to create a 3-D volume, which is repeated at many time points. Thus, many volumes are created in fast succession to create a time series set of images. Analysis of this dataset can then reveal both regional changes in brain activity and temporal changes in response to stimulation.

17.3 fMRI Image Data and Processing

The majority of MRI systems will store 3-D images using the DICOM format. As mentioned, a range of other formats exist, and many analysis tools require the format to be Analyze 7.5. The DICOM file format consists of just a single file, with fields describing the procedure and image recorded in a header within the file that itself is followed by the image data. DICOM files are usually saved with the ".dcm" file extension. The Analyze file format consists of two separate files: one being the header, the other being the image data file. An Analyze image file will be stored using the ".img" file extension, whereas the corresponding header file will have a ".hdr" extension. Conveniently, a series of 3-D volumes can be recorded as a 4-D file using the Analyze format. Many tools exist for converting between DICOM and Anlyze image formats.

When considering image data files, another term you need to be aware of is "endian" and associated phrases "big endian" and "little endian." Big endian and little endian are words that merely describe the order in which a sequence of bytes is stored in memory (a little like left to right or right to left). The origin of the term is not so high-tech and actually comes from *Gulliver's Travels,* describing the way two warring factions crack open their hard-boiled eggs. Your average PC (and modern Macs) are little endian, so why should we care? Simply because some medical image binary data files may be stored as big endian and some of the R functions we will use in this chapter request the "endian-ness" of the data input file. Because PCs are little endian, the default for endian status in these functions is "little," but if you are aware that images are stored as big endian, you will need to change the option.

fMRI data from a 3-D volume is comprised of elements called voxels (from volumetric and pixel). Whereas a pixel is a data measurement in 2-D space, a voxel is a data measurement in 3-D space. When visualizing a 3-D image of an MRI scan, the voxels are rendered. When statistically analyzing fMRI data, the intensity scores within each voxel are assessed.

A raw fMRI dataset must be preprocessed before you can analyze it statistically. Image and signal processing techniques are applied to the raw data to reduce noise and artifacts, and improve the validity and power of the analysis. Patient undergoing scanning might move their head, which means the position of the brain within the functional images will vary over time. Thus, motion correction would be applied to the data to align a voxel's time series so that it refers to the same point within the brain. Time slice correction may be applied next to counter the fact that every slice in a volume is captured at a slightly different time. Spatial filtering should be applied to the dataset to reduce noise level while retaining the underlying signal. Finally, the data should undergo an intensity normalization stage to compensate for variations of signal intensity over time, either within an fMRI session or between sessions. This chapter will not focus on preprocessing fMRI data because this is not something carried out using R, but readers should be aware of the necessary steps involved and apply these steps on their dataset before performing statistical analysis.

17.4 R Packages for Medical Imaging

Currently, only a handful of R packages are dedicated to medical imaging. There is a package available for simulation and reconstruction of PET images and the DICOM package for importing imaging data using the DICOM standard. The other packages, AnalyzeFRMI and fmri, are used for the statistical analysis of fMRI datasets. Although few in number, these packages are incredibly useful. AnalyzeFMRI is dedicated to the analysis of data in the Analyze format, whereas fmri allows for the import of other data formats, including DICOM. The combination of just these packages with the 3-D visualization package misc3d actually provides us with a powerful suite of tools for analysis and visualization of fMRI data. Packages used in this chapter are shown in List 17.2.

List 17.2 R packages used for analysis and visualization of fMRI data.

- DICOM
- fmri
- AnalyzeFMRI
- misc3d

If you have not already done so, download the packages in List 17.2:

```
>install.packages(c("DICOM", "fmri", "AnalyzeFMRI", "misc3d",
"rgl"))
```

Create a working directory in the R folder called "imaging," and change the working directory to the "imaging" folder:

```
> setwd("imaging")
```

Now load the libraries for these packages (Code 17.1).

```
> library(DICOM)
> library(AnalyzeFMRI)
> library(fmri)
> library(misc3d)
> library(rgl)
```

Code 17.1.

17.5 Importing Medical Images with the DICOM Package

The DICOM package allows you to load either a single DICOM file or multiple files into an object. We will import a single DICOM file using the `dicom.info()` function, which creates a list structure to contain both the header and image data. We will import the DICOM file "dicom.dcm" from the imaging directory into an object called `dcm.list`:

```
> dcm.list <- dicom.info("dicom.dcm",flipud=FALSE)
```

The second parameter in this function, called `flipud`, is set to FALSE so that the function does not rotate the image by 180°.

The `summary` function can show us what is held within the list in our `dcm.list` object (Code 17.2).

```
> summary(dcm.list)
    Length Class      Mode
hdr    6   data.frame list
img 3339   -none-     numeric
```

Code 17.2.

The `hdr` list in `dcm.list` contains six vectors, which house the metadata for the image. There are 54 items (fields) of metadata stored in the header for our image (just type "dcm.list$hdr" to see the entire data frame). To view one example of a DICOM field from the header, we can just print a single row from `dcm.list` (Code 17.3) and see that the field name is called Manufacturer and the value for this field (i.e., the manufacturer's name) is ezDICOM.

```
> dcm.list$hdr[17,]
   group element         name code length    value
17  0008    0070 Manufacturer   LO      8 ezDICOM
```

Code 17.3.

To view the values in all of the fields, we can just print the field name and the value instead of returning a complete output of vectors from the metadata list (Code 17.4).

```
> dcm.list$hdr[,c(3,6)]
    name                        value
1   MetaElementGroupLength      70
2   FileMetaInformationVersion  <NA>
3   TransferSyntaxUID           1.2.84.10008.1.2.1
4   ImplementationClassUID      2.16.84.1.113662.5
5   IdentifyingGroupLength      300
6   ImageType                   ORIGINAL PRIMARY OTHER
7   SOPClassUID                 1.2.84.10008.5.1.4.1.1.4
8   SOPInstanceUID              2.16.84.1.1136621
9   StudyDate                   280307
```

```
10 SeriesDate                         280307
11 StudyTime                          1030
12 SeriesTime                         1030
13 AcquisitionTime                    1035
14 ImageTime                          1035
15 AccessionNumber                    1
16 Modality                           MR
17 Manufacturer                       ezDICOM
18 InstitutionName                    ANONYMIZED
19 InstitutionAddress                 ANONYMIZED
20 StudyDescription                   FUNCTIONAL
21 PatientGroupLength                 30
22 PatientName                        John Smith
23 PatientID                          0001
24 AcquisitionGroupLength             72
25 MRAcquisitionType                  2D
26 SliceThickness                     3.0
27 SpacingBetweenSlices               0.0
28 ReconstructionDiameter             159.0
29 PatientPosition                    HFS
30 ImageGroupLength                   170
31 StudyInstanceUID                   1.1
32 SeriesInstanceUID                  1.1
33 StudyID                            1
34 SeriesNumber                       1
35 ImageNumber                        1
36 ImagePositionPatient               -79.5 -94.5 0.0
37 ImageOrientationPatient            1.0 0.0 0.0 0.0 1.0 0.0
38 SliceLocation                      0.0
39 ImagePresentationGroupLength       160
40 SamplesPerPixel                    1
41 PhotometricInterpretation          MONOCHROME2
42 NumberOfFrames                     4
43 FrameIncrementPointer
44 Rows                               63
45 Columns                            53
46 PixelSpacing                       3.00 3.00
47 BitsAllocated                      16
48 BitsStored                         16
49 HighBit                            15
50 PixelRepresentation                0
51 RescaleIntercept                   0.00
52 RescaleSlope                       0.02392651
53 PixelDataGroupLength               26724
54 PixelData                          <NA>
```

Code 17.4.

The usefulness of DICOM becomes apparent when you view the individual fields and realize the extent of information related to an image that can be stored in a single file. The Rows and Columns fields tell us the dimension of our image, which is 53 by 63 voxels. Other fields tell us when, where, and with what equipment our image was captured. We can view the actual image in R by parsing the dcm.list$img object to the image() function in the graphics package:

```
> image(dcm.list$img, col=grey(0:127/128), axes=FALSE, xlab="",
lab="")
```

You can use the dicom.table() function to create a data frame of valid DICOM fields:

```
> x<-dicom.table(dcm.list, dcm.list$hdr[,3])
```

This data frame can then be stored in an ASCII text file using the write.dicom.hdr() function:

```
> write.dicom.hdr(x, "DICOMheader.txt")
```

17.6 Reading fMRI Data in the Analyze Format

With the Analyze format being a popular choice for storing fMRI data, you will need to have image files stored in this format to use a combination of both dedicated data analysis packages in R. AnalyzeFMRI will only import files in Analyze format, although the fmri package will allow you to import other formats, including Analysis of Functional Neuro-Images (AFNI), DICOM, and Neuroimaging Informatics Technology Initiative (NIFTI). One of the fMRI example datasets that we will work with throughout the rest of the chapter is a simple, artificial finger-tapping experiment. The experiment involved a participant tapping a finger for two seconds when given a signal. This two-second period was followed by a further nine two-second rest periods to make up an epoch. The epoch was repeated a further six times to yield a 4-D dataset comprised of 70 image volumes. All volumes reside in one image file, and you should ensure that the example files, called "finger.hdr" and "finger.img," are in the working directory. AnalyzeFMRI has two useful functions to extract information from header files (hdr) and image files (img) in the Analyze format.

The f.analyze.file.summary() function extracts voxel and volume dimensional data for an image. If we apply this function to the finger-tapping image data, we can see that the voxel dimensions are 3.75 × 3.75 × 4 mm (Code 17.5). The size of the x and y planes is 64 voxels, whereas the height of the volume is 25 voxels. The number of voxels in the z dimension also equals the number of 2-D images, so we know that each three-dimensional image volume is made up of 25 slices.

```
> x<-f.analyze.file.summary("finger.img")

       File name: finger.img
  Data Dimension: 4-D
     X dimension: 64
     Y dimension: 64
     Z dimension: 25
  Time dimension: 70 time points
Voxel dimensions: 3.75  x 3.75  x 4
       Data type: signed short (16 bits per voxel)
```

Code 17.5.

A second function called f.read.analyze.header() can load the corresponding header file and display the most important information contained within it (Code 17.6).

```
> x<-f.read.analyze.header("finger.hdr")
> x
$swap
```

```
[1] 0

$file.name
[1] "finger.img"

$dim
[1]   4 64 64 25 70  0  0  0

$vox.units
[1] ""

$cal.units
[1] ""

$datatype
[1] 4

$data.type
[1] "signed short"

$bitpix
[1] 16

$pixdim
[1] 0.00 3.75 3.75 4.00 0.00 0.00 0.00 0.00
attr(,"Csingle")
[1] TRUE

$glmax
[1] 0

$glmin
[1] 0
```

Code 17.6.

The `AnalyzefMRI` package has a number of functions to retrieve image data at a number of levels, including whole 4-D volume set, single time point volumes, volume slices at a single or at all time points, and time series data from a particular 3-D position given *x*, *y*, and *z* coordinates.

An entire 3-D or 4-D dataset comprised of single or multiple volumes, respectively, can be loaded using the `f.read.analyze.volume()` function and parsing the name of the image file:

```
> data <-f.read.analyze.volume("finger.img")
```

Now if you are interested in a particular region in the brain, you may be eager to visually inspect the intensity values from the relative voxels within a slice, volume, or perhaps across the time series. The data object we have just created is a 64 × 64 × 25 × 70 array, and you would have to be pretty sharp to work out where those precious voxels lie in 7,168,000 data points (yes, that is more than seven million pieces of information!). That is not to diminish the value of this function, as we would need to create a 4-D data object for statistical analysis, as we shall see.

On a smaller scale, we can load data for a single volume using `f.read.analyze.tpt`, which allows you to specify a particular time point. Code 17.7 shows the voxel data for the first nine *x* dimension values and twenty *y* dimension values (in the first slice) of the tenth volume of our finger-tapping experiment.

```
> a<-f.read.analyze.tpt("finger.img",10)
> a
       [,1] [,2] [,3] [,4] [,5] [,6] [,7] [,8] [,9]
 [1,]    0    0    0    0    0    0    0    0    0
 [2,]   23    4   18   10    8   30   17    5   16
 [3,]   17   13    7   20   14   10   14   10    1
 [4,]   12    8   14    7    4    9   14   11   10
 [5,]   15    4   16   12    5   10   30    9    4
 [6,]   20    9   12    6    7   24    4    5   15
 [7,]    5   28   19   13   21    4    9   22    5
 [8,]   10    1    5   14   20   10   16    6    9
 [9,]    9    8   20    4    6    5   22   27   22
[10,]    2   15   13    5    3    4    7    3   17
[11,]    8   19   12   10   25   13   16   11   24
[12,]    7   11    8   16   18   14   20   15    6
[13,]   17   16   24    4   21   29   13   39   28
[14,]    9   18   24    1    2    4    7   17   23
[15,]   19   10   11    6   25   17   34   18   20
[16,]   21   12   35   64   29   47   39   44   26
[17,]   29   28   39   49   23   17   42    8   68
[18,]   24   21   18   28   27   19   34   44  247
[19,]   21   26   27   37   27    5   57  146   89
[20,]   23   21   21   28   19   40  206  143   35
```

Code 17.7.

We can retrieve data for single slices, too, using the f.read.analyze.slice() function. Code 17.8 shows data from the fifth slice of the tenth volume.

```
> a<-f.read.analyze.slice("finger.img",5,10)
> a
       [,1] [,2] [,3] [,4] [,5] [,6] [,7] [,8] [,9]
 [1,]    0    0    0    0    0    0    0    0    0
 [2,]   14   11    2    6   12   16   21   22   13
 [3,]   13    8   23    7   14   11   16   25   15
 [4,]   14    7    9   18   25    7    8   18   20
 [5,]    7    6   12    9   16    8   21   16   12
 [6,]   12   21    9    8   17   22    8    9   25
 [7,]    3   17    9   17   24   10   11    9   10
 [8,]   10    7    4   13    8   24    2   18    3
 [9,]   18   16   15   15   11    3   12   21   10
[10,]   13   27   21   13   13   13    3   17   18
[11,]   25   11   16   34   11   10   21   30   24
[12,]   15    6   33   12   29   13   31   11    2
[13,]   17   28   33   23   15   27   38   11    7
[14,]   21   27   17   28   21   15   27   30   25
[15,]   48   40   15   21   20   10   22   32   19
[16,]   58   25   19   22   11   29   13   17   39
[17,]   16   13   47   27   10   18    9   55   74
[18,]   41   43   24   33   22   14   13   41   65
[19,]   20   13   40   15   36   22   32  122   25
[20,]    7   34   12   25    9   12   98   67   13
```

Code 17.8.

Data can also be retrieved for a specified slice at all time points using the f.read.analyze.slice.at.all.timepoints() function.

At the finest level, we can retrieve data at a single point across all volumes in the time series using f.read.analyze.ts(). The parameters for the function simply state the *x*, *y*, and *z* coordinates (Code 17.9).

```
> a<-f.read.analyze.ts("finger.img", 32, 32, 12)
> a
 [1] 580 567 580 565 574 570 578 577 586 580 585 587 583
[14] 592 614 603 603 611 601 617 620 603 596 605 599 607
[27] 618 617 618 627 602 624 625 629 614 623 617 624 612
[40] 627 632 629 625 630 624 651 614 634 621 623 620 627
[53] 621 638 618 627 636 619 621 615 629 624 639 632 629
[66] 630 626 609 620 626
```

Code 17.9.

Combining f.read.analyze.ts() and plot(), we can visualize the voxel values over the time series. As an example, the following line of code creates a line graph of the voxel values at a point in an active area in our finger-tapping experiment (Figure 17.1):

```
> plot(f.read.analyze.ts("finger.img", 34, 18, 6), type="l",xlab =
"Time",ylab="Response")
```

Notice how the voxel values peak just after every tenth time point, corresponding to activation in this part of the brain roughly six seconds after the repetitive finger-tapping events.

Because we will be using the fmri package for statistical analysis, we also need to know how to load in Analyze format files to create a useable object. The

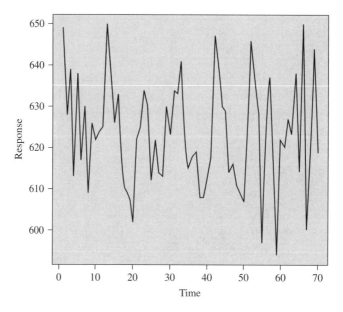

FIGURE 17.1   A line graph of the voxel values.

read.ANALYZE() function allows you to read in a series of image files in a given folder. If you look closely at the function parameters, you will see that you have to specify a prefix and a postfix for the names of the files so that they will be recognized as a sequence. This is useful if all of the volumes in an experiment are saved as individual 3-D datasets, as opposed to a 4-D file like "finger.img." The function, therefore, assumes that your file names all start with exactly the same suffix, contain a number of the form "001" as part of an ordered series, and possibly end with a string after the number. An example for a file list with a postfix could be "series_001_finger.img," "series_002_finger.img," "series_003_finger.img," etc. Be careful supplying a prefix, however, as you need to account for the variable number of zeros in a series of file name numbers. For instance, typing:

```
> data <- read.ANALYZE(prefix ="series_, numbered = TRUE, postfix
= "_finger", picstart = 1, numbpic = 70)
```

to read in 70 image files that are named "series_001_finger.img" to "series_070_finger.img" would not work because the prefix before numbers one to nine is "series_00" and thereafter is "series_0." Therefore, you need to create a vector of prefixes for each image name, such as:

```
> names<-c(paste("series_00", 1:9, sep = ""), paste("series_0",
10:70, sep = ""))
```

Once this is done, we can load our 70 volumes into the object to hold the data, remembering to set the postfix parameter and the numbered parameter to TRUE:

```
> finger.data <- read.ANALYZE(prefix = names, numbered = TRUE,
postfix = "_finger", picstart = 1, numbpic = 70)
```

Alternatively, an example where there is no postfix would be "series_001.img," "series_002.img," "series_003.img," etc.

```
> finger.data <- read.ANALYZE(prefix = names, numbered = TRUE,
postfix = "", picstart = 1, numbpic = 70)
```

If the volumes are held in a 4-D dataset, such as the finger-tapping example "finger.img," we simply call the same function but set the prefix to the name of the file (i.e., "finger"), numbered to FALSE, no postfix, and both picstart and numbpic parameters to 1:

```
> finger.data <- read.ANALYZE(prefix = "finger", numbered = FALSE,
postfix = "", picstart = 1, numbpic = 1)
```

The finger.data object that we have loaded our images into is of the fmridata class and contains a number of lists (Code 17.10):

```
> names(finger.data)
[1] "ttt"     "format"  "delta"   "origin"  "orient"  "dim"      "weights"
"header"
[9] "mask"
```

Code 17.10.

The `ttt` list contains the 4-D data matrix. The `header` list contains values for many fields, such as the number of dimensions as well as the endian status (Code 17.11).

```
> names(finger.data$header)
 [1] "sizeofhdr"    "endian"      "datatype1" "dbname"    "extents"
 [6] "sessionerror" "regular"     "hkey"      "dimension"   "unused"
[11] "datatype"     "bitpix"      "dimun0"    "pixdim"     "voxoffset"
[16] "funused"      "calmax"      "calmin"    "compressed"  "verified"
[21] "glmax"        "glmin"       "describ"   "auxfile"      "orient"
[26] "originator"   "generated"   "scannum"   "patientid"  "expdate"
[31] "exptime"      "histun0"     "views"     "voladded"  "startfield"
[36] "fieldskip"    "omax"        "omin"      "smax"        "smin"

> finger.data$header[c(2,9)]
$endian
[1] "little"

$dimension
[1]  4 64 64 25 70  0  0  0
```

Code 17.11.

17.7 2-D and 3-D Visualization of fMRI Data

Visualizing our fMRI data in two and three dimensions allows us to assess areas of brain activation in different ways. Interactive viewing in 3-D lets us rotate a volume at a particular time point to reveal activation in potential regions of interest. By viewing sequential volumes, we can track changes in these regions and, more than anything, get a good overall feel of temporal activation in our experiment. Volumes can also be viewed by slice, where each slice is plotted sequentially one after the other. This allows for inspection of activated voxels at a finer level.

Two-dimensional volume slice images can be produced by plotting the `fmridata` object. To do this for the `finger.data` object into which we loaded our finger-tapping example data, we enter:

```
> plot.fmridata(finger.data, type = "slice")
```

Figure 17.2 displays images of the 25 slices with a scale of voxel intensity. The `plot.fmridata()` function also allows you to produce 2-D images of t-statistics and p-values derived from statistical parametric mapping (SPM), as we will see.

Three-dimensional imaging of volumes may be called using the `contour3D()` function in the `misc3d` package. First, we revert back to using the data object created using the `f.read.analyze.volume()` function in the `AnalyzeFMRI` package (see previous section). Next, we create a new matrix for one of the volumes in this dataset:

```
> volume <- data[,,,11]
```

Then we call the `contour3d()` function, parsing the `volume` object as a 3-D matrix along with the x, y, and z dimensions as a sequence; a three-number vector for the

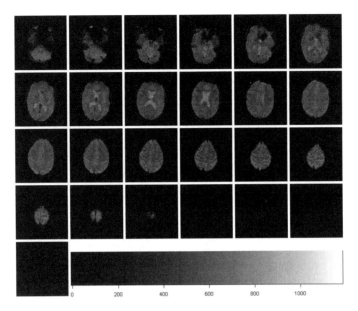

FIGURE 17.2   Image of 25 slices with scale of voxel intensity.

levels at which to construct contour surfaces; an alpha channel vector; and a vector for the color scheme to use:

```
> contour3d(volume, 1:64, 1:64, 1.5*(1:25), lev=c(500, 700, 800),
alpha = c(0.2, 0.5, 1), color = c("white", "red", "green"))
```

A screenshot of the 3-D image produced is shown in Figure 17.3, where we are viewing the brain from above (looking down the z-axis). For now, you can accept the default values for the function, but it is worth experimenting with the level parameters and the color scheme to create visually effective plots. The 3-D window is interactive, and the image may be rotated using the mouse. The levels and colors we have selected

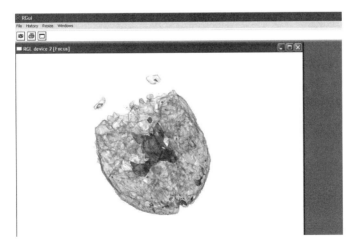

FIGURE 17.3   3D image using the contour3d() function.

for plotting show good contrast between active areas in this volume and those areas of the brain that have not been stimulated. The significance of voxel intensity within these active areas will be tested in Section 17.11.

17.8 Quality Assessment of fMRI Response Data

fMRI data is noisy, and even after preprocessing, you would want to assess the dataset for potential artifacts that could invalidate or reduce the power of the statistical analysis. A function called f.spectral.summary() provides a useful technique for carrying out a quick quality control of the data by revealing whether artifacts exist at single frequencies.

This function works by calculating a periodogram for each voxel time series in the dataset. A periodogram can be described here as an estimate of the spectral density of a voxel time series. Take another look at the voxel time series in Figure 17.1, which, as you recall, is in an area of activation after finger tapping. Although we can see a pattern of high peaks with a periodicity, the curve shows noise too. In a perfect world, the time series plot would be a nice sine-shaped curve peaking roughly every 10 seconds. f.spectral.summary() calculates the periodicity for all voxels across all time points, and the quantiles of normalized periodogram ordinates are then plotted for each frequency (Figure 17.4). Every voxel will vary in signal over time, so the minimum periodicity that could be calculated is two, and the maximum periodicity would be half the number of time points (think about this!). For our dataset, the maximum periodicity is 35 because we have 70 time points. Partial textual output for the finger-tapping experiment is shown in Code 17.12. A quick look at the distribution of fre-

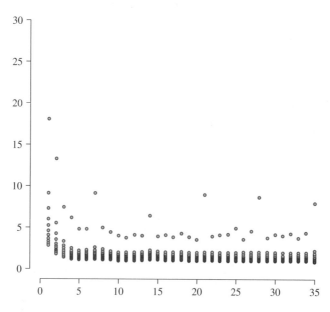

FIGURE 17.4 Normalized periodogram ordinates for each frequency.

quencies in Figure 17.4 shows an interesting pattern. There is a regular pattern of activity, with a frequency of seven, consistent with the periodicity of the seven finger-tapping events that occurred in our experiment. Note also that there is an increase in activity at periods that are multiples of seven, suggesting either potential artifacts or other periodic activity triggered by the finger-tapping event.

```
> f.spectral.summary("finger.img", mask.file=FALSE, ret.flag=TRUE)
Processing slices... [1] [2] [3] [4] [5] [6] [7] [8] [9] [10] [11] [12] [13]
[14] [15] [16] [17] [18] [19] [20] [21] [22] [23] [24] [25]
         [,1]      [,2]      [,3]      [,4]      [,5]
50%   507.9415  2.749968  1.704772  1.363275  1.109877
55%   522.2102  3.024452  1.860144  1.465709  1.196062
60%   536.2598  3.319359  2.023286  1.577771  1.283357
65%   550.2696  3.653869  2.209193  1.703604  1.376978
70%   565.3195  4.059034  2.430626  1.841837  1.484467
75%   581.3014  4.532157  2.697535  1.998565  1.601249
80%   600.4132  5.134843  3.046322  2.189012  1.727369
85%   622.1056  5.975501  3.519776  2.421473  1.883092
90%   649.1529  7.236717  4.246808  2.732506  2.103670
95%   693.6312  9.119967  5.493657  3.233810  2.458819
100%  982.1763 18.048448 13.238980  7.412540  6.111643
         [,6]      [,7]      [,8]      [,9]     [,10]
50%   1.011470  1.045086  1.123036  1.051072  0.9885595
55%   1.071125  1.118995  1.211654  1.126839  1.0346535
60%   1.149765  1.202594  1.304396  1.212059  1.1007263
65%   1.233467  1.284060  1.408102  1.304407  1.1804597
70%   1.324581  1.378364  1.514535  1.400117  1.2713437
75%   1.428860  1.478413  1.634230  1.513718  1.3587541
80%   1.543145  1.598907  1.775135  1.641300  1.4680042
85%   1.675013  1.740811  1.946329  1.799894  1.5935272
90%   1.865118  1.920487  2.171193  1.994789  1.7590132
95%   2.160519  2.205813  2.535863  2.328012  2.0204041
100%  4.768781  4.752230  9.078462  4.942271  4.3739172
```

Code 17.12.

17.9 An Overview of fMRI Statistical Analysis

Statistical analysis of fMRI data allows identification of functionally active regions of the brain after stimulation. Differences in brain activity over a specified time need to be calculated within each individual voxel. Given that an fMRI time series often involves more than 100 time points, each voxel can show considerable variation in response values, and the applied statistical procedure must be robust to identify true activity. The approach described in this chapter is based on statistical parametric mapping (SPM), and is the most familiar method for determining disease-related change and functional anatomy within the brain.[30] Furthermore, the outlined procedure applies adaptive spatial smoothing to the SPM to reduce noise and improve signal detection.[31] We will avoid immersing ourselves too deeply into the technical details and math of the methodologies, which are described in the fmri package help files and Worsley et al.[32] The analysis procedure using the fmri package, however, is quite straightforward and can be broken down into individual steps (List 17.3).

List 17.3. Steps involved in analysis of fMRI data.

- Create an expected BOLD response
- Create a linear model design matrix
- Estimate linear model parameters
- Create a statistical parametric map (SPM)
- Apply adaptive smoothing to the SPM
- Calculate p-values for the smoothed SPM

Similar to other statistical approaches applied to fMRI data, the method we will apply tests specific hypotheses about the expected BOLD response for each voxel. We can think of the expected BOLD response as an estimate of activity or signal within each voxel post-stimulus over time. Given that many voxels and time points must be considered, most approaches use the (general) linear model to generate an expected BOLD response. The linear model is essentially the weighted sum of multiple hypothesis effects. The expected BOLD response is defined formally as a convolution between a hemodynamic response function (hrf) and an external stimulus function (a convolution is the mathematical term for blending two functions). In simpler terms, the expected BOLD response is derived by combining information about what the experimenter believes should be a neural response during the experiment (external stimulus function) and the predicted delay in neural response due to slow time lag between stimulation of nerve cells and blood flow. The expected BOLD response for our study is shown in Figure 17.5.

The linear model upon which we will measure our time series response data is specified by a design matrix. Once we have created our expected BOLD response, the next step in the analysis is to create this design matrix to contain the expected BOLD

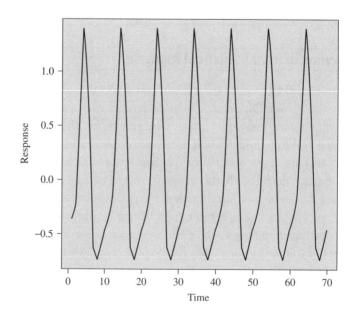

FIGURE 17.5 Expected BOLD response.

response values. The design matrix also contains information about possible varying drift in voxel signal and other external effects. The drift is removed by adding polynomial covariates in the time series to a specified degree. The number of columns in the design matrix will account for the expected BOLD response and the number of polynomial covariates.

The third step in the analysis is to generate the SPM by estimating the model parameters and variances. These parameters are weights that represent the estimated brain response (effect) over time calculated for each voxel from a set of regressors. The regressors are derived using the hypothesized brain responses. These hypothesized effects will be compared to an error term that represents the variability across the time series not accounted for by hypothesized effects.

Once the SPM is created in this way, it is smoothed by structure adaptive smoothing so as to reduce noise and improve the sensitivity of signal detection. Smoothing also has the benefit of reducing the problem of multiple testing because the method reduces the number of independent decisions made when detecting signals.

A test statistic may then be derived for each voxel by comparing the size of the hypothesized effect to the error term. P Values are calculated for each voxel, and it is these P Values that tend to be plotted when we visualize an SPM to highlight significant change in neural activity over time.

The functions in the fmri package essentially automate the whole process just described, and the breakdown of tasks shown in List 17.3 are each handled by a single function. The next few sections give a step-by-step guide to generating an SPM for our finger-tapping dataset using these functions.

17.10 Creating an Expected BOLD Response and Linear Model (SPM)

All of the steps in this section use functions from the fmri package. A detailed paper on the fmri package and implemented approaches may be found in the R news archives (http://cran.r-project.org/doc/Rnews/Rnews_2007-2.pdf). First, a function called fmri.stimulus() can be used to create a vector of expected BOLD responses, given the parameters of the task. fmri.stimulus() takes as parameters the total number of time points, the expected response time points, the repetition time (TR), and the scan duration:

```
> hrf <- fmri.stimulus(70, c(1, 11, 21, 31, 41, 51, 61), 2, 2)
```

This line tells us that we are creating a BOLD response vector called hrf based on 70 scans predicting a response at time points 1, 11, 21, 31, 41, 51, and 61 based on a stimulus of two seconds (TR) and a time between scans of two seconds.

We can view the expected BOLD response curve by plotting the hrf vector as a line graph:

```
> plot(hrf, type="l",xlab = "Time",ylab="Response")
```

Figure 17.5 shows the periodicity of the predicted response. Note that because of the time lag between stimulation of nerve cells and blood flow, the hemodynamic response peaks approximately five or six seconds after the expected response times. We

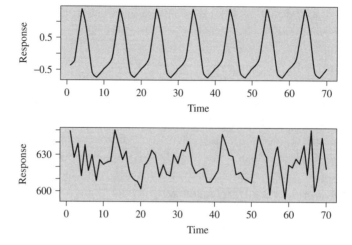

FIGURE 17.6 Voxel time series response and expected response data.

can quite easily compare voxel time series response data to the expected response data (Figure 17.6). First, set the plot window to hold two graphs:

```
> par(mfrow = c(2, 1))
```

and then plot the expected BOLD response curve:

```
> plot(hrf, type="l",xlab = "Time",ylab="Response")
```

Then simply load the activated area voxel data (with dimensions $x = 34$, $y = 18$, $z = 6$), as we did in Section 17.6 within the plot function, and plot the time series underneath the expected response curve:

```
> plot(f.read.analyze.ts("finger.img", 34, 18, 6), type="l",xlab = "Time",ylab="Response")
```

The plot does actually reveal that despite being quite "noisy," there is a clear pattern of expected periodicity in response for this particular voxel across the time points.

With the expected BOLD response vector in hand, we can now create the design matrix for our study:

```
> design.matrix <- fmri.design(hrf,order = 2)
```

Here we have parsed the expected BOLD response and the number of polynomial drift terms up to the order specified, which, in this case, is two (the default). We can visualize the BOLD response and drift terms as a series of plots (Code 17.13).

```
> par(mfrow=c(2,2))
> for (i in 1:4) plot(design.matrix [,i],type="l")
```

Code 17.13.

The design matrix is then taken as input, along with the raw dataset, to create a linear model. Using these inputs, the function fmri.lm() estimates the parameters of the model and then creates the model (Code 17.14).

```
> finger.model <- fmri.lm(finger.data, design.matrix)
fmri.lm: entering function
fmri.lm: calculating AR(1) model
0% . 10% . 20% . 30% . 40% . 50% . 60% . 70% . 80% . 90% .
fmri.lm: recalculating linear model with prewithened data
0% . 10% . 20% . 30% . 40% . 50% . 60% . 70% . 80% . 90% .
fmri.lm: calculating spatial correlation
fmri.lm: determining df: 23.35230
fmri.lm: exiting function
```

Code 17.14.

We have now created an object of class `fmrispm` called `finger.model`. An `fmrispm` object has a number of lists associated with it that all hold information about the model, including the parameter estimates (Code 17.15).

```
> names(finger.model)
[1] "beta"      "cbeta"     "var"    "res"   "arfactor" "rxyz"     "scorr"
[8] "weights"   "vwghts"    "mask"   "dim"   "hrf"      "resscale"
[14] "bw" "df"
```

Code 17.15.

The `finger.model` object is our statistical parametric map, which now requires smoothing before we calculate significant voxel signal across time.

17.11 Creating a Smoothed SPM and Determining Signal

To smooth our SPM, we call the `fmri.smooth()` function to reduce noise and improve signal detection. We parse as parameters the SPM object `finger.model` and the voxel bandwidth on which to smooth (Code 17.16).

```
> finger.smooth <- fmri.smooth(finger.model, hmax = 4)
fmri.smooth: entering function
fmri.smooth: smoothing the statistical parametric map
0.076% . 0.17% . 0.29% . 0.44% . 0.62% . 0.85% . 1.1% . 1.5% . 2% . 2.5% . 3.2%
. 4.1% . 5.2% . 6.6% . 8.3% . 10% . 13% . 17% . 21% . 26% . 33% . 41% . 51% .
64% . 80% . 100% .
Vaws3D: first variance estimation
Vaws3D: smooth the residuals
Vaws3D: estimate correlations
Vaws3D: final variance estimation

fmri.smooth: determine local smoothness
fmri.smooth: exiting function
```

Code 17.16.

Once smoothed, we can calculate *t*-statistics for each voxel time series and define P Values using the `fmri.pvalue()` function (Code 17.17).

```
> finger.pvalue <- fmri.pvalue(finger.smooth)
fmri.pvalue: entering function
fmri.pvalue: calculate treshold and p-value method: basic
fmri.pvalue: thresholding
fmri.pvalue: exiting function
```

Code 17.17.

The `fmri.pvalue()` function creates an object of the `fmripvalue` class, and although we have not included these options in the previous line, can calculate different types of P Values and permits the user to specify a minimum value for signal detection.

17.12 Visualizing the Effects of Smoothing the SPM and P Values

With the signal P Values calculated, we can now determine activated regions of interest by visualizing the distribution of significant P Values. First, however, we will have a look at the signal of the smoothed SPM we created and compare that to the raw signal in Figure 17.2:

```
> plot(finger.smooth)
```

By viewing the SPM in Figure 17.7, it is now much easier to hypothesize about where potential regions of activity lie. By plotting the P Values (Figure 17.8), we can view exactly which voxels have significant signal:

```
> plot(finger.pvalue)
```

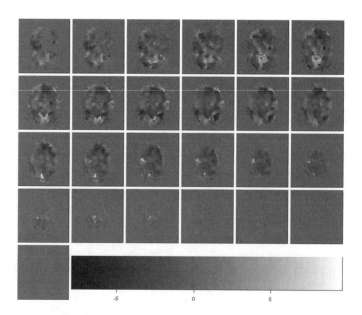

FIGURE 17.7   Images of smoothed SPM.

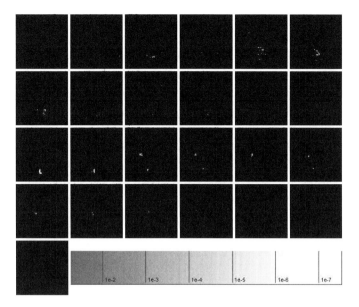

FIGURE 17.8   Images of significant voxels.

Significant activation signal occurs in a number of regions in the brain, including the occipital, parietal, and frontal lobes. This is only a hypothetical dataset, so it is futile to draw inferences from these results, but this exercise has hopefully demonstrated that once the concepts are understood, it is extremely easy to produce and visualize statistical parametric maps using R.

17.13 Examining fMRI Data with Independent Components Analysis

The statistical parametric mapping approach for determining regions of activity within the brain from fMRI data is a typical example of a hypothesis-driven method of analysis. By applying this approach, we needed to make prior assumptions about the data to generate an expected BOLD response and estimate a number of parameters for the model. If you were presented with an fMRI dataset without any metadata or experimental details, you would obviously not be able to apply a hypothesis-driven approach. You would, however, still be able to hypothesize activated regions of the brain by using data-driven, exploratory approaches that are becoming more widespread in the analysis of neurological time series data. Trying to determine the structure of fMRI data using a data-driven method is particularly useful if the expected time course of brain activation is difficult to stipulate a priori.

The data-driven method most frequently applied to fMRI data is Independent Components Analysis (ICA). Formally, we can define ICA as a statistical method used to discover maximally independent hidden features from a set of observed measurements.[33] Discovery of these independent features from data is often referred to as blind signal separation. The term "blind" refers to the fact that ICA tries to "unmix" different signals from a mixed bag of signals (e.g., an fMRI time series dataset) without any prior knowledge. A full introduction and technical discussion of ICA applied to fMRI data is beyond the scope of this book, but a good introduction to the subject and references therein may be found in Calhoun and Aldari[33] and McKeown et al.[34]

ICA as a tool for fMRI data analysis is still under development and is very much an area of continuing research. ICA applied to fMRI data may be used to separate either spatially or temporally independent sources of signal, hence the terms "spatial ICA" (sICA) and "temporal ICA" (tICA) encountered in the literature. The AnalyzeFMRI package in R has a function to apply sICA to fMRI data, so only sICA will be considered here. The multivariate nature of sICA reveals relationships between voxels either proximal or distal to each other. This gives the potential to create activation maps and also generate hypotheses about co-activation within different regions of the brain. To complete this chapter, we will simply scratch the surface of how sICA can be applied in R to fMRI datasets. We will begin by demonstrating sICA applied to the finger-tapping dataset before trying to predict activated regions from data derived from a more complex experiment.

17.14 Spatial ICA on fMRI Data Using R

Performing sICA on an fMRI dataset in R is straightforward. The author of the AnalyzeFMRI package has implemented sICA in one single function called f.ica.fmri(). In its simplest use (as we will see here), the user only has to parse the dataset name and number of components to return, and that is it. The function then decomposes the data into the specified number of components using an algorithm called fastICA. Using fastICA and the C language to carry out all computations is a clever approach by the package author. The function returns some matrices into an object, and although these can be explored by the user, all we need to do to see a result is plot the object.

Before we begin, you will need to install the fastICA package:

```
>install.packages("fastICA")
```

To simply demonstrate the use of sICA, we will first apply the function to the finger-tapping dataset that we have already analyzed using statistical parametric mapping. Remember that this experiment was a simple finger-tapping paradigm over a relatively small number of epochs, and you would not necessarily be inclined to perform sICA on such a dataset. However, we can make a comparison between predicted regions of activation by sICA and the statistical parametric map we generated earlier. We parse the file name "finger.img" and the number of components for decomposition (Code 17.18). Here, for simplicity, we have only asked for eight components to be extracted, but in reality you would seek to extract a much larger number. Two issues to be addressed when applying ICA to fMRI data are:

- How many components should be extracted?

- What is the order of the components in relation to the variance that they explain?

Because there are no functions in AnalyzeFMRI to provide the answers to these questions, we would have to hard-code some functions ourselves. Therefore, we will use a simplistic approach of assessing the quality of components to retain in terms of periodicity and not concern ourselves with the importance of order of retained components.

```
> ica<-f.ica.fmri("finger.img", 8)
Reading in dataset
Making mask
Mask size = 21388
Creating data matrix
Running ICA
Centering
Whitening
Symmetric FastICA using logcosh approx. to neg-entropy function
Iteration 1 tol=0.570846
Iteration 2 tol=0.944197
Iteration 3 tol=0.836856
Iteration 4 tol=0.481388
Iteration 5 tol=0.485264
Iteration 6 tol=0.306091
Iteration 7 tol=0.143688
Iteration 8 tol=0.083511
Iteration 9 tol=0.050716
Iteration 10 tol=0.033394
Iteration 11 tol=0.022803
Iteration 12 tol=0.015949
Iteration 13 tol=0.011811
Iteration 14 tol=0.008697
Iteration 15 tol=0.006735
Iteration 16 tol=0.005173
Iteration 17 tol=0.004054
Iteration 18 tol=0.003190
Iteration 19 tol=0.002495
Iteration 20 tol=0.001983
Iteration 21 tol=0.001547
Iteration 22 tol=0.001233
Iteration 23 tol=0.000959
Iteration 24 tol=0.000764
Iteration 25 tol=0.000595
Iteration 26 tol=0.000474
Iteration 27 tol=0.000368
Iteration 28 tol=0.000294
Iteration 29 tol=0.000229
Iteration 30 tol=0.000182
Iteration 31 tol=0.000142
Iteration 32 tol=0.000113
Iteration 33 tol=0.000088
```

Code 17.18.

The various steps of the algorithm, including data preprocessing, are listed in the output as the function chugs along. The process is iterative, although the number of iterations can be controlled using the `maxit` parameter on the function (e.g., `maxit = 1000`).

To view each of the eight returned independent component (activation) maps (spatial mapping of signal), we use the `f.plot.ica.fmri()` function (Code 17.19).

```
> f.plot.ica.fmri(ica, 1)
> f.plot.ica.fmri(ica, 2)
> f.plot.ica.fmri(ica, 3)
...
> f.plot.ica.fmri(ica, 8)
```

Code 17.19.

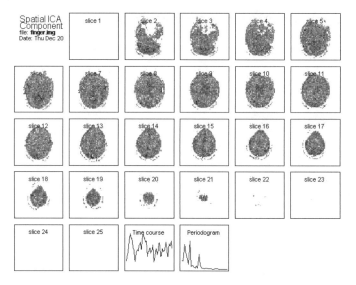

FIGURE 17.9   ICA component activation map.

If you are not fond of the default orange color scheme, you can parse other color palettes or, as I have done for my plotting, a grayscale palette:

```
> f.plot.ica.fmri(ica, 2,  col=gray(0:8/8))
```

The `f.plot.ica.fmri()` function returns a series of plots for the component, revealing an activation map across each slice along with a time course plot and periodogram. As mentioned, the periodogram is useful for assessing reliable frequency of activation as opposed to noise.

By applying the `f.plot.ica.fmri()` function for each component and assessing the activation maps with periodograms, we can soon discover how many components we may reliably keep. One important point to mention is that if you repeat the analysis, the components returned each time will differ in order and often in fine spatial mapping of signal. If sICA is suitable for the dataset, however, you should get repeatability of retainable component number and, of course, patterns of activation.

Out of our eight components returned, one in particular showed an interesting activation map (Figure 17.9) with respectable periodicity in terms of noise. You can compare the activation map to the mapping of smoothed SPM and subsequent P Values in Figures 17.7 and 17.8, respectively. The regions of activation predicted by this sICA component match closely, with significant regions of activation predicted by SPM in the occipital and parietal lobes. Indeed, sICA has added useful information to the results obtained by SPM, suggesting observed signal represents co-activation in the occipital and parietal lobes.

17.15 sICA of a Speech fMRI Dataset

Now that we are familiar with the `f.ica.fmri()` function, we will move to applying this function to a dataset that is more complex in design than the finger-tapping experiment. This dataset (kindly provided by Steffanie Brassen) was derived from a

young, healthy female in a recent fMRI experiment conducted to examine neural cor-
relates underlying the successful encoding of words that can be freely recalled or recognized but not recalled.[35] This high-profile study was exceedingly well planned with
a comprehensive analysis of activated regions using spatial parametric mapping.

The dataset is a series of 408 volumes, where each volume consists of 46 slices of
dimension 53×46 voxels. The subject was presented with 120 single words, where
the duration of each word was two seconds (hence, we will refer to the data as the
"word" dataset). Words were comprised of 40 neutral, 40 positive, and 40 negative ad-
jectives in a randomized order. The time between images (TR) was 2.41 seconds. The
subject was instructed to view each word for the entire display time. Each word was
then replaced with a rating screen, and the subject had to indicate an emotional rating
by pressing one of three buttons (for positive, neutral, or negative emotional response).
The study found that successful encoding of words that were recognized but not recalled relies on a subset of regions of the brain activated during successful encoding
of freely recalled words. The statistical analysis in the study revealed activation (depending on successful encoding of recognized words) in ventrolateral and dorsolateral
prefrontal cortex, the hippocampus, inferior temporal gyrus, and the cerebellum. Activation was bilateral in each region. Activation in these regions was dependent on the
ability of the subject to recognize and recall words.

The word data has been preprocessed by image registration, normalization, and
smoothing. In applying sICA to this dataset, we will make no prior assumptions about
the data and not take the previous statistical results into account. In other words, we will
carry out the analysis blind and purely explore the data. We call the `f.ica.fmri()`
function and parse the "word.img" file that houses the 4-D dataset and state that we
want to decompose 50 components with a maximum of 100 iterations (Code 17.20).
Again, in practice, you may want to extract more components.

```
> ica<-f.ica.fmri("word.img", 50, maxit=1000)
Reading in dataset
Making mask
Mask size = 79823
Creating data matrix
Running ICA
Centering
Whitening
Symmetric FastICA using logcosh approx. to neg-entropy function
Iteration 1 tol=0.428584
Iteration 2 tol=0.716126
Iteration 3 tol=0.753930
Iteration 4 tol=0.443528
Iteration 5 tol=0.687427
Iteration 6 tol=0.765650
Iteration 7 tol=0.823128
Iteration 8 tol=0.706111
Iteration 9 tol=0.188495
Iteration 10 tol=0.015873
Iteration 11 tol=0.009713
Iteration 12 tol=0.008846
Iteration 13 tol=0.009836
Iteration 14 tol=0.011401
Iteration 15 tol=0.010866
Iteration 16 tol=0.007721
Iteration 17 tol=0.006076
```

```
Iteration 18 tol=0.004964
Iteration 19 tol=0.003637
Iteration 20 tol=0.002877
Iteration 21 tol=0.002578
Iteration 22 tol=0.002263
Iteration 23 tol=0.002000
Iteration 24 tol=0.001826
Iteration 25 tol=0.001766
Iteration 26 tol=0.001711
Iteration 27 tol=0.001702
Iteration 28 tol=0.001665
Iteration 29 tol=0.001572
Iteration 30 tol=0.001418
Iteration 31 tol=0.001216
Iteration 32 tol=0.000993
Iteration 33 tol=0.000774
Iteration 34 tol=0.000625
Iteration 35 tol=0.000547
Iteration 36 tol=0.000474
Iteration 37 tol=0.000404
Iteration 38 tol=0.000340
Iteration 39 tol=0.000282
Iteration 40 tol=0.000240
Iteration 41 tol=0.000234
Iteration 42 tol=0.000231
Iteration 43 tol=0.000231
Iteration 44 tol=0.000237
Iteration 45 tol=0.000255
Iteration 46 tol=0.000275
Iteration 47 tol=0.000297
Iteration 48 tol=0.000321
Iteration 49 tol=0.000345
Iteration 50 tol=0.000369
Iteration 51 tol=0.000392
Iteration 52 tol=0.000411
Iteration 53 tol=0.000427
Iteration 54 tol=0.000438
Iteration 55 tol=0.000446
Iteration 56 tol=0.000447
Iteration 57 tol=0.000440
Iteration 58 tol=0.000427
Iteration 59 tol=0.000409
Iteration 60 tol=0.000385
Iteration 61 tol=0.000358
Iteration 62 tol=0.000328
Iteration 63 tol=0.000297
Iteration 64 tol=0.000266
Iteration 65 tol=0.000270
Iteration 66 tol=0.000277
Iteration 67 tol=0.000281
Iteration 68 tol=0.000281
Iteration 69 tol=0.000277
Iteration 70 tol=0.000269
Iteration 71 tol=0.000260
Iteration 72 tol=0.000249
Iteration 73 tol=0.000238
Iteration 74 tol=0.000224
Iteration 75 tol=0.000210
Iteration 76 tol=0.000194
Iteration 77 tol=0.000178
Iteration 78 tol=0.000163
Iteration 79 tol=0.000148
Iteration 80 tol=0.000133
Iteration 81 tol=0.000119
Iteration 82 tol=0.000106
Iteration 83 tol=0.000093
```

Code 17.20.

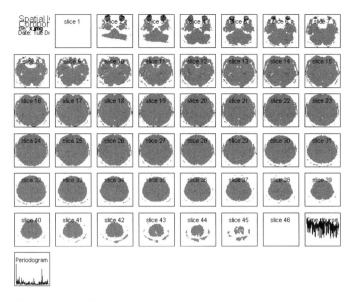

FIGURE 17.10 Activation map of first component.

As with the finger-tapping analysis, we can proceed to visualize the independent activation maps for each component using the f.plot.ica.fmri() function. Of the 50 components in the analysis, 4 components showed activation maps of interest when accounting for periodicity.

The first component retained (Figure 17.10) is not that informative, as the activation map involves just the eyes, as seen in the first five or six slices. The activation map of the second retained component (Figure 17.11) shows potential co-activation in the frontal lobe (slices 27 to 30) and parietal lobe (slices greater than 26). A clear peak exists within the periodogram, although there is also clear evidence of noise. The third

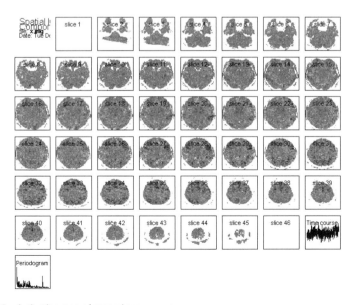

FIGURE 17.11 Activation map of second component.

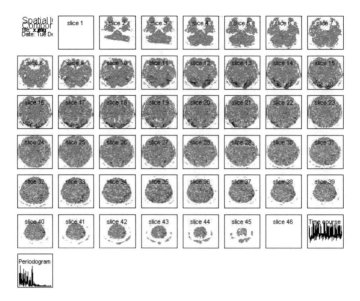

FIGURE 17.12 Activation map of third component.

component (Figure 17.12) shows activation in the occipital lobe (approximate slices 11 to 21). The fourth component (Figure 17.13) shows an activation map with distinct signal in small regions of the parietal lobe (slices 29 to 44).

As an exercise in exploring the data, sICA has provided hypotheses as to where areas of activation and co-activation occur in word recognition and recall given the criteria of the original experiment. More interestingly, sICA has predicted areas of activation that overlap with those areas predicted in the original study using SPM. One example is the area of activation in the prefrontal cortex we saw in the second component extracted (Figure 17.11).

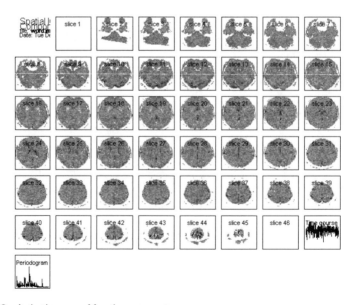

FIGURE 17.13 Activation map of fourth component.

17.16 Summary

We have only touched the surface of how fMRI data can be statistically analyzed using univariate or data-driven exploratory methods. The approaches described here have been basic and simply served to demonstrate that R has some powerful packages for the analysis of fMRI data. Analysis can be more rigorous, and the interested reader should explore the literature to get a good feel for the variations in the way fMRI data is analyzed. Principal components analysis is sometimes used in the preprocessing stage for data reduction. Variations exist in the construction of SPMs and, using algorithms that already exist in R, the skilled user could implement novel approaches from the literature (for instance, the sICA approach developed for the AnalyzeFMRI package relies on the fastICA package already in R). Different algorithms also exist for ICA, and you may use ICA for analysis of groups of subjects. Although there are only two dedicated packages for fMRI analysis in R, together they are quite powerful if you want to analyze data in the Analyze format.

17.17 Questions

1. What does fMRI measure?

2. What four R packages, used in this chapter, are useful for processing and analyzing fMRI data?

3. What function from the DICOM package allows us to import images in DICOM format?

4. An object of what class must be parsed to the `image()` function to view a DICOM image?

5. What is a voxel?

6. What R function can we use to view 3-D fMRI image data?

7. What is a BOLD response?

8. What does the P Value relate to exactly in the function `fmri.pvalue()`?

9. Why do we apply independent components analysis to fMRI data?

10. What argument can we parse to the `f.ica.fmri()` function to control the number of iterations?

18 Retrieving Public Microarray Datasets

18.1 Background

In the next three chapters, we will turn our attention to a topic that really demands an entire book to itself. Indeed, microarray gene expression analysis already has many excellent books detailing the technologies, how data is generated, processed, and, of course, analyzed. R is already a leader in terms of microarray analysis software mainly due to the Bioconductor project.[36]

Bioconductor is an open-source development and academic project in its own right (http://www.bioconductor.org/). Like R, Bioconductor is a well-managed, controlled project with a core list of developers. Based mainly on R, the development contains a number of packages that may be used for the analysis of genomic data. The reader is encouraged to thoroughly research the Bioconductor project to gain insight into the many functions available.

In this chapter, we will begin to explore how we can obtain gene expression data from a public repository. Obtaining data from such a repository allows you to explore previous studies of a disease of interest. In Chapter 19 we will then look at how we can perform quality control on raw data and create a list of genes that are differentially expressed between disease and control samples. Finally, in Chapter 20, we will take our gene list and annotate each gene to find out as much information as possible using online databases. There is much more to do with microarray data than just what is detailed in these three chapters. We will literally get as far as annotating a gene list, but using skills learned in previous chapters, you could go on to find patient groups or clusters of genes that have similar expression patterns across samples. With sufficient data, you can even apply classification methods to model gene expression patterns that could prove to be useful biomarkers for diagnosis or prognosis.

This chapter begins with a brief introduction to DNA microarray technologies, including the Affymetrix platform and the type of data produced.

18.2 DNA Microarray Technologies

Microarray technologies used for gene expression profiling come in a variety of flavors. A DNA microarray is generally a chip housing a collection of microscopic spots of DNA covalently bound to a suitable matrix. The DNA spots usually represent single genes. There are many variations in technologies, but when considering carrying out a microarray experiment, we generally consider one of two types: spotted microarrays or oligonucleotide arrays.

Spotted microarrays, also referred to as two-channel microarrays, have oligonucleotide or cDNA probes spotted onto the microarray surface. Typically, cDNA corresponding to mRNA from two different samples labeled with two different fluorophores (normally red and green) is hybridized to the array. Often one sample will be cDNA derived from disease tissue and the other sample cDNA from normal tissue. The fluorophores most commonly used for labeling are Cy3 (green) and Cy5 (red). The labeled cDNA from both samples is then mixed and hybridized to a single microarray, and cDNA representing mRNA from genes being expressed in each sample will bind to the array probes. The microarray is then scanned using scanning technology that presents an image of the fluorescence from the two fluorophores. For each spot (i.e., gene), the fluorescence intensities can be measured for both fluorophores to give a ratio of expression for the gene in each sample. Thus, the intensity ratios reveal the patterns of up-regulation or down-regulation for genes between two samples. The intensities, however, cannot be used to predict absolute expression values for a gene.

Oligonucleotide microarray technologies differ in that only a single sample is hybridized to the array. This type of technology is often referred to as "Affymetrix" or "Affy," due to Affymetrix, Inc., being a major player in the market with their GeneChip® technology. It is worth pointing out that oligonucleotide chips are also produced by other companies, including Agilant Technologies. Generally, oligonucleotide probes matching regions of mRNA sequences from known or predicted genes are bound to an acrylamide surface, and an array can contain hundreds of thousands of spots. RNA from a sample may be reverse-transcribed to cDNA, in vitro transcribed to cRNA, and tagged with a biomolecule such as biotin. The labeled cRNA is then hybridized to the array and the chip scanned after necessary processing. Absolute expression values for genes may then be estimated by reading intensities during scanning of the chip.

In this chapter we will focus on data generated using the Affymetrix GeneChip Human Genome U133 Array Set (HG-U133A), which is an oligonucleotide-type platform. Although we will only concentrate on one platform type, the reader should be aware that there are R packages for dealing with different stages of processing and analysis of both images and data generated from spotted arrays and other microarray technologies.

18.3 A Primer on Affymetrix GeneChip Technology

An Affymetrix GeneChip array is divided into many tiled areas, called cells, each containing many copies of a unique probe. Each of these tiles is composed of smaller tiles of probes, 25 bases long, arranged in pairs that consist of a perfect match (PM) probe and a mismatch (MM) probe. In these probe pairs, the only difference between PM and MM probes is that the central base in the MM probe has been altered, reducing the complementary nature of the sequence (hence the term mismatch). These collections of probes are called probe sets, and the probe sequences in a probe set are specific for a transcript. Older-generation Affymetrix arrays, such as the HuGeneFL, contain probe sets of 20 probe pairs, whereas newer arrays, including the Human U133 series, have as few as 11 pairs. Depending on the array type, probe pairs in a probe set can be either spread across the array or arranged contiguously.

In many circumstances, researchers will have a service provider perform the microarray experiment and will have a dataset returned containing the expression values of genes for each of their samples. For a study utilizing Affymetrix technology, the dataset is often produced by algorithms housed in Affymetrix-licensed software, such as Microarray Suite (e.g., Mas5.0) or the more recent GeneChip Operating Software (GCOS). The probe set plays a critical role in deciding the concentration of a transcript that is specific to the probes and how reliably the transcript has been detected. The Statistical Expression Algorithm in the Affymetrix Mas5.0 software suite uses each probe set to calculate an average of the logged difference between PM and MM values. The significance of the differences between PM and MM values is tested using a one-sided Wilcoxon's signed rank test that generates a P Value. This P Value is then compared to user-defined thresholds to determine a PMA detection call for that particular transcript. Quite simply, a transcript may be called Present (if P Value is less than threshold), Absent (if P is greater than or equal to threshold), or Marginal (if the signal is at the limit of detection), hence the acronym PMA. The reader should be aware that other methods of deriving expression values from oligonucleotide arrays exist, such as RMA, GC-RMA, and PLIER. A typical dataset returned to the user from the microarray service provider will provide information for each transcript detailing the probe ID, expression value, detection call, and corresponding P Value.

A good introduction to expression calls is given by Pepper et al.[37] and a comparison of methods by Seo and Hoffman.[38]

18.4 Affymetrix Data Files

A gene expression study using one of the Affymetrix GeneChip technology platforms produces a number of different file types. Raw image data from a scanned array is contained in a DAT file. Generally, researchers keen to explore their expression data ignore raw image data and turn instead to the data contained in the CEL file, which contains information about each probe that is on the chip. This is the information extracted from the image data using the dedicated analysis software such as Mas 5.0 or GCOS. Crucially, the CEL file contains information describing the position of each probe on the chip and the intensity values for each probe. As mentioned, a transcript is represented by a number of different probes positioned across the array collectively making up a probe set. It is intensity readings from across a probe set that give a single expression value for a particular gene. The CEL file by itself, however, does not provide information as to which probe set a probe belongs. This information is stored in a separate file called a CDF library file.

In the next chapter we will learn by example the usefulness of studying intensity data at the probe level using a Bioconductor package called `simpleAffy`. Usually, though, a researcher will simply be interested in using the expression measurements produced by the analysis software, so prior to using simpleAffy, we will run through a typical analysis of an Affymetrix dataset. For our example, we will obtain a series of gene expression data from a study that is housed in the Gene Expression Omnibus (GEO) repository at the National Center for Biotechnology Information (NCBI).[39]

18.5 Gene Expression Omnibus Repository

GEO is a public repository that allows the scientific community to submit high-throughput experimental data for free access by other researchers. Data types in GEO include single- and dual-channel microarray, measuring mRNA, miRNA, genomic DNA (including arrayCGH and SNP), and protein abundance. Public datasets are available for nearly 4000 platforms, more than 170,000 individual samples, and nearly 7000 collections of samples for species including human, mouse, and rat. The website, however, is more than just a data storage facility; it also has a collection of web-based tools for querying and downloading of datasets. Luckily for the R user, a package called GEOquery is available as part of the Bioconductor suite that allows us to download datasets from GEO directly to our desktops. We will look at GEOquery later on, but first let us take a quick tour of the GEO website to find a dataset for our example. In your web browser, navigate to http://www.ncbi.nlm.nih.gov/geo/.

GEO data is organized in different record types, where the basic types are platform, sample, series, dataset, and profile. A **Platform** record includes information relative to the technology used, and each platform record will have a unique GEO accession number (GPLxxx). **Sample** records contain information submitted by experimenters detailing the handling conditions and manipulations that a sample underwent. A Sample record is a unique GEO accession number with the format GSMxxx. Related Sample records are linked together by **Series** records, supplied by the submitter, which give a description of the whole study. Furthermore, Series records may also contain extracted data, analysis results, and interpretations of findings. Series record accession numbers are in the format GSExxx. The fourth record type is the GEO DataSet, which is the submitted Series records reassembled by GEO staff and given the accession number GDSxxx. A GDS record set is a powerful collection of GEO sample information, as the data in these record sets may be compared biologically and statistically using data analysis tools (including those provided by the GEO site). By exploring the GEO site, a researcher may view examples of each record type.

18.6 Finding Datasets in GEO

For our example, let us query GEO to find a microarray dataset from an experiment related to the disease polycystic ovary syndrome (List 18.1).

List 18.1 Querying GEO for a microarray data set.

1. Navigate to the main page on the GEO site.

2. Enter the search term "polycystic ovary syndrome" in the DataSets field under the GEO navigation tab.

3. Click the corresponding Go button.

4. At the time of this writing, eight datasets were returned. Each dataset may be identified by the GDS record number found adjacent to the dataset title.

5. Navigate to the dataset entitled "Polycystic ovary syndrome: adipose tissue," which has the record accession number GDS2084. Below the title, you will find a summary of the study followed by further information that includes a

link to download all the CEL files made available from the study (under Supplementary Files) and individual links for information describing each sample (under Samples).

6. Click the "GDS2084 record" hyperlink to the DataSet record for this particular study.

Now let us look at the information contained in the DataSet record page. The GDS Summary tab displays information relating to the study background. The Title ("Polycystic ovary syndrome: adipose tissue"), Summary, and Platform sections tell us the study focused on analysis of omental adipose tissues of morbidly obese patients with polycystic ovary syndrome (PCOS), which is a common hormonal disorder among women of reproductive age, characterized by hyperandrogenism and chronic anovulation. The study was carried out using the Affymetrix GeneChip Human Genome U133 Array Set (HG-U133A) with the GEO accession number GPL96.

Note that the Series record number for this particular study is GSE5090, as we will require this number later on. The Subset and Sample Info tab details subsets of eight samples derived from patients with polycystic ovary syndrome and seven samples from control cases. Underneath this tab you will find hyperlinks to information about each sample used in the study. Indeed, by following these links, you can download the CEL file for each sample. Alternatively, to obtain the CEL files for the whole study, navigate back to the GEO DataSets results page that you generated for your search term and follow the hyperlink for Supplementary Files. By clicking the link, you will be directed to an FTP site with a further link to a zipped TAR file called "GSE5090_RAW.tar" that contains the complete set of CEL files for the study. You may download this file for later use, but we will also download it directly from R later on. In fact, the GEOquery package in R can save you the bother of navigating the GEO site to extract and view information for a microarray study.

18.7 Gathering Information About Your Data—The GEOquery Package

As we saw in the previous section, GEO DataSets can contain a wealth of information relevant to an experiment of interest, and this information can be captured using the GEOquery package in R. In this section we will use GEOquery to explore in detail the PCOS dataset held in GEO.

```
> source("http://bioconductor.org/biocLite.R")
> biocLite("GEOquery")
```

Code 18.1.

Download the GEOquery package from Bioconductor (Code 18.1).

To begin, let us load the GEOquery package:

```
>library(GEOquery)
```

We can download the PCOS GEO DataSet into a data structure called gds using the getGEO() function:

```
>gds <- getGEO("GDS2084")
```

R will then connect to a GEO FTP site, open the URL, download the data, and parse the data subsets into the `gds` data structure from the GEO DataSet (Code 18.2).

```
trying URL 'ftp://ftp.ncbi.nih.gov/pub/geo/DATA/SOFT/GDS/GDS2084.soft.gz'
ftp data connection made, file length 875237 bytes
opened URL
downloaded 854Kb

File stored at:
C:\DOCUME~1\SCSLEWPD\LOCALS~1\Temp\RtmpKisSsb/GDS2084.soft.gz
parsing geodata
parsing subsets
ready to return
```

Code 18.2.

We can then view the GEO DataSet summary and sample information for the study (Code 18.3).

```
> Meta(gds)
[[1]]
[1] "able_begin" "able_end"

$channel_count
[1] "1"

$description
[1] "Analysis of omental adipose tissues of morbidly obese patients with
polycystic ovary syndrome (PCOS). PCOS is a common hormonal disorder among
women of reproductive age, and is characterized by hyperandrogenism and chronic
anovulation. PCOS is associated with obesity."

$feature_count
[1] "22283"

$order
[1] "none"

$platform
[1] "GPL96"

$platform_organism
[1] "Homo sapiens"

$platform_technology_type
[1] "in situ oligonucleotide"

$pubmed_id
[1] "17062763"

$reference_series
[1] "GSE5090"

$sample_count
[1] "15"

$sample_organism
[1] "Homo sapiens"
```

```
$sample_type
[1] "RNA"

$title
[1] "Polycystic ovary syndrome: adipose tissue"

$type
[1] "gene expression array-based"

$update_date
[1] "Mar 21 2007"

$value_type
[1] "count"
```

Code 18.3.

As you can see, much of the output is the same as that displayed on the GEO DataSet Record page for this particular study.

We can display individual expression values for each probe across our samples using the Table() function. The table in Code 18.4 shows the expression values for the first 10 probes for the first 3 chips in the dataset.

```
> Table(gds)[1:10, 1:5]
      ID_REF IDENTIFIER GSM114841 GSM114844 GSM114845
1   1007_s_at       DDR1     222.6   252.700     219.3
2    1053_at       RFC2      35.5    24.500      23.4
3     117_at      HSPA6      41.5    53.300      31.3
4     121_at       PAX8     229.8   419.600     274.5
5   1255_g_at     GUCA1A     14.3    13.000      29.6
6    1294_at      UBE1L     150.8   116.000      89.9
7    1316_at       THRA      29.7    35.400      53.0
8    1320_at     PTPN21      35.1    44.800      28.8
9   1405_i_at       CCL5     47.8    53.200      41.7
10   1431_at     CYP2E1      22.5    24.900      38.7
```

Code 18.4.

The first column (ID_REF) gives the probe ID assigned by Affymetrix for each probe. The second column (IDENTIFIER) is an ID for the corresponding transcript. The remaining columns are the expression values returned for each array—here three arrays are listed (GSM114841, GSM114844, and GSM114845). As an example we would read the expression value for the paired box gene 8 (PAX8) gene transcript with probe ID 121_at as 229.8 for sample GSM114841.

When we typed "Meta(gds)," R returned a stream of useful output about this microarray experiment, which also told us that there were 15 samples of tissue from PCOS cases and controls. We are interested in establishing which genes differ for expression value between cases with and without PCOS, so we need to know the disease status associated with each sample. This information can be extracted from the gds data object by using the Columns() function (for clarity, we will just output the first two columns) (Code 18.5).

```
> Columns(gds)[1:2]
        sample          disease.state
1  GSM114841                  control
2  GSM114844                  control
3  GSM114845                  control
4  GSM114849                  control
5  GSM114851                  control
6  GSM114854                  control
7  GSM114855                  control
8  GSM114834 polycystic ovary syndrome
9  GSM114842 polycystic ovary syndrome
10 GSM114843 polycystic ovary syndrome
11 GSM114847 polycystic ovary syndrome
12 GSM114848 polycystic ovary syndrome
13 GSM114850 polycystic ovary syndrome
14 GSM114852 polycystic ovary syndrome
15 GSM114853 polycystic ovary syndrome
```

Code 18.5.

The first seven samples listed are described as control, and the latter eight are PCOS. A description of each sample can also be viewed in the output, telling us that the source of the sample is "Omental adipose tissue" for each case (simply type "Columns(gds)" at the prompt).

The level of information for each sample can also be examined in more detail by viewing each Sample record. Let us have a look at the GSM data structure for our first sample, GSM28370. To do this, we can again use getGEO(), which we used to retrieve the DataSet record (Code 18.6).

```
> gsm <- getGEO("GSM114841")
trying URL
'http://www.ncbi.nlm.nih.gov/geo/query/acc.cgi?targ=self&acc=GSM114841&form=
text&view=full'
Content type 'geo/text' length unknown
opened URL
downloaded 603Kb

File stored at:
C:\DOCUME~1\SCSLEWPD\LOCALS~1\Temp\RtmpKisSsb/GSM114841.soft
```

Code 18.6.

The information for this sample is now stored in the gsm object. We can apply the same functions to the gsm object to extract information about our sample as we did to the gsd object containing the DataSet. Metadata for the sample can be retrieved using the Meta() function (Code 18.7).

```
> Meta(gsm)
$biomaterial_provider_ch1
[1] "Ramón y Cajal Hospital, Madrid, Spain"

$channel_count
[1] "1"
```

```
$characteristics_ch1
[1] "Morbidly obese control subject"

$contact_address
[1] "ARTURO DUPERIER"

$contact_city
[1] "MADRID"

$contact_country
[1] "Spain"

$contact_email
[1] "bperal@iib.uam.es"

$contact_fax
[1] "34 91 5854401"

$contact_institute
[1] "INSTITUTO DE INVESTIGACIONES BIOMEDICAS, CSIC-UAM"

$contact_name
[1] "BELEN,,PERAL"

$contact_phone
[1] "34 91 5854478"

$contact_state
[1] "MADRID"

$`contact_zip/postal_code`
[1] "28029"

$data_processing
[1] "MAS 5.0, scaled to 100 and RMA"

$data_row_count
[1] "22283"

$description
[1] "Total RNA was extracted from omental  adipose tissue from a control
subject"

$geo_accession
[1] "GSM114841"

$label_ch1
[1] "Biotin"

$last_update_date
[1] "Jun 16 2006"

$molecule_ch1
[1] "total RNA"

$organism_ch1
[1] "Homo sapiens"

$platform_id
[1] "GPL96"

$series_id
[1] "GSE5090"

$source_name_ch1
[1] "Omental adipose tissue"

$status
[1] "Public on Jun 17 2006"
```

```
$submission_date
[1] "Jun 16 2006"

$supplementary_file
[1] "file:///samples/GSM114841/GSM114841.CEL.gz"
[2] "file:///samples/GSM114841/GSM114841.EXP.gz"

$title
[1] "EP3_adipose_control"

$type
[1] "RNA"
```

Code 18.7.

In addition to details about the researcher and the institution that carried out this experiment, we see more specific details about this sample, such as that this individual was classified as a "morbidly obese control subject." What the data-processing subheading in the metadata also tells us in this example is that the expression values were generated for this sample using Affymetrix Mas 5.0 software with the RMA algorithm. We can also see that expression values were "scaled to 100." Now, let us take a closer look at the expression data available for this sample. We know from using the `Table()` function on the `gds` object that expression data was recorded for each sample, but let us look at what additional data was made available by the researcher in the `gsm` object. We can quickly see the extent of available information using the `Columns()` function (Code 18.8).

```
> Columns(gsm)
              Column
1             ID_REF
2              VALUE
3           ABS_CALL
4  Detection p-value

Description
1
2 Signal intensity - MAS 5.0, scaled to 100 and RMA
3 Presence/absence of gene transcript in sample; the call in an absolute
analysis that indicates if the transcript was present (P), absent (A), marginal
(M), or no call (NC)
4 p-value that indicates the significance level of the detection call
```

Code 18.8.

Each column is described by a header, and there is also a description of each column given. Immediately, we can see that the researcher has provided output from the Mas 5.0 analysis software for each transcript, including the probe ID (ID_REF), expression value (VALUE), Detection Call (ABS_CALL), and a P Value (Detection P Value) indicating the significance level of the detection call. Remember that the detection call tells us whether the transcript is called Present, Marginal, or Absent, and is determined by the corresponding P Value. We can look at the complete data for our sample using the `Table()` function, but for a quick look, let us retrieve data for 10 transcripts in the dataset (Code 18.9).

```
> Table(gsm)[500:510,]
          ID_REF VALUE ABS_CALL Detection.p.value
500      33307_at  47.8      P         0.039365
501      33304_at  35.3      A         0.339558
502      33197_at  58.4      P         0.017001
503      33148_at  21.4      P         0.019304
504      33132_at  15.8      A         0.189687
505      32837_at 274.0      P         0.000959
506      32836_at 319.8      P         0.006532
507      32811_at 213.3      P         0.024711
508      32723_at  35.0      P         0.004863
509    32699_s_at  43.9      A         0.162935
510      32625_at  67.1      A         0.117160
```

Code 18.9.

The expression values are more meaningful when the PMA detection call is also present. The probe 33304_at has a low expression value of 35.3, and the analysis software has determined a detection P Value of 0.339558, strongly suggesting that this transcript is not present (i.e., the gene is not being expressed) in this sample. Conversely, the transcript with probe ID 32837_at has an expression value of 274.0 and a detection P Value of 0.000959, and we may consider this gene as being expressed and retain it for further analysis.

In our example, the study involved 15 arrays, but often, a study will be based on perhaps 50 arrays or more. Earlier, I mentioned a third GEO data structure called the Series record. The Series record contains lists for the GPL platform object and the GSM Sample record object. By using the `getGEO()` and `GSMList()` functions, we can retrieve the GSE data structure for our example and output all of the information contained within the GSE object in one go (Code 18.10).

```
> gse <- getGEO("GSE5090")
trying URL
'ftp://ftp.ncbi.nih.gov/pub/geo/DATA/SOFT/by_series/GSE5090/GSE5090_family.soft
.gz'
ftp data connection made, file length 5750718 bytes
opened URL
downloaded 5615Kb

File stored at:
C:\DOCUME~1\SCSLEWPD\LOCALS~1\Temp\RtmpKisSsb/GSE5090.soft.gz
Parsing....
^PLATFORM = GPL96
^SAMPLE = GSM114834
^SAMPLE = GSM114841
^SAMPLE = GSM114842
^SAMPLE = GSM114843
^SAMPLE = GSM114844
^SAMPLE = GSM114845
^SAMPLE = GSM114847
^SAMPLE = GSM114848
^SAMPLE = GSM114849
^SAMPLE = GSM114850
```

```
^SAMPLE = GSM114851
^SAMPLE = GSM114852
^SAMPLE = GSM114853
^SAMPLE = GSM114854
^SAMPLE = GSM114855
```

Code 18.10.

We now have an object called `gse`, which contains the information we saw for each GSM Sample record, not just for one, but for all samples in the study. Using the `GSMList()` function, we would see the same data returned collectively for sample GSM114841 as we did when we applied the `Meta()`, `Columns()`, and `Table()` functions separately on the single GSM object for this sample:

```
> GSMList(gse)[[1]]
```

The usefulness of the GSE object can be demonstrated by combining the `Table()` function with the `GSMList()` function so that we can view expression values, PMA detection calls, and P Values for each sample without having to download this information separately for individual GSM objects per sample (Code 18.11).

```
> Table(GSMList(gse)[[1]])[1:5, ]
          ID_REF VALUE ABS_CALL Detection.p.value
1 AFFX-TrpnX-M_at  0.6       A          0.932322
2 AFFX-TrpnX-5_at  1.6       A          0.645547
3 AFFX-TrpnX-3_at  0.5       A          0.891021
4  AFFX-ThrX-M_at  2.0       A          0.672921
5  AFFX-ThrX-5_at  1.7       A          0.814869
> Table(GSMList(gse)[[2]])[1:5, ]
          ID_REF VALUE ABS_CALL Detection.p.value
1 AFFX-TrpnX-M_at  1.3       A          0.937071
2 AFFX-TrpnX-5_at  1.6       A          0.544587
3 AFFX-TrpnX-3_at  0.8       A          0.843268
4  AFFX-ThrX-M_at  5.2       A          0.588620
5  AFFX-ThrX-5_at  1.1       A          0.772364
> Table(GSMList(gse)[[3]])[1:5, ]
          ID_REF VALUE ABS_CALL Detection.p.value
1 AFFX-TrpnX-M_at  1.3       A          0.963431
2 AFFX-TrpnX-5_at  2.6       A          0.672921
3 AFFX-TrpnX-3_at  0.5       A          0.910522
4  AFFX-ThrX-M_at  4.3       A          0.631562
5  AFFX-ThrX-5_at  1.9       A          0.897835
```

Code 18.11.

18.8 Creating a Simple Expression Dataset Using GEOquery

It is easy to see now how we could simply extract expression data from a GSE object to create a new data object that could be input for statistical analysis. Again, by combining GEOquery functions, we can quickly do this (Code 18.12).

```
>arraydata <- do.call("cbind", lapply(GSMList(gse), function(x) {
tab <- Table(x)
return(tab$VALUE)
}))

>rownames(arraydata) <- Table(GPLList(gse)[[1]])$ID
>colnames(arraydata) <- names(GSMList(gse))

>arraydata[1:5, ]

          GSM114834 GSM114840 GSM114841 GSM114842 GSM114843 GSM114844 GSM114845
1007_s_at       0.6       1.3       1.3       1.0       1.8       1.2       5.0
1053_at         1.6       1.6       2.6       1.6       2.1       2.0       6.1
117_at          0.5       0.8       0.5       0.5       0.4       0.5       4.1
121_at          2.0       5.2       4.3       2.0       6.3       2.5       9.6
1255_g_at       1.7       1.1       1.9       2.1       3.4       2.2      10.1
          GSM114846 GSM114847 GSM114848 GSM114849 GSM114850 GSM114851 GSM114852
1007_s_at       1.1       1.2       1.4       0.9       0.5       1.2       0.5
1053_at         2.4       1.5       1.6       1.2       1.0       1.2       0.6
117_at          0.3       0.3       0.8       0.6       0.5       0.3       0.4
121_at          4.2       4.0       3.0       4.4       3.4       1.8       3.9
1255_g_at       1.3       3.3       1.3       2.9       2.6       1.5       2.7
          GSM114853 GSM114854 GSM114855
1007_s_at       1.4       1.0       0.8
1053_at         0.7       1.3       1.6
117_at          0.2       0.7       0.4
121_at          5.3       1.3       2.1
1255_g_at       2.2       2.2       1.4
```

Code 18.12.

First, we create a matrix to hold the expression data that is extracted from each sample using the `do.call()` function. Once the table has been created, we assign the probe ID and sample ID names to the row and column headers, respectively, using the `rownames()` and `colnames()` functions. Now you have a data matrix that can be used to analyze the data to look for differential gene expression between groups of samples or to reveal potential subgroupings of genes or patients according to expression patterns.

18.9 Summary

In this chapter we used functions from the `GEOquery` package to retrieve an Affymetrix gene expression dataset from the Gene Expression Omnibus. We were also able to extract relevant information about the dataset and present that. More importantly, we were able to use additional `GEOquery` functions to extract the raw expression data. Before we can analyze this data, we need to perform quality control, which is the basis of the next chapter.

18.10 Questions

1. What are the benefits to medical research if one can access publicly available microarray datasets?

2. What is the relationship between R and Bioconductor?

3. What exactly do we parse to the `getGEO()` function from the `GEOquery` package?

4. What are the different types of GEO records?

5. What type of GEO information does a Series record contain?

6. What is the purpose of the `do.call()` function?

19 | Working with Microarray Data

19.1 Background

In the previous chapter, we learned how to find and view data from a microarray study in the Gene Expression Omnibus (GEO) repository using the GEO data structures and the Bioconductor `GEOquery` package. We were able to extract expression data, view sample and metadata for the study, and compile the expression data into a matrix.

Following on from that work, we would now be at the point where we would like to analyze the data. However, what if we were the researchers producing our own microarray data? What if our data was simply presented to us as raw CEL files or we were only able to obtain the CEL files? What if we wished to generate expression data using a different method to that, say, in the Mas 5.0 algorithm? We may want to generate our own PMA detection calls. There are many circumstances in which we may want to start an analysis from scratch, working directly with raw microarray data.

In this chapter, we will take a look at a couple of Bioconductor packages called `affy`[40] and `simpleAffy`[41] that allow us to process Affymetrix raw data as well as carry out quality control and simple data analysis.

To begin, download the `affy` and `simpleAffy` packages (Code 19.1).

```
> source("http://bioconductor.org/biocLite.R")
> biocLite("affy")
> biocLite("simpleaffy")
```

Code 19.1.

19.2 Getting Some Raw Affymetrix Data

When exploring the GEO website for the polycystic ovary syndrome (PCOS) study in the last chapter, you may have already downloaded the CEL files by following the Supplementary Files hyperlink. If not, do this now to obtain the study CEL files that we will use next. Alternatively, you may download them directly using the `download.file()` function (Code 19.2).

```
>download.file(url="ftp://ftp.ncbi.nih.gov/pub/geo/DATA/supplementary/series/
GSE5090/GSE5090_RAW.tar", destfile="arraydata.tar")
trying URL 'ftp://ftp.ncbi.nih.gov/pub/geo/DATA/supplementary/series/GSE5090/
GSE5090_RAW.tar'
```

```
ftp data connection made, file length 63569920 bytes
opened URL
downloaded 62080Kb
```

Code 19.2.

Note that we are downloading a zipped ".tar" file that contains all of the CEL files and storing this file as "arraydata.tar." You should change the R working directory to a folder of your choice before downloading. You can unzip the individual CEL files, each of which is also zipped in a ".tar" file, into a destination folder of your choice. I have created a folder called "affy files" within the main R folder. You also need to create a file called "covdesc" that contains a list of the CEL files and a simple description about the experimental condition (in this case, smoking status). For our example, the content of the file could look like that shown in Code 19.3.

```
        disease
GSM114841.cel  control
GSM114844.cel  control
GSM114845.cel  control
GSM114849.cel  control
GSM114851.cel  control
GSM114854.cel  control
GSM114855.cel  control
GSM114834.cel  polycystic_ovary_syndrome
GSM114842.cel  polycystic_ovary_syndrome
GSM114843.cel  polycystic_ovary_syndrome
GSM114847.cel  polycystic_ovary_syndrome
GSM114848.cel  polycystic_ovary_syndrome
GSM114850.cel  polycystic_ovary_syndrome
GSM114852.cel  polycystic_ovary_syndrome
GSM114853.cel  polycystic_ovary_syndrome
```

Code 19.3.

The file thus contains only two columns: a column with no header listing the names of the CEL files and a second column with the disease status as described by the researcher. The covdesc file should be placed in the same folder as the CEL files. With the CEL files and descriptor file in place, we are now ready to start processing raw data from our example.

Load the `simpleAffy` library:

```
>library(simpleaffy)
```

The `simpleaffy` package uses another package called `affy`, which should have been installed alongside `simpleaffy`. The `affy` package has a range of functions, allowing you to get from probe-level data from the CEL files to expression measures for each gene on the arrays. The functions fall into four basic types (List 19.1).

List 19.1 `simpleaffy` **function types.**

- Quality control (QC)
- RNA degradation assessment

- Probe-level normalization/background correction
- Conversion of probe-level data to expression measurements

The package also contains some useful plotting functions. The `simpleaffy` package is really an extension to `affy` that provides more functionality. We will use `simpleaffay` to provide us with additional functions for gene filtering using the PMA detection calls and gene filtering by differential expression.

Before we begin, remember to change the working directory to the folder containing the CEL files. Now we can read in the raw CEL file data to an object called `data.raw` using a function called `read.affy()`:

```
> data.raw <- read.affy("covdesc")
```

The `read.affy()` function reads in all CEL files from the same folder and also reads the covdesc file to gain some information as to the number of sample groups we have. The `data.raw` object is an instance of the `AffyBatch` class, which is a lot more than just a means to hold the data. Many functions are available to view information about the probe-level data within the arrays as well as to preprocess the data. To find out more about this class, type:

```
> ?AffyBatch
```

You will notice that the `AffyBatch` class extends another class called `eSet` from Bioconductor's Biobase package that is a container for microarray experimental and metadata. Here you will find a whole host of further functions to explore your data with. You will also see that `AffyBatch` or `eSet` class structures hold data in something called a slot. To access data contained within a slot, you need to use the "@" operator. As an example, our `data.raw` object is an instance of the `AffyBatch` class that has a slot called `cdfName`, which tells us the Affymetrix array platform used:

```
> data.raw@cdfName
[1] "HG-U133A"
```

Another slot called `nrow` can tell us the number of rows in the Affymetrix chip used for our example study:

```
> data.raw@nrow
[1] 712
```

The HG-U133A array has 712 rows and 712 columns, which equates to 506,944 individual spots per chip. We can even obtain probe-level data from each individual spot using the `AffyBatch` class function `exprs()` (Code 19.4).

```
> intensity(data.raw)[506940:506944,1]
 506940  506941  506942  506943  506944
 118.5 12505.8   251.8 12583.0   217.3
```

Code 19.4.

Here, we have merely extracted probe-level intensity data for the last four spots on the first array of the `data.raw` object. We will return a bit later to see how some additional functions in the `simpleaffy` package can be used to gather information about expression-level data.

19.3 How Good Is Your Affy Data? Normalizing and Quality Control

Before we process and analyze our data, let us have a quick check on how "good" the raw data actually is by carrying out some quality control. There are a number of reasons why we would need to check the quality of our data, but we are concerned primarily with the quality of a chip or perhaps the mRNA from our samples. Bioconductor provides a number of QC methods that we can perform on our data, many of which use plots such as density plots of PM (perfect match/mismatch) intensities, RNA degradation plots, or MvA plots.

The MvA plot is particularly useful to assess how similar probe-level data is between replicate samples or samples in the same subgroup, such as the controls in our example. What is an MvA plot? These plots allow the detection of any bias in intensities between samples by assessment of the curve produced. The data plotted can differ, depending on the microarray technology used, but generally, for Affymetrix arrays, for any probe, the "M" represents log fold change between two samples and "A" represents the mean absolute log expression between the samples. M is plotted on the *y*-axis and A on the *x*-axis. MvA plots can be generated using the `MAplot()` function in the `affy` package. So, let us carry out a quick QC on some of our PCOS and control samples using the MvA plotting function available in `affy`. We will look at raw data from the first and fourth control samples (GSM114841 and GSM114849):

```
> MAplot(data.raw[c(1,4)], pairs=TRUE)
```

The plot in Figure 19.1 shows the distribution of points for each transcript along the line representing zero-fold change. Hence, points plotted above and below this line represent transcripts that differ in concentration between the two samples for the range of mean log expression values. As an exercise, you could create MvA plots for all pairwise comparisons of samples to see how they differ. The distribution is summarized by a Loess curve, showing how the distribution differs from zero-fold change. The shape of the plot suggests a generally symmetrical distribution around the zero-fold change line. The Loess curve does, however, suggest a slight trend whereby as the average probe intensity between the two samples increases, the fold change slightly decreases. This slight variation may exist due to technical or systematic error that occurred during array processing.

At this point, you may want to normalize your array data to correct for systematic variation. To demonstrate the effect of normalizing Affymetrix data, we can apply the `normalize()` function in `affy`:

```
> data.norm <- normalize(data.raw)
```

Now if we generate an MvA plot (Figure 19.2) for the normalized data for samples GSM114841 and GSM114849, we see that the Loess curve is no longer distinguishable from the line where fold change is equal to zero:

```
> MAplot(data.norm[c(1,4)], pairs=TRUE)
```

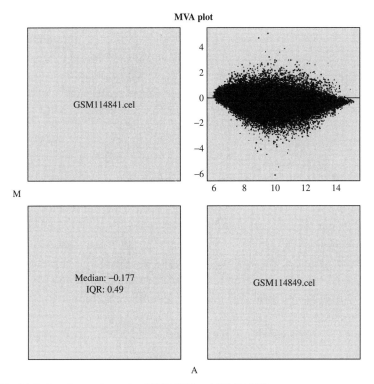

FIGURE 19.1   MvA plot of control samples GSM114841 and GSM114849.

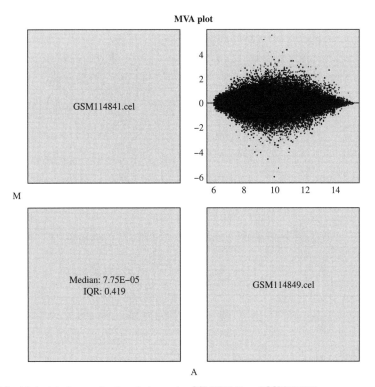

FIGURE 19.2   MvA plot of normalized control samples GSM114841 and GSM114849.

So we can see that normalizing the data may improve the quality of the dataset and remove bias due to systematic error. More importantly, normalizing microarray data allows for comparability of gene expression values across a range of samples. We have spoken of the Mas 5.0 algorithm as being widely applied to generate expression-level data from probe-level data, and as part of this process, Mas 5.0 actually normalizes the data. The normalization step in Mas 5.0 involves scaling by adjusting the average signal value of all arrays to a common value.

19.4 From Probe-Level Data to Expression Data

Now that we have covered the concept of normalization, let us move on to the next step of processing the raw probe-level data obtained from each CEL file to generate transcript expression levels and corresponding detection calls. As part of this process, we can also carry out some more quality control on our dataset. To do this, we will use the Mas 5.0 algorithm option in a function called `call.exprs()` in the `simpleaffy` package to generate the expression levels from the `data.raw` object we created earlier.

To create an expression-level dataset type:

```
> data.mas5 <- call.exprs(data.raw,"mas5")
```

The `data.mas5` object is a new instance of the `exprSet` class described in Bioconductor's Biobase package and is actually the class for which `AffyBatch` is extended. The object now contains normalized expression summaries for all probes in all arrays and will form the input into functions we will use to filter genes later. One of the advantages of using the Mas 5.0 algorithm to generate our expression-level data is that it can also generate the PMA detection calls and associated P Values. As we will see in the next section, PMA calls can be invaluable during the process of filtering genes to create a list of truly useful genes differentially expressed between samples. We can generate the PMA detection calls and P Values using `simpleaffy`'s `detection.p.val()` function:

```
> sample1<-detection.p.val(data.raw[,1],calls=TRUE)
```

The function has calculated PMA calls and P Values for all probes in the first array in our dataset and put the information into the `sample1` object. We can view the PMA calls and P Values separately for the first 10 probes (Code 19.5).

```
> sample1$call[1:10]
 [1] "P" "P" "P" "P" "A" "P" "P" "A" "P" "M"
> sample1$pval[1:10]
 [1] 0.003066739 0.027859596 0.013091780 0.014936519 0.250723677 0.006531992
 [7] 0.004180367 0.302547472 0.003815252 0.060419342
```

Code 19.5.

Before we move on to find a list of genes differentially expressed between the PCOS and control samples in the dataset, we will continue with our QC assessment using some additional functions implemented in `simpleaffy`. The `qc` function can generate commonly used quality control metrics, including average background, scale factor, number of genes called present, and 3′ to 5′ end ratios of the beta-actin and GADPH genes. To apply the `qc` function to our example dataset, type:

```
> qc <- qc(data.raw, data.mas5)
```

The quality control information in the `qc` object can all be displayed in a single plot:

```
> plot(qc)
```

This plot (Figure 19.3) may appear slightly daunting at first, but is really quite easy to interpret. We can see on the left side the list of arrays in our example dataset followed by a pair of numbers and a chart. Each array is separated by a gray dotted line. The upper adjacent number shows the percentage of probes called "present" for that array. There is good agreement across all arrays in the study, with this number varying by less than 10%. The bottom number tells us the average background, and again there is good consistency across nearly every array, with the exception of GSM114845, which has a considerably higher average background. The chart gives information of the scaling applied by Mas 5.0 and the 3′:5′ ratios of the beta-actin and GADPH genes. We can see that the scale on the chart ranges from −3 through zero to 3. Each array has a horizontal line originating at 0 showing the scale factor that Mas 5.0 applied. A scale factor line to the left corresponds to down-scaling, whereas a line

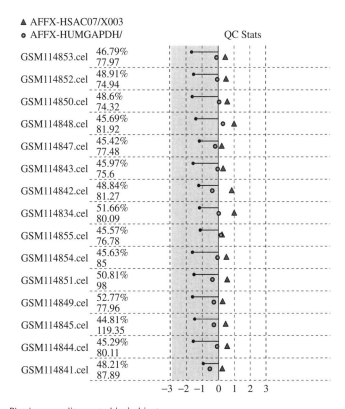

FIGURE 19.3 Plotting a quality control (qc) object.

to the right of zero corresponds to up-scaling. Affymetrix recommends that scale factors should always fall within the 3-fold range. Large fluctuations in scaling factors signal potential problems, such as systematic error, poor sample quality, or differences in the amount of starting material. The scale factors for the example dataset all fall within 2-fold of one another. The circles and triangles plotted show the 3′:5′ ratios of beta-actin and GADPH. Affymetrix suggests that the GADPH ratio should approximate 1, but in practice it is often less, as is the case with our example arrays. The recommended ratio for beta-actin is less than 3, and again this is the case in this study. Therefore, all the quality control metrics appear acceptable for our example dataset, giving us a good degree of confidence to go ahead and analyze our data for differential gene expression.

19.5 Creating a List of Differentially Expressed Genes

We learned previously that our `data.mas5` object, which holds our expression data, is an instance of the `AffyBatch` class. This means we can apply `AffyBatch` functions to this object to extract expression data and metadata. We can use the `exprs()` function to return expression-level data for different probes (Code 19.6).

```
> exprs(data.mas5)[1:10,1:4]
            GSM114841.cel GSM114844.cel GSM114845.cel GSM114849.cel
1007_s_at        7.799031      7.982501      7.936151      8.017302
1053_at          5.149059      4.618321      4.996079      4.973884
117_at           5.376999      5.736024      6.646779      5.426191
121_at           7.845280      8.713759      8.313314      7.828122
1255_g_at        3.842650      3.705601      4.138562      4.029924
1294_at          7.237356      6.858445      6.873898      6.967967
1316_at          4.894248      5.146691      5.526283      5.393529
1320_at          5.133438      5.484950      5.576856      5.313660
1405_i_at        5.578731      5.733813      6.576959      6.869568
1431_at          4.493393      4.637869      3.530006      3.901150
```

Code 19.6.

Here we have returned the expression levels computed by the Mas 5.0 algorithm for the first 10 probes on the array for the control samples GSM114841, GSM114844, GSM114845, and GSM114849. This is the data we can now use to determine a list of genes differentially expressed between the PCOS and control samples. Anyone familiar with the literature on finding significantly differentially expressed genes will be aware that there are a number of statistical ways of doing this, many of which have variations and modifications to improve the robustness of the test. Investigating these different methods is beyond the scope of this book. The reader is encouraged to explore the literature on methods contained within Bioconductor and to apply different methods to the example dataset. To generate our gene list, we will use the functions implemented in `simpleaffy` that the authors of that package included for a quick analysis of your data. The function `pairwise.comparison()` will generate fold changes, *t*-tests, and means for pairs of experimental groups and is ideal for the PCOS dataset.

The `pairwise.comparison()` function takes as input an `exprSet` object, so we may apply it to the `data.mas5` object created earlier:

```
> pair <- pairwise.comparison(data.mas5, "disease",
c("control","polycystic_ovary_syndrome"), data.raw)
```

Other arguments we have included as input to the function are the group label "disease," labels for members of the group "control" and "polycystic_ovary_ syndrome," and a spots argument referencing the `data.raw` object that allows the function to generate PMA calls and corresponding P Values. The `pair` object now holds all the results of the tests applied and is known as a pairwise comparison object, which is an instance of `simpleaffy`'s `PairComp` class. Before we learn how to extract information from the analysis, let us take a look at another `simpleaffy` function that we can apply to filter the data according to some sensible criteria. For instance, because we are only interested in generating a list of genes significantly differentially expressed between the PCOS and control samples, there is probably no point in keeping fold change or *t*-test results for genes that are not being expressed (i.e., called Absent) across all the samples by the Mas 5.0 algorithm. Therefore, there is a function called `pairwise.filter()` in `simpleaffy` that allows you to carry out filtering of probes. You can filter by expression level, PMA detection call, fold change, and *t*-test statistic, either individually or collectively. To filter by expression, you need to pass a minimum expression level as an argument to `pairwise.filter`, and there is also the possibility of passing an optional argument to state how many arrays must have the minimum expression value (where all arrays would be the default). Similarly, you can pass as arguments the minimum number of arrays in which a gene is called Present across all groups or by group. Thresholds may also be set by argument for minimum fold change and maximum *t*-test P Value for a gene to be returned.

For our example dataset, we will filter the results of the pairwise comparison using the following criteria: Genes must be called Present in at least three samples in one of the groups, the *t*-test P Value must be less than 0.01, and fold change between means of the two groups must be at least 1.5 (Code 19.7).

```
> pair.filt <- pairwise.filter(pair, min.present.no= 3,
present.by.group=T, fc=log2(1.5),tt= 0.05)
[1] "Checking member control in group: ' disease '"
[1] "Checking member polycystic_ovary_syndrome in group: ' disease '"
```

Code 19.7.

The `pair.filt` object we created is also an instance of the `PairComp` class. A quick check on the `simpleaffy` documentation reveals that there are a number of slots associated with the `PairComp` class (Table 19.1).

These slots are useful because they allow us to organize our gene list according to expression level or statistical analysis information. After creating a list of differentially expressed genes, the first question we ask is exactly how many genes have been returned given the criteria passed to the `pairwise.filter()` function. This can be

Table 19.1 `PairComp` **Slots.**

`means`	Mean values for each of the experimental factors
`fc`	Fold change between the means
`tt`	P-score between the factors
`calls`	Detection p-values for each probeset on each array
`group`	Name of the factor that was compared
`members`	List containing the two levels compared between
`pData`	phenoData for the members that were compared
`calculated.from`	Original expression set that was being compared

determined easily using `nrow()` and counting the instances in one of the slots (such as the `means` slot) (Code 19.8).

```
> nrow(pair.filt@means)
[1] 67
```

Code 19.8.

Thus, we have a list of 67 genes significantly differentially expressed between the PCOS and normal samples. We can now explore the contents of our `pair.filt` object by viewing some of the slots contained within the class. We can start by viewing the expression-level means of a selection of probes in each of the two experimental groups in the dataset using the `means` slot (Code 19.9).

```
> pair.filt@means[1:10,1:2]
              control polycystic_ovary_syndrome
200832_s_at 10.333091                 11.004764
200951_s_at  4.417044                  5.088299
200974_at   10.134294                  9.469463
200978_at    9.624845                 10.284560
201211_s_at  7.136844                  6.529505
201468_s_at  8.909398                  9.756174
201496_x_at  6.486854                  5.147633
201497_x_at  9.179405                  8.083944
202040_s_at  6.770394                  7.417757
202274_at    6.554181                  5.574886
```

Code 19.9.

We can also view the PMA detection calls for these genes using the `calls` slot (Code 19.10).

```
> pair.filt@calls[1:10,]
            GSM114841.cel.present GSM114844.cel.present GSM114845.cel.present
200832_s_at "P"                   "P"                   "P"
200951_s_at "A"                   "A"                   "A"
200974_at   "P"                   "P"                   "P"
```

```
200978_at    "P"                      "P"                      "P"
201211_s_at  "P"                      "P"                      "P"
201468_s_at  "P"                      "P"                      "P"
201496_x_at  "P"                      "P"                      "P"
201497_x_at  "P"                      "P"                      "P"
202040_s_at  "P"                      "P"                      "P"
202274_at    "P"                      "P"                      "P"
             GSM114849.cel.present GSM114851.cel.present GSM114854.cel.present
200832_s_at  "P"                      "P"                      "P"
200951_s_at  "A"                      "A"                      "A"
200974_at    "P"                      "P"                      "P"
200978_at    "P"                      "P"                      "P"
201211_s_at  "P"                      "P"                      "P"
201468_s_at  "P"                      "P"                      "P"
201496_x_at  "P"                      "P"                      "M"
201497_x_at  "P"                      "P"                      "M"
202040_s_at  "P"                      "P"                      "P"
202274_at    "P"                      "P"                      "A"
             GSM114855.cel.present GSM114834.cel.present GSM114842.cel.present
200832_s_at  "P"                      "P"                      "P"
200951_s_at  "A"                      "P"                      "P"
200974_at    "P"                      "P"                      "P"
200978_at    "P"                      "P"                      "P"
201211_s_at  "P"                      "P"                      "P"
201468_s_at  "P"                      "P"                      "P"
201496_x_at  "P"                      "P"                      "P"
201497_x_at  "P"                      "P"                      "P"
202040_s_at  "P"                      "P"                      "P"
202274_at    "P"                      "P"                      "A"
             GSM114843.cel.present GSM114847.cel.present GSM114848.cel.present
200832_s_at  "P"                      "P"                      "P"
200951_s_at  "A"                      "A"                      "A"
200974_at    "P"                      "P"                      "P"
200978_at    "P"                      "P"                      "P"
201211_s_at  "P"                      "P"                      "A"
201468_s_at  "P"                      "P"                      "P"
201496_x_at  "P"                      "A"                      "P"
201497_x_at  "P"                      "P"                      "P"
202040_s_at  "P"                      "P"                      "P"
202274_at    "P"                      "A"                      "A"
             GSM114850.cel.present GSM114852.cel.present GSM114853.cel.present
200832_s_at  "P"                      "P"                      "P"
200951_s_at  "A"                      "A"                      "P"
200974_at    "P"                      "P"                      "P"
200978_at    "P"                      "P"                      "P"
201211_s_at  "A"                      "A"                      "A"
201468_s_at  "P"                      "P"                      "P"
201496_x_at  "P"                      "P"                      "A"
201497_x_at  "P"                      "P"                      "A"
202040_s_at  "P"                      "P"                      "P"
202274_at    "A"                      "P"                      "A"
```

Code 19.10.

Fold change between the two groups for each gene can be accessed using the `fc` slot (Code 19.11).

```
> pair.filt@fc[1:10]
200832_s_at 200951_s_at   200974_at    200978_at 201211_s_at 201468_s_at
 -0.6716723  -0.6712554   0.6648302   -0.6597152   0.6073387  -0.8467760
201496_x_at 201497_x_at 202040_s_at    202274_at
  1.3392202   1.0954607  -0.6473627    0.9792949
```

Code 19.11.

Finally, we can also view the *t*-test P Values for our selected genes to show the level of significance for the differences between groups (Code 19.12).

```
> pair.filt@tt[1:10]
200832_s_at 200951_s_at   200974_at   200978_at 201211_s_at 201468_s_at
0.036249247 0.045360948 0.035356311 0.009561160 0.044191895 0.033220095
201496_x_at 201497_x_at 202040_s_at   202274_at
0.027175288 0.036267591 0.001434368 0.017101796
```

Code 19.12.

With all this information on board, we have now arrived at the point where researchers new to microarray studies get excited without realizing, as we will see in the next chapter, that the hard work is only just about to begin.

19.6 Summary

In this chapter we used our knowledge of obtaining raw data for an Affymetrix microarray study based on polycystic ovary syndrome. We were able to use functions from the `affy` and `simpleaffy` packages to perform quality control on the raw data and extract expression-level information from probe-level data. We were then able to create a list of genes significantly differentially expressed between disease and control samples. The gene list is, of course, meaningless unless it has biological meaning. The next step then is to trawl biological databases and published literature to gather as much information as possible on each gene in that list.

19.7 Questions

1. What three main tasks does the the `simpleAffy` package allow you to perform?

2. What is an Affymetrix CEL file?

3. What R function can we use to produce an MvA plot?

4. What quality control metric can the `qc()` function yield?

5. What statistical test does the `pairwise.comparison()` function perform?

6. How can we filter differentially expressed genes using components of objects from the `pairwise.comparison` class?

20 Annotating Microarray Gene Lists

20.1 Background

In this last chapter devoted to gene expression data, we will look at how we can annotate a gene list. The annotation step is the one that makes or breaks a microarray study and leaves the researchers anxious as to whether they have truly discovered meaningful biological differences between test and control samples. Annotation of gene lists is tedious, despite the ever-growing mass of information available for each gene represented on arrays. It is important to be able to determine gene identities, chromosomal location, and gene ontology (GO) data about pathway involvement of the encoded proteins. Even a small gene list of about 20 genes would allow the researcher to go back to the laboratory and confirm the findings by other techniques.

Any tool that speeds up this process is gratefully received. Little wonder then that packages are being developed as part of the Bioconductor to assist in the annotation process.

This chapter looks at how we can take our gene list and use three Bioconductor packages to gather publicly available information on each gene, including ontology pathway information. Although we talk in terms of a "gene" list, we actually work with probe IDs.

To begin, download the appropriate packages listed in Code 20.1.

```
> source("http://bioconductor.org/biocLite.R")
> biocLite("annotate")
> biocLite("hgu133a")
```

Code 20.1.

20.2 The hgu133a Package for Annotating Genes

The hgu133a package contains functions that allow you to map information from the HG-U133A array's probesets to annotation data stored within the package. This package was built using another Bioconductor package called AnnBuilder. AnnBuilder allows you to build annotation data packages for a given set of genes that can be mapped to GenBank accession numbers, UniGene identifiers, Image identifiers, or Entrez Gene identifiers.

We know that the HG-U133A chip has more than 22,000 probesets, and we check the completeness of annotation data held within the package rda data file using the hgu133a() function (Code 20.2).

```
> hgu133a()
Quality control information for hgu133a
Date built: Created: Mon Apr 23 12:08:27 2007

Number of probes: 22283
Probe number mismatch: None
Probe mismatch: None
Mappings found for probe-based rda files:
        hgu133aACCNUM found 22283 of 22283
        hgu133aCHR found 21453 of 22283
        hgu133aCHRLOC found 20527 of 22283
        hgu133aENZYME found 2739 of 22283
        hgu133aENTREZID found 21822 of 22283
        hgu133aGENENAME found 21469 of 22283
        hgu133aGO found 19489 of 22283
        hgu133aMAP found 21362 of 22283
        hgu133aOMIM found 16092 of 22283
        hgu133aPATH found 6322 of 22283
        hgu133aPFAM found 21201 of 22283
        hgu133aPMID found 21293 of 22283
        hgu133aPROSITE found 21201 of 22283
        hgu133aREFSEQ found 21105 of 22283
        hgu133aSYMBOL found 21469 of 22283
        hgu133aUNIGENE found 21113 of 22283
Mappings found for non-probe-based rda files:
        hgu133aCHRLENGTHS found 25
        hgu133aENZYME2PROBE found 748
        hgu133aGO2ALLPROBES found 8046
        hgu133aGO2PROBE found 5855
        hgu133aPATH2PROBE found 190
        hgu133aPMID2PROBE found 136367
```

Code 20.2.

Just scanning the list, we can see that there is excellent quality annotation data for virtually every resource. Retrieving annotation data for genes in our list is straightforward because there is a function in hgu133a to achieve this for each resource.

20.3 Creating an Ordered Gene List

When annotating a gene list, researchers often prefer to order the list from the most significant genes to the least. In the last chapter, we created a gene list and stored it in a PairComp object called pair.filt. We can order the elements in the pair.filt object slots using the order() function in the objects index:

```
> pair.filt.sorted <- pair.filt[order(pair.filt@tt) , ]
```

Note that the elements are ordered according to the tt slot representing the p-values.

The next step is to retrieve the probe ID values from the sorted list. We can create a new list called "gene.list" by extracting the row names from the means slot of our sorted PairComp object:

```
> gene.list<-row.names(pair.filt.sorted@means)
```

The first thing you would want to do now is view the list of probe IDs so you can begin to annotate each gene in order (Code 20.3).

```
> pair.filt.sorted@tt
          215207_x_at                    221473_x_at
          0.0002053879                   0.0003035978
          206788_s_at                    202040_s_at
          0.0007563956                   0.0014343685
          219237_s_at                    203134_at
          0.0031571109                   0.0039280305
          210935_s_at                    214140_at
          0.0051927148                   0.0055196189
          202305_s_at                    221246_x_at
          0.0057344527                   0.0059346585
          215698_at                      209474_s_at
          0.0060619611                   0.0075607193
          204840_s_at                    212629_s_at
          0.0076537919                   0.0088280626
          204873_at                      200978_at
          0.0095364828                   0.0095611596
          218999_at                      215694_at
          0.0110288367                   0.0112021844
          221496_s_at                    209788_s_at
          0.0118014027                   0.0118604182
          219789_at                      215404_x_at
          0.0119851192                   0.0135982219
          217239_x_at                    204731_at
          0.0144194641                   0.0157245619
          212249_at                      202274_at
          0.0158429469                   0.0171017961
          207434_s_at                    216288_at
          0.0181238421                   0.0201027966
          216657_at                      205467_at
          0.0206706210                   0.0213540306
          222372_at                      219957_at
          0.0217905826                   0.0234892733
          202554_s_at                    210892_s_at
          0.0243546496                   0.0257458831
          217566_s_at                    205187_at
          0.0260493725                   0.0265474041
          201496_x_at                    209909_s_at
          0.0271752876                   0.0272246403
          210916_s_at                    214916_x_at
          0.0273601390                   0.0277689848
          203213_at                      212382_at
          0.0285195499                   0.0314237835
          207815_at                      201468_s_at
          0.0322404834                   0.0332200945
          206098_at                      200974_at
          0.0347978956                   0.0353563105
          214336_s_at                    200832_s_at
          0.0359843293                   0.0362492469
          201497_x_at                    222020_s_at
          0.0362675909                   0.0365621204
          202927_at                      214600_at
          0.0386202430                   0.0390166172
          205302_at                      220187_at
          0.0391835921                   0.0416602671
          212806_at                      208733_at
          0.0417694423                   0.0419415445
          220392_at   AFFX-HUMISGF3A/M97935_5_at
          0.0422510242                   0.0430068251
          201211_s_at                    217793_at
```

```
          0.0441918949              0.0442295620
             204602_at                 222330_at
          0.0444085624              0.0447227993
           200951_s_at               204939_s_at
          0.0453609479              0.0463718762
           215646_s_at               204938_s_at
          0.0463827423              0.0468811145
           216207_x_at
          0.0475708203
```

Code 20.3.

20.4 Retrieving Gene Names, Identifiers, and Chromosome Data

To fetch annotation data for a particular resource, we tap into a series of R environments provided by the hgu133a package. As well as returning information about a gene, some of these data resources allow you to map the probe ID to gene identifiers in public resources. As an example, let us return some data for the probe ID at the 20th position in our gene list (Code 20.4).

```
> gene.list[20]
[1] "209788_s_at"
```

Code 20.4.

This probe ID, 209788_s_at, can be sent to one of the environments holding data for different types of information, such as gene name and chromosome location. To return the desired information, we use the get() function, parsing the gene.list object index and specifying the environment to use.

The first piece of information we will retrieve for probe ID 209788_s_at is the gene name. The environment that provides mappings between the manufacturer's probe ID and gene name is called "hgu133aGENENAME." Therefore, we just need to set the env argument of get() to hgu133aGENENAME (Code 20.5).

```
> get(gene.list[20], env = hgu133aGENENAME)
[1] "type 1 tumor necrosis factor receptor shedding aminopeptidase regulator"
```

Code 20.5.

We can also get the gene name abbreviation by setting the env argument to hgu133aSYMBOL (Code 20.6).

```
> get(gene.list[20], env = hgu133aSYMBOL)
[1] "ARTS-1"
```

Code 20.6.

The gene name abbreviation returned is ARTS-1. We can also use the `getSYMBOL()` function in the `annotate` package to return all the gene name abbreviations in our gene list (Code 20.7).

```
> getSYMBOL(gene.list, data = "hgu133a")
           215207_x_at                221473_x_at
           "LOC729148"                  "SERINC3"
           206788_s_at                202040_s_at
               "CBFB"                   "JARID1A"
           219237_s_at                 203134_at
             "DNAJB14"                   "PICALM"
           210935_s_at                 214140_at
                "WDR1"                  "SLC25A16"
           202305_s_at                221246_x_at
               "FEZ2"                      "TNS1"
            215698_at                 209474_s_at
             "JARID1A"                   "ENTPD1"
           204840_s_at                212629_s_at
                "EEA1"                      "PKN2"
            204873_at                  200978_at
                "PEX1"                      "MDH1"
            218999_at                  215694_at
             "TMEM140"                  "SPATA5L1"
           221496_s_at                209788_s_at
                "TOB2"                    "ARTS-1"
            219789_at                 215404_x_at
                "NPR3"                     "FGFR1"
           217239_x_at                 204731_at
           "LOC652102"                   "TGFBR3"
            212249_at                  202274_at
               "PIK3R1"                    "ACTG2"
           207434_s_at                 216288_at
               "FXYD2"                   "CYSLTR1"
            216657_at                  205467_at
                "ATXN3"                   "CASP10"
            222372_at                 219957_at
               "MAGI1"                    "RUFY2"
           202554_s_at                210892_s_at
               "GSTM3"                    "GTF2I"
           217566_s_at                 205187_at
                "TGM4"                     "SMAD5"
           201496_x_at                209909_s_at
               "MYH11"                    "TGFB2"
           210916_s_at                214916_x_at
                "CD44"                       "IL8"
            203213_at                  212382_at
                "CDC2"                      "TCF4"
            207815_at                 201468_s_at
                "PF4V1"                     "NQO1"
            206098_at                  200974_at
               "ZBTB6"                    "ACTA2"
           214336_s_at                200832_s_at
                "COPA"                       "SCD"
           201497_x_at                222020_s_at
               "MYH11"                      "HNT"
            202927_at                  214600_at
                "PIN1"                     "TEAD1"
            205302_at                  220187_at
              "IGFBP1"                   "STEAP4"
            212806_at                  208733_at
             "KIAA0367"                   "RAB2A"
            220392_at  AFFX-HUMISGF3A/M97935_5_at
                "EBF2"                     "STAT1"
           201211_s_at                 217793_at
               "DDX3X"                   "RAB11B"
```

```
            204602_at              222330_at
              "DKK1"                 "PDE3B"
          200951_s_at            204939_s_at
             "CCND2"                  "PLN"
          215646_s_at            204938_s_at
              "VCAN"                  "PLN"
          216207_x_at
          "IGKV1D-13"
```

Code 20.7.

Different public resources have different identifier numbers for genes. Having these unique identifiers at hand saves a lot of time navigating the corresponding online resources.

To retrieve the ENTREZ gene identifier, set the `env` argument to `hgu133aENTREZID` (Code 20.8).

```
> get("209788_s_at", env = hgu133aENTREZID)
[1] 51752
```

Code 20.8.

Similarly, we can return the Unigene identifier (Code 20.9).

```
> get(gene.list[20], env = hgu133aUNIGENE)
[1] "Hs.436186"
```

Code 20.9.

To determine the chromosome in which the gene resides, as well as the actual location and cytoband reference, you can set the `env` arguments to `hgu133aCHR`, `hgu133aCHRLOC`, and `hgu133aMAP`, respectively (Code 20.10).

```
> get(gene.list[20], env = hgu133aCHR)
[1] "5"
> get(gene.list[20], env = hgu133aCHRLOC)
        5          5
-96135945 -96122270
> get(gene.list[20], env = hgu133aMAP)
[1] "5q15"
```

Code 20.10.

20.5 Mapping Gene Ontology Data

The Gene Ontology (GO) encompasses three ontologies that describe genes and their products.[42] The three ontologies are cellular component (CC), molecular function (MF), and biological process (BP).

We can map GO data to our probe IDs using functions in the `annotate` package. To retrieve GO identifiers for the gene list, we can parse the list as an argument to the `getGO()` function:

```
> gene.GO<-getGO(gene.list, data = "hgu133a")
```

We have created a list called `gene.GO` to contain the data. To return details on information held for any probe ID in the list, such as 200978_at, we can just index it (Code 20.11).

```
> gene.GO[[20]][[1]]
$GOID
[1] "GO:0001525"

$Evidence
[1] "TAS"

$Ontology
[1] "BP"
```

Code 20.11.

Now that we have the GO ID number held within the list, we can use this as an argument to other functions to obtain specific GO data about a gene. Using the function `getGOdesc()`, we can retrieve GO terms for the first gene in our list (Code 20.12).

```
> getGOdesc(gene.GO[[20]][[1]][["GOID"]], "ANY")
$'GO:0001525'
GOID = GO:0001525

Term = angiogenesis

Definition = Blood vessel formation when new vessels emerge from the
      proliferation of preexisting blood vessels.

Ontology = BP
```

Code 20.12.

We specified "ANY" to tell the function to return information on any of the three GO categories. In this case, biological process information was given for angiogenesis.

Let us look at another way to map GO category data to a gene. In the previous section, we determined the gene name (ARTS-1) abbreviation and ENTREZ ID (51752) for the 20th gene in our list. We can use the ENTREZ ID to retrieve GO terms for this gene using the `getOntology()` function and specifying one of the ontology category codes.

Get ontology terms for the ARTS-1 gene and put them into `arts1.GO` object (Code 20.13).

```
> arts1.mf<-getOntology(GOENTREZID2GO$"51752", ontology= "MF")
> arts1.bp<-getOntology(GOENTREZID2GO$"51752", ontology= "BP")
> arts1.cc<-getOntology(GOENTREZID2GO$"51752", ontology= "CC")
```

Code 20.13.

We have actually created three lists to hold terms for the MF, BP, and CC categories. To view the terms, you can parse each list to a function called `getGOTerm()` (Code 20.14).

```
> getGOTerm(arts1.mf)
$MF
GO:0004178
"leucyl aminopeptidase activity"
GO:0004179
"membrane alanyl aminopeptidase activity"
GO:0004239
"methionyl aminopeptidase activity"
GO:0005138
"interleukin-6 receptor binding"
GO:0005151
"interleukin-1, Type II receptor binding"
GO:0005515
"protein binding"
GO:0008237
"metallopeptidase activity"
GO:0008270
"zinc ion binding"
GO:0046872
"metal ion binding"

> getGOTerm(arts1.bp)
$BP

GO:0001525
"angiogenesis"
GO:0006509
"membrane protein ectodomain proteolysis"          GO:0008217
"blood pressure regulation"
GO:0019885
"antigen processing and presentation of endogenous peptide antigen via MHC
class I"
GO:0045088
"regulation of innate immune response"
GO:0045444
"fat cell differentiation"
GO:0045766
"positive regulation of angiogenes

> getGOTerm(arts1.cc)
$CC
GO:0005576
"extracellular region"
GO:0005737
"cytoplasm"
GO:0005783
"endoplasmic reticulum"
GO:0005788
"endoplasmic reticulum lumen"
GO:0005829
"cytosol"
GO:0016021
"integral to membrane"
```

Code 20.14.

20.6 Mapping Information from the Literature

There is another useful function called `pm.getabst()` in the `annotation` package that allows you to parse a gene list to query PubMed for instances of the gene in listed abstracts. The function downloads the abstracts, creating an object of class `pubMedAbst`, which is a list of lists. Code 20.15 shows partial output for such a query.

```
> gene.list.abstracts <- pm.getabst(gene.list, "hgu133a")
Read 1649 items
Read 6573 items
Read 5876 items
Read 2886 items
Read 4243 items
...
```

Code 20.15.

To determine the number of abstracts returned for each gene, we can apply the `length()` function to the `gene.list.abstracts` object. Code 20.16 shows the abstract count for the first five genes in the list.

```
> lapply(gene.list.abstracts[1:5], length)
$'215207_x_at'
[1] 1

$'221473_x_at'
[1] 6

$'206788_s_at'
[1] 26

$'202040_s_at'
[1] 17

$'219237_s_at'
[1] 3
...
```

Code 20.16.

To retrieve details of each abstract, including the title, authors, and journal, you just index the corresponding elements in the list (Code 20.17).

```
> gene.list.abstracts[[20]][1:5]
[[1]]
An object of class 'pubMedAbst':
Title: Construction and characterization of a full-length-enriched and a
5'-end-enriched cDNA library.
PMID: 9373149
Authors: Y Suzuki, K Yoshitomo-Nakagawa, K Maruyama, A Suyama, S Sugano
```

```
Journal: Gene
Date: Oct 1997

[[2]]
An object of class 'pubMedAbst':
Title: Prediction of the coding sequences of unidentified human genes. IX. The
complete sequences of 100 new cDNA clones from the brain, which can code for
large proteins in vitro.
PMID: 9628581
Authors: T Nagase, K Ishikawa, N Miyajima, A Tanaka, H Kotani, N Nomura,
O Ohara
Journal: DNA Res
Date: Feb 1998

[[3]]
An object of class 'pubMedAbst':
Title: Molecular cloning of adipocyte-derived leucine aminopeptidase highly
related to placental leucine aminopeptidase/oxytocinase.
PMID: 10220586
Authors: A Hattori, H Matsumoto, S Mizutani, M Tsujimoto
Journal: J Biochem (Tokyo)
Date: May 1999

[[4]]
An object of class 'pubMedAbst':
Title: Characterization of recombinant human adipocyte-derived leucine
aminopeptidase expressed in Chinese hamster ovary cells.
PMID: 11056387
Authors: A Hattori, K Kitatani, H Matsumoto, S Miyazawa, T Rogi,
N Tsuruoka, S Mizutani, Y Natori, M Tsujimoto
Journal: J Biochem (Tokyo)
Date: Nov 2000

[[5]]
An object of class 'pubMedAbst':
Title: Genomic organization of the human adipocyte-derived leucine
aminopeptidase gene and its relationship to the placental leucine
aminopeptidase/oxytocinase gene.
PMID: 11481040
Authors: A Hattori, K Matsumoto, S Mizutani, M Tsujimoto
Journal: J Biochem (Tokyo)
Date: Aug 2001
```

Code 20.17.

The titles of each abstract in the list can also be returned separately using the pm.titles() function (Code 20.18).

```
> pm.titles(gene.list.abstracts[20])
[[1]]
 [1] "Construction and characterization of a full-length-enriched and a 5'-end-
enriched cDNA library."
 [2] "Prediction of the coding sequences of unidentified human genes. IX. The
complete sequences of 100 new cDNA clones from the brain, which can code for
large proteins in vitro."
 [3] "Molecular cloning of adipocyte-derived leucine aminopeptidase highly
related to placental leucine aminopeptidase/oxytocinase."
 [4] "Characterization of recombinant human adipocyte-derived leucine
aminopeptidase expressed in Chinese hamster ovary cells."
 [5] "Genomic organization of the human adipocyte-derived leucine
aminopeptidase gene and its relationship to the placental leucine
aminopeptidase/oxytocinase gene."
 [6] "Identification of 33 polymorphisms in the adipocyte-derived leucine
aminopeptidase (ALAP) gene and possible association with hypertension."
```

[7] "Identification of ARTS-1 as a novel TNFR1-binding protein that promotes TNFR1 ectodomain shedding."
[8] "ERAAP customizes peptides for MHC class I molecules in the endoplasmic reticulum."
[9] "An IFN-gamma-induced aminopeptidase in the ER, ERAP1, trims precursors to MHC class I-presented peptides."
[10] "The ER aminopeptidase ERAP1 enhances or limits antigen presentation by trimming epitopes to 8-9 residues."
[11] "Generation and initial analysis of more than 15,000 full-length human and mouse cDNA sequences."
[12] "An aminopeptidase, ARTS-1, is required for interleukin-6 receptor shedding."
[13] "The secreted protein discovery initiative (SPDI), a large-scale effort to identify novel human secreted, and transmembrane proteins: a bioinformatics assessment."
[14] "Shedding of the type II IL-1 decoy receptor requires a multifunctional aminopeptidase, aminopeptidase regulator of TNF receptor type 1 shedding."
[15] "Distribution of adipocyte-derived leucine aminopeptidase (A-LAP)/ ER-aminopeptidase (ERAP)-1 in human uterine endometrium."
[16] "Regulation of the human leukocyte-derived arginine aminopeptidase/endoplasmic reticulum-aminopeptidase 2 gene by interferon-gamma."
[17] "Concerted peptide trimming by human ERAP1 and ERAP2 aminopeptidase complexes in the endoplasmic reticulum."
[18] "The oxytocinase subfamily of M1 aminopeptidases."
[19] "The ER aminopeptidase, ERAP1, trims precursors to lengths of MHC class I peptides by a \"molecular ruler\" mechanism."
[20] "Diversification of transcriptional modulation: large-scale identification and characterization of putative alternative promoters of human genes."
[21] "Extracellular TNFR1 release requires the calcium-dependent formation of a nucleobindin 2-ARTS-1 complex."
[22] "Expression of endoplasmic reticulum aminopeptidases in EBV-B cell lines from healthy donors and in leukemia/lymphoma, carcinoma, and melanoma cell lines."
[23] "Processing of a class I-restricted epitope from tyrosinase requires peptide N-glycanase and the cooperative action of endoplasmic reticulum aminopeptidase 1 and cytosolic proteases."

Code 20.18.

Finally, there is a function called pmAbst2HTML() that allows you to create an HTML file of the titles and publication date:

```
> pmAbst2HTML(gene.list.abstracts[[20]], filename="abstracts.html")
```

The file will be saved to the working directory as "abstracts.html" and can be opened using a standard web browser (Figure 20.1).

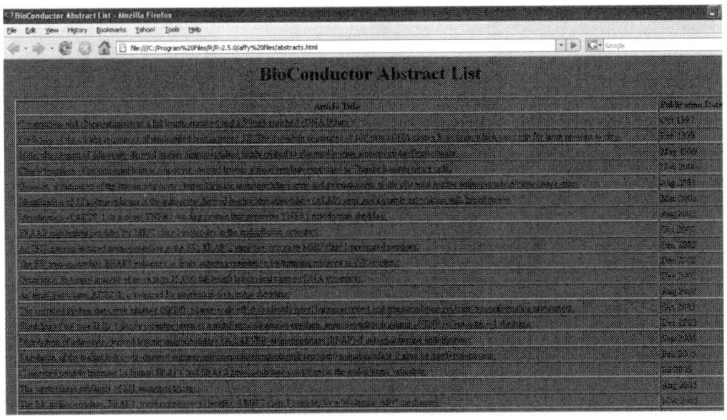

FIGURE 20.1 HTML file for abstract list created by pmAbst2HTML().

20.7 Summary

In this last chapter devoted to microarray gene expression data analysis, we have looked at ways of mapping information from Affymetrix chips to public resource data and Gene Ontology terms. The annotation process of a microarray experiment is a long and tedious process, but using the `annotate` and `hgu177a` packages can speed up this process.

20.8 Questions

1. Why do you think that annotating gene lists from a microarray experiment might be the most time-consuming stage of the entire process?

2. Can you describe three important pieces of information that are supplied as values to the `env` argument of the `get()` function in the `hgu133a` package?

3. What information can the `hgu133a()` function provide us with?

4. What does gene ontology refer to, and what R function can we use to retrieve this information for a gene list derived from the `hgu133a` chip?

5. How can we retrieve information for a gene from PubMed?

6. What type of information can be searched in PubMed for a gene in our gene list?

21 Array CGH Analysis

21.1 Background

There is evidence that most human cancers demonstrate genetic instability at one or more levels. Some cancers exhibit genetic instability at a level where just one (point mutation) or a few nucleotides become substituted, lost, or gained, changing the sequence of a gene and ultimately the structure or function of the protein that the gene encodes. An examination of cells from many cancer types, however, would reveal that instability occurs at a level that witnesses loss or gain of large portions of chromosomes, if not entire chromosomes.[43]

The key stages of initiation and progression that define the tumorigenesis pathway in different cancer types may be characterized by combinations of genetic instabilities. Consider loss of heterozygosity (LOH) whereby on a chromosome pair, one copy of a gene is lost due to a gross deletion event but the other, "good" copy of the gene is still active within the cell. If a point mutation occurs in this gene at a later time, the encoded protein might become functionally inactivated. Such a sequence of events was originally proposed (and now accepted by many) to be the mechanism that leads to the inactivation of tumor suppressor genes.[44] Indeed, it was quickly established that the characterization of chromosomal losses and gains could quickly lead to the identification of tumor suppressor gene loci.

The mapping of genetic gains and losses in cytogenetics laboratories has been paralleled by the quest to develop and improve the necessary technologies to perform such studies. In the 1990s, chromosome banding techniques gave way to methods that could detect genomic structural change using fluorescently labeled probes. Nowadays, techniques such as fluorescent in situ hybridization (FISH) are commonplace in the diagnostic setting. Comparative genomic hybridization (CGH) is a related fluorescent-based technique that was first proposed by Kallioniemi et al.[45]

CGH allows for a comprehensive and genome-wide screening of losses and gains across all chromosomes in a tumor sample. In a CGH experiment, one sample of DNA is obtained from disease tissue (test) and another sample is obtained from normal tissue from the same patient.[46] Disease DNA is labeled using a green fluorochrome and the normal reference DNA labeled with a red fluorochrome. Both labeled DNA samples are then mixed and hybridized to normal human metaphase spreads, where the differentially labeled disease and normal DNA compete to hybridize to the complementary loci on the individual chromosomes. By measuring the fluorescence intensities along each chromosome using a fluorescent microscope (and image analysis software), one can quickly establish the green-to-red ratio and determine positions of genetic gain or loss in that tumor sample.

When using metaphase spreads, CGH is limited by resolution, where deletions can be detected from about 5 to 10 Mb and amplifications from about 1 to 2 Mb.[47] To circumvent this problem, microarray-based CGH (array-CGH) has been developed where the target sequences for the labeled DNA are mapped genomic clones as opposed to metaphase chromosomes. Array-based CGH technology can now detect copy number changes in a chromosome by as little as 5 to 10 Kb, providing extremely high-resolution mapping. The purpose of this chapter is to provide an introduction to the processing and analysis of array-CGH data in R using a well-defined bladder cancer dataset.

21.2 array-CGH Technologies

There are currently two array-based CGH technologies used for screening chromosomal alterations in human disease.[48] The first of these methods to be introduced involves production of an array using bacterial artificial chromosome (BAC) clones containing human genomic DNA as the target sequence. BAC clones are carefully selected from a physical map of the genome for coverage as wide as possible. The DNA from these clones is then spotted in replicate onto a solid support, such as a glass slide. Tumor and normal DNA samples, labeled using different fluorochromes, may then be hybridized to the array in similar fashion to CGH. This method is quite expensive and labor-intensive. It is also felt that the limit of BAC array-CGH resolution has been met and so a new technology has emerged to overcome this. Oligonucleotide array-CGH, based on the production of synthetic probes, not only has a better resolution, but also allows greater flexibility in probe design and greater all-around coverage of the genome.[48]

Although oligonucleotide array-CGH appears to be the technology of the future, currently, the majority of available data exists for BAC-based array-CGH experiments. This chapter will focus on the analysis of a BAC array-CGH dataset. However, the reader should be reassured that the knowledge gained here using R will stand them in good stead for analysis of other types of array-CGH data.

21.3 array-CGH Data

Before we begin learning how to analyze array-CGH data, let us indulge in a (very) brief primer on how chromosomal loss and gain data is generated using BAC array-CGH technology. Once the hybridization step is complete, images of the array may be captured manually or using a microarray scanner. Usually, an image is captured for each fluorescent dye and the background signal intensity calculated to adjust fluorochrome intensities in image pixels. The next step is to generate signal data from each spot on the array by calculating the fluorescence intensity ratio. This would be done either using proprietary software or noncommercial software, such as the SPOT 2.0 software package, which is freely available (http://cancer.ucsf.edu/array/index.php) for academic use and runs on Windows platforms.[49] Once this is done for all the spots on the array, ratios from replicate spots can be averaged. Again, using SPOT 2.0, this process can be automated by a program in SPOT called SPROC, which will produce normalized ratios for spots and output this information, as well as a great deal of other information, in a standard text file.

Essential data variables for BAC array-CGH data analysis include log(2) ratio for individual clones (represented by spots), the clone ID, and the chromosomal position of each clone.

21.4 Packages Required

A number of R packages have been developed for the processing and analysis of array-CGH data. The `Manor` package, for example, is devoted to the normalization of array-CGH data, whereas the `Glad` package is used to detect breakpoints in chromosomes and assign normal, gain, or loss status to these regions.

There is another extremely useful package, called `aCGH`, that was first introduced in Bioconductor as a major upgrade to the 1.4 release.

The great thing about the `aCGH` package is that it provides a broad spectrum of functions, allowing you to read, analyze, and generate plots for array-CGH data. As you will see, the `aCGH` package also deals quite seamlessly with output files from SPOT and SPROC.

21.5 An Example Dataset

Many array-CGH studies have been published over the last few years reporting chromosomal gains and losses in a range of different cancer types. In this chapter we will analyze a single, publicly available bladder cancer array-CGH dataset.[50] The complete dataset contains 57 SPROC output text files for 57 bladder cancer samples. Each file contains log(2) ratio, gain/loss/normal, and outlier status for 2385 clones. The files can be downloaded at http://microarrays.curie.fr/publications/oncologie_moleculaire/bladder_TCM/, where you will find the link to the zipped ".tar" file "CGH_raw_data.tar.bz2."

Click this link, and you will be directed to a download page. Download the file, and after opening it using an unzipping tool, you should find a second ".tar" file called "CGH_raw_data.tar." Within this zipped file, you will find all 57 SPROC files; these can be extracted to a folder that you can use as your working directory for this project in R. In addition to the SPROC files are two ".csv" files called "filenames_samples.csv" and "Clinical_parameters.csv," and both of these should also be extracted to the working directory. I have called my working directory "cgh data," but any folder name will do.

The SPROC output file names correspond to the sample ID numbers from the experiment. Before we get started with the analysis, we need to do a little bit of housekeeping with the file names.

1. If you open the "filenames_samples.csv" file (in Excel or Notepad), you will see that the numbers in the "Sample" column are shorter than the corresponding file names in the "Filename" column but that the sample number is actually contained within the file name. As an example, look at the first patient in row 2, which has the sample name 1033. Compare this to the corresponding file name (in your working directory folder, you will find the corresponding SPROC file

"021007FR1033sproc.txt"). Unfortunately, the aCGH package will adopt the file name as the sample name when reading in these text files, and such long-winded names can make plots and output look ugly in R. Therefore, rename the SPROC text files according to the sample names in "Clinical_parameters.csv."

2. In your working directory, delete the following files if they exist:

HA20_clonepos_Jul03_v1_1.txt

HA20_spotclone_Jul03_v1_1.txt

021021FR520sproc.txt

021103FR1533-10sproc.txt

021017FR1533-13sproc.txt

These files either represent duplicate samples or are unworkable later in the analysis.

3. Open each SPROC file and scroll across the column headers until you reach the "KB" column header. Replace the term "KB" with "KB.POSITION."

4. Navigate your web browser to http://www.jbpub.com/biology/bioinformatics, click this book, locate the files for this chapter, and download the file "clones.info.ex.dat." This file holds the clone ID and chromosome information for each clone in four columns: "Clone," "Target," "Chromosome," and "KB."

5. At the same URL, download the file "phenotype.dat." This file contains phenotypic data for each tumor sample in nine columns: "ID," "sample," "patient," "primary," "stage," "grade," "nodes," "metastasis," and "gender."

6. Finally, download the two R code files "plotfs.R" and "plotgenome.R" (more about these later).

At this point, my working directory contains the files in List 21.1.

List 21.1 SPROC files listed in working directory.

- 1033sproc.txt
- 10871sproc.txt
- 1210sproc.txt
- 12112sproc.txt
- 1212sproc.txt
- 1333sproc.txt
- 1343sproc.txt
- 1352sproc.txt
- 1379sproc.txt
- 1382sproc.txt
- 1410sproc.txt
- 1412sproc.txt
- 1413sproc.txt
- 1448sproc.txt

- 1484sproc.txt
- 1498sproc.txt
- 1512-6sproc.txt
- 1528-7sproc.txt
- 1533-1sproc.txt
- 1541-4sproc.txt
- 195sproc.txt
- 2259-1sproc.txt
- 2297sproc.txt
- 2307sproc.txt
- 2335sproc.txt
- 2380sproc.txt
- 2505sproc.txt
- 2541sproc.txt
- 2542sproc.txt
- 2571sproc.txt
- 259sproc.txt
- 2691sproc.txt
- 2816sproc.txt
- 2817sproc.txt
- 2821sproc.txt
- 3031sproc.txt
- 3046sproc.txt
- 3161sproc.txt
- 338sproc.txt
- 3395sproc.txt
- 3554sproc.txt
- 407sproc.txt
- 413sproc.txt
- 447sproc.txt
- 448sproc.txt
- 506sproc.txt
- 813-1sproc.txt
- 824sproc.txt
- 848sproc.txt
- 877sproc.txt
- 908sproc.txt
- 910sproc.txt
- 924-1sproc.txt

- 943sproc.txt
- phenotype.dat
- clones.info.ex.dat
- plotfs.R
- plotgenome.R

21.6 The aCGH Package

In an array-CGH experiment, once the researcher has signal ratios in her hand, she can determine exactly which regions of the chromosomes are gained or lost within her disease DNA samples. The main step involved in this process has recently been the focus of much research. The goal is to segment the clone signal ratios across each chromosome into distinct states, where all the clones in a state have the same copy number. The best way to think of a state is as representing a region in a chromosome that has been lost or gained. Segmentation of clone signal ratios is not trivial, and a number of different methods, which are beyond the scope of this book, have been proposed.

The aCGH package uses an unsupervised hidden Markov model approach to determine states from raw log(2) ratio values and refers to predicted gains and losses as genomic events. aCGH also allows you to generate a range of plots for your data to visualize log(2) ratio-based genomic events for each sample, predicted states across each chromosome in each sample, and clustering of samples by genomic events. Another powerful aspect of aCGH is the ability to compare samples for genomic events according to phenotype class and compare these classes using statistical methods. The example analysis that follows will focus on using aCGH to generate a range of plots for the bladder cancer dataset, as well as investigate differences in genomic states for patients grouped according to tumor stage.

The fundamental component of aCGH is the aCGH class. Objects of the aCGH class store all the data necessary for a full analysis. The benefit of this is that you only have to work with one object throughout the analysis if you so wish.

Download the aCGH package and load the library (Code 21.1).

```
> source("http://bioconductor.org/biocLite.R")
> biocLite("aCGH")
library(aCGH)
```

Code 21.1.

Load the additional R code files sitting in your working directory (Code 21.2).

```
> source("plotfs.R")
> source("plotgenome.R")
```

Code 21.2.

21.7 Reading array-CGH Data Files

We have the SPROC output data for all of our samples in our working directory, so the first step is to load the data held within each file into an aCGH object (make sure you have the working directory set to the folder where you placed your array-CGH files). We will call our aCGH object pcos, and all the files can be loaded in one go using the aCGH.read.Sprocs function (Code 21.3). The function will attempt to load any text file held within the working directory, so ensure you have no other files in that directory with the extension ".txt," or R will throw an error. When we call aCGH.read.Sprocs note that we have set parameters to allow the function to search for text files where the full file path is not read in as a sample name and only data chromosomes 1 through 23 are read.

```
> pcos <- aCGH.read.Sprocs(dir(getwd(), pattern = "txt", full.names = FALSE),
chrom.remove.threshold = 23)
Trying to read  1033sproc.txt
Trying to read  10871sproc.txt
Trying to read  1210sproc.txt
Trying to read  12112sproc.txt
Trying to read  1212sproc.txt
Trying to read  1333sproc.txt
Trying to read  1343sproc.txt
Trying to read  1352sproc.txt
Trying to read  1379sproc.txt
Trying to read  1382sproc.txt
Trying to read  1410sproc.txt
Trying to read  1412sproc.txt
Trying to read  1413sproc.txt
Trying to read  1448sproc.txt
Trying to read  1484sproc.txt
Trying to read  1498sproc.txt
Trying to read  1512-6sproc.txt
Trying to read  1528-7sproc.txt
Trying to read  1533-1sproc.txt
Trying to read  1541-4sproc.txt
Trying to read  195sproc.txt
Trying to read  2259-1sproc.txt
Trying to read  2297sproc.txt
Trying to read  2307sproc.txt
Trying to read  2335sproc.txt
Trying to read  2380sproc.txt
Trying to read  2505sproc.txt
Trying to read  2541sproc.txt
Trying to read  2542sproc.txt
Trying to read  2571sproc.txt
Trying to read  259sproc.txt
Trying to read  2691sproc.txt
Trying to read  2816sproc.txt
Trying to read  2817sproc.txt
Trying to read  2821sproc.txt
Trying to read  3031sproc.txt
Trying to read  3046sproc.txt
Trying to read  3161sproc.txt
Trying to read  338sproc.txt
Trying to read  3395sproc.txt
Trying to read  3554sproc.txt
Trying to read  407sproc.txt
Trying to read  413sproc.txt
Trying to read  447sproc.txt
Trying to read  448sproc.txt
```

```
Trying to read   506sproc.txt
Trying to read   813-1sproc.txt
Trying to read   824sproc.txt
Trying to read   848sproc.txt
Trying to read   877sproc.txt
Trying to read   908sproc.txt
Trying to read   910sproc.txt
Trying to read   924-1sproc.txt
Trying to read   943sproc.txt

Averaging duplicated clones
CTB-102E19       920 921 922
CTB-142O24       1974 1975
CTB-339E12       1968 1969
CTB-36F16        1510 1511
GS1-124N22       828 829
GS1-20208        882 883
RP11-105K5       1076 1077
RP11-10G10       1117 1118
RP11-119J20      589 590
RP11-11M9        449 450
RP11-123F4       880 881
RP11-12N7        295 296
RP11-138E20      857 858
RP11-13C20       276 277
RP11-172D2       1080 1081
RP11-176L20      291 292
RP11-17D23       211 212
RP11-204M16      1035 1036
RP11-20K4        1503 1504
RP11-221P7       1074 1075
RP11-224H3       271 272
RP11-238H10      1108 1109
RP11-261B20      633 634
RP11-268N2       1067 1068
RP11-284E5       308 309
RP11-295J19      157 158
RP11-30M1        310 311
RP11-31B6        2246 2247
RP11-32C8        195 196
RP11-368K23      248 249
RP11-416H1       258 259
RP11-442L5       268 269
RP11-443B20      173 174
RP11-496P1       161 162
RP11-498O20      260 261
RP11-4K20        155 156
RP11-541A15      184 185
RP11-560C7       169 170
RP11-72C6        1279 1280
RP11-83O14       1071 1072
RP11-94M13       1137 1138
RP11-99M6        1078 1079
```

Code 21.3.

Now that we have the SPROC output data in our `pcos` object, we can do a quick check to make sure all our samples have been read in. We could look at the output after calling `aCGH.read.Sprocs`, but by using `summary(pcos)`, we can view a summary of the extent of data held within the object (Code 21.4). Immediately we can see that all 54 arrays were read in and that data exists for 2318 clones across our samples.

We can ignore the lines that follow for now, as they refer to further processing of the data, which we have yet to do.

```
> summary(pcos)
aCGH object
Call: aCGH.read.Sprocs(dir(getwd(), pattern = "txt", full.names = TRUE),
    chrom.remove.threshold = 23)

Number of Arrays 54
Number of Clones 2318
Imputed data does not exist
HMM states are not assigned
samples standard deviations are not computed
genomic events are not assigned
phenotype does not exists
```

Code 21.4.

21.8 A Look Inside Our aCGH Object

With the SPROC data safely read in, we can now explore the pcos object a little further and add other information that we will need for the analysis. An aCGH object (pcos in our case) contains a number of lists and functions. When you first create an aCGH object, it will be five lists; to view the names and order of these, we can simply call the names function (Code 21.5).

```
> names(pcos)
[1] "log2.ratios"       "clones.info"       "phenotype"
[4] "qual.rep"          "bad.quality.index"
```

Code 21.5.

The help documentation for the aCGH object (viewed by typing ?aCGH in the aCGH package) actually refers to these lists as slots, but be aware that they are indeed lists and data within them cannot be accessed using the "@" operator.

The first list, called "log2.ratios," is mandatory in the sense that to process your CGH data further and utilize package functions, you must have filled this with log(2) ratios from your data. Because everything has run smoothly so far, we can be sure that this list contains log(2) ratio data from our samples, but what does this data look like? The first 10 clone log(2) ratios for each of the 54 bladder cancer samples is shown in Code 21.6. A close inspection of these values reveals that some data is missing for different clones and declared as NA. This is par for the course in a CGH experiment—some clones will not yield a valid signal. Also note that values can be positive or negative, reflecting the fact that a positive log(2) value results from a ratio of tumor to normal DNA greater than 1, whereas a negative value means the ratio was less than 1. Thus, for each clone, the higher the logged value, the greater the copy number gain. Similarly, the more negative the number, the greater the copy number loss.

```
> pcos2$log2.ratios[1:10,]
                1033sproc.txt 10871sproc.txt 1210sproc.txt 12112sproc.txt
RP11-82D16       -0.102988     -0.134150     -0.145369     -0.246308
RP11-62M23        0.056675     -0.029314      0.040022      0.035205
RP11-11105        0.170740      0.079844      0.098052      0.175996
RP11-51B4         0.159399      0.106763      0.130184      0.166643
RP11-60J11        0.180978      0.119105      0.099000      0.197605
RP11-813J5       -0.003498      0.052311     -0.011150      0.083762
RP11-19901        0.050194      0.016932      0.057358      0.088063
RP11-188F7       -0.111300     -0.099679     -0.104899     -0.016987
RP11-178M15       0.029541     -0.014219     -0.054160      0.021911
RP11-219F4        0.151537      0.119420      0.119638      0.151911
                1212sproc.txt 1333sproc.txt 1343sproc.txt 1352sproc.txt
RP11-82D16       -0.270396     -0.261510     -0.107072     -0.143339
RP11-62M23       -0.082273     -0.036287     -0.221575     -0.015615
RP11-11105        0.064376      0.083855            NA      0.105267
RP11-51B4         0.039715      0.081370     -0.285926      0.005680
RP11-60J11        0.119975      0.130800     -0.329361      0.083792
RP11-813J5       -0.085532     -0.013100     -0.103742     -0.073492
RP11-19901       -0.038589      0.038027            NA     -0.012311
RP11-188F7       -0.224717     -0.105731     -0.200228     -0.108160
RP11-178M15      -0.026075     -0.019455     -0.283969     -0.057417
RP11-219F4        0.025388      0.127956     -0.471008      0.041803
                1379sproc.txt 1382sproc.txt 1410sproc.txt 1412sproc.txt
RP11-82D16       -0.267549     -0.071657     -0.148236     -0.151478
RP11-62M23       -0.152543      0.135268     -0.046211     -0.040757
RP11-11105        0.120711      0.199935      0.099823      0.062141
RP11-51B4        -0.090654      0.219807      0.055271      0.084568
RP11-60J11        0.097544      0.239827      0.118751      0.093567
RP11-813J5       -0.116325      0.076882     -0.021093      0.015374
RP11-19901       -0.024502      0.151565     -0.011377      0.036427
RP11-188F7       -0.054214     -0.002408     -0.096791     -0.079449
RP11-178M15      -0.003446      0.057665      0.042846     -0.007273
RP11-219F4        0.050192      0.155553      0.120423      0.122684
                1413sproc.txt 1448sproc.txt 1484sproc.txt 1498sproc.txt
RP11-82D16       -0.257931      0.109043     -0.124225      0.343663
RP11-62M23       -0.137868      0.169271     -0.003616      0.418612
RP11-11105        0.100001      0.321358      0.096927      0.560688
RP11-51B4        -0.034965      0.276938      0.131016      1.019456
RP11-60J11        0.078429      0.335916      0.133152      0.611537
RP11-813J5       -0.021630      0.115817     -0.021946      0.067857
RP11-19901       -0.095283      0.185380     -0.053061      0.094554
RP11-188F7       -0.192856     -0.130185     -0.221261     -0.014445
RP11-178M15      -0.045529      0.077672     -0.108521      0.051084
RP11-219F4        0.056806      0.283387      0.004269      0.169950
                1512-6sproc.txt 1528-7sproc.txt 1533-1sproc.txt
RP11-82D16       -0.169332     -0.367707     -0.136839
RP11-62M23       -0.064259     -0.191847      0.061630
RP11-11105        0.114951     -0.048024      0.066893
RP11-51B4        -0.000359     -0.100509      0.142827
RP11-60J11        0.156800      0.003564      0.101414
RP11-813J5        0.000355     -0.065820      0.057433
RP11-19901        0.037755     -0.054117      0.034362
RP11-188F7       -0.132610     -0.141882     -0.076967
RP11-178M15      -0.047984     -0.120600     -0.032009
RP11-219F4        0.090065     -0.188188      0.150903
                1541-4sproc.txt 195sproc.txt 2259-1sproc.txt
RP11-82D16             NA     -0.235128            NA
RP11-62M23        0.149054     -0.081231      0.100920
RP11-11105        0.123227     -0.031676      0.093661
RP11-51B4         0.152905      0.013072      0.028193
RP11-60J11        0.104463      0.071914      0.001155
RP11-813J5        0.057663     -0.078869     -0.003213
RP11-19901        0.118454     -0.038800            NA
RP11-188F7        0.156086     -0.194608      0.072586
RP11-178M15            NA     -0.136513      0.040837
RP11-219F4        0.073952      0.000864     -0.064605
```

	2297sproc.txt	2307sproc.txt	2335sproc.txt	2380sproc.txt
RP11-82D16	-0.223979	-0.300080	-0.127626	-0.241090
RP11-62M23	-0.034073	-0.023541	-0.000726	0.063666
RP11-11I05	0.025096	0.040337	0.114261	0.090721
RP11-51B4	0.040310	0.126351	0.113525	0.115795
RP11-60J11	0.066655	-0.009915	0.200532	0.124596
RP11-813J5	0.011075	0.022456	-0.044942	0.118230
RP11-199O1	0.000899	0.038032	-0.006606	0.079768
RP11-188F7	-0.071728	-0.035510	-0.186480	-0.001022
RP11-178M15	-0.002625	-0.015468	-0.075392	0.051791
RP11-219F4	0.144198	0.058841	0.152588	0.165578

	2505sproc.txt	2541sproc.txt	2542sproc.txt	2571sproc.txt
RP11-82D16	0.276176	-0.260319	0.111028	0.114783
RP11-62M23	0.378122	-0.102464	0.048755	0.110600
RP11-11I05	0.497866	-0.079348	0.218874	0.237668
RP11-51B4	0.593539	-0.025095	0.146614	0.231768
RP11-60J11	0.552578	-0.024027	0.188847	0.284808
RP11-813J5	0.146355	-0.166031	0.063075	0.132859
RP11-199O1	0.145066	-0.138064	0.169949	0.161481
RP11-188F7	0.020438	-0.277516	0.004830	0.080295
RP11-178M15	0.035987	-0.190472	0.089377	0.090824
RP11-219F4	0.232658	-0.077152	0.248816	0.291732

	259sproc.txt	2691sproc.txt	2816sproc.txt	2817sproc.txt
RP11-82D16	-0.082924	-0.066435	-0.063779	0.084733
RP11-62M23	-0.037742	0.000577	0.001631	0.102874
RP11-11I05	-0.001080	0.132360	0.152306	0.071720
RP11-51B4	0.157170	0.080785	0.157760	0.195755
RP11-60J11	0.008019	0.157039	0.167136	0.239519
RP11-813J5	0.053046	0.007743	-0.018354	0.088421
RP11-199O1	-0.016631	0.007103	0.005507	0.115464
RP11-188F7	-0.079678	-0.153710	-0.162955	-0.049127
RP11-178M15	-0.077211	-0.050622	-0.041456	-0.045237
RP11-219F4	0.022310	0.095374	0.110924	0.066843

	2821sproc.txt	3031sproc.txt	3046sproc.txt	3161sproc.txt
RP11-82D16	-0.113889	0.294455	0.191126	-0.065639
RP11-62M23	-0.050295	0.368315	0.257202	0.085537
RP11-11I05	0.051197	0.497524	0.379045	0.056406
RP11-51B4	0.048631	0.435510	0.374111	0.182273
RP11-60J11	0.084701	0.536968	0.439617	0.120826
RP11-813J5	-0.011646	0.021726	0.261055	0.134142
RP11-199O1	-0.006766	0.001938	0.271432	0.095751
RP11-188F7	-0.161304	-0.088810	0.086380	0.064905
RP11-178M15	-0.038658	-0.003741	0.139073	0.007990
RP11-219F4	0.080920	0.121991	0.301505	0.102011

	338sproc.txt	3395sproc.txt	3554sproc.txt	407sproc.txt
RP11-82D16	-0.123159	NA	-0.223872	-0.105761
RP11-62M23	0.000049	-0.023423	-0.112772	0.081726
RP11-11I05	0.107186	-0.034639	0.065356	0.047431
RP11-51B4	0.078043	NA	-0.046704	0.148164
RP11-60J11	0.105582	-0.110258	0.136275	0.049537
RP11-813J5	-0.032685	-0.119867	-0.053187	0.081476
RP11-199O1	0.030169	NA	-0.023591	0.133686
RP11-188F7	-0.158954	-0.164797	-0.140862	0.031147
RP11-178M15	-0.050580	-0.157078	-0.028967	0.029249
RP11-219F4	0.096785	-0.202826	0.056561	0.084002

	413sproc.txt	447sproc.txt	448sproc.txt	506sproc.txt
RP11-82D16	-0.149651	-0.107256	-0.148617	-0.055142
RP11-62M23	0.009301	0.062776	0.004249	-0.017021
RP11-11I05	0.106068	0.084008	0.108167	0.095456
RP11-51B4	0.106338	0.112549	0.105432	-0.358694
RP11-60J11	0.156561	0.119602	0.151878	0.108011
RP11-813J5	0.056196	0.108418	-0.011086	-0.027976
RP11-199O1	0.063269	0.084048	0.070836	-0.022293
RP11-188F7	-0.004022	-0.038337	-0.110526	-0.177734
RP11-178M15	0.017471	0.011291	-0.023999	-0.081388
RP11-219F4	0.145335	0.128659	0.100582	0.103910

	813-1sproc.txt	824sproc.txt	848sproc.txt	877sproc.txt
RP11-82D16	-0.191470	-0.385479	0.006610	-0.232471

```
RP11-62M23      0.012128    -0.264554    0.146441    -0.061412
RP11-11105      0.090663    -0.200755    0.040701     0.000349
RP11-51B4       0.082392    -0.130898    0.251278     0.053201
RP11-60J11      0.116161    -0.196217    0.141245     0.029462
RP11-813J5     -0.013678    -0.162827    0.118767    -0.008938
RP11-19901      0.026051    -0.229264    0.131573    -0.028857
RP11-188F7     -0.106094    -0.312793    0.024426    -0.153554
RP11-178M15    -0.057690    -0.246832    0.015706    -0.114420
RP11-219F4      0.063604    -0.023646    0.222551     0.005912
             908sproc.txt 910sproc.txt 924-1sproc.txt 943sproc.txt
RP11-82D16     -0.189508     0.009917    -0.144784     0.052742
RP11-62M23     -0.004537     0.117676    -0.102761     0.271530
RP11-11105      0.107900     0.260341     0.116346           NA
RP11-51B4       0.150940     0.282159     0.009560     0.057473
RP11-60J11      0.146627     0.263415     0.184981     0.022990
RP11-813J5     -0.016099     0.196523     0.037236     0.062084
RP11-19901      0.072395     0.184090    -0.015783     0.078624
RP11-188F7     -0.127963     0.050253    -0.072956     0.089071
RP11-178M15     0.014582     0.162310    -0.030734     0.068809
RP11-219F4      0.086847     0.298191     0.103953    -0.111746
```

Code 21.6.

The second aCGH list containing useful information at this stage is clones.info (shown for the first 10 clones in Code 21.7). In particular, this list details the name of each clone, the target chromosome, and the position of the target sequence on that chromosome.

```
> pcos2$clones.info[1:10,]
        Clone          Target Chrom    kb
2   RP11-82D16   HumArray2H11_C9    1    1008
3   RP11-62M23   HumArray2H10_N30   1    3767
4   RP11-11105   HumArray2H10_B18   1    4738
5   RP11-51B4    HumArray2H10_Q30   1    6346
6   RP11-60J11   HumArray2H10_T30   1    7621
7   RP11-813J5   HumArray2H10_B19   1   10417
8   RP11-19901   HumArray2H10_W30   1   11074
9   RP11-188F7   HumArray2H9_C14    1   13178
10  RP11-178M15  HumArray2H9_F14    1   14564
11  RP11-219F4   HumArray2H9_I14    1   15607
```

Code 21.7.

21.9 Adding More Data to Our aCGH Object

The third useful list in the pcos object is "phenotype," which can be accessed by typing "pcos2$phenotype." For now, R would return NULL because we have yet to add any phenotypic data to the pcos object for our samples.

In a disease study, phenotype data could include clinical, pathological, or perhaps demographic data about the patient. In terms of data analysis, as you will see in the next few sections, phenotype data could be anything that forms categories of interest subdividing the patients for statistical comparison.

Sitting in our working directory is a file called "phenotype.dat," which holds all phenotypic data for each sample. We can load this data into the pcos object using a combination of the read.table() function, with the header parameter set to TRUE, and the phenotype() function provided in the aCGH package (Code 21.8). The phenotype data can then be displayed by calling the phenotype() function by itself.

```
> phenotype(pcos)<-read.table("phenotype.dat", header=T)
> phenotype(pcos)
     ID sample patient primary stage grade nodes metastasis gender
1    1033   P1033     yes  T2:4    G3    N2       M0      M
2   10871   P1087     yes  T2:4    G3    N1       M0      M
3    1210   P1210      no  T2:4    G2    N1       M0      M
4   12112   P1207     yes  T2:4    G2    N0       M0      M
5    1212   P1212      no    T1    G3    N0       M0      F
6    1333   P1333     yes    Ta    G1    N0       M0      M
7    1343   P1343     yes  T2:4    G3    Nx       Mx      M
8    1352   P1352     yes    Ta    G1    N0       M0      M
9    1379   P1379     yes    Ta    G1    N0       M0      M
10   1382   P1382     yes  T2:4    G3    N0       M0      M
11   1410   P1343     yes  T2:4    G3    Nx       Mx      M
12   1412   P1412     yes    Ta    G2    N0       M0      M
13   1413   P1413     yes    Ta    G1    N0       M0      M
14   1448   P1448     yes  T2:4    G3    N2       M0      M
15   1484   P1484     yes  T2:4    G3    Nx       Mx      F
16   1498   P1498     yes    Ta    G3    N0       M0      M
17   1512   P1512     yes  T2:4    G3    N2       M0      F
18   1528   P1528      no  T2:4    G3    N2       M1      M
19   1533   P1533      no  T2:4    G3    Nx       Mx      M
20   1541   P1541     yes  T2:4    G3    N0       M0      M
21    195    P195     yes  T2:4    G3    N0       M0      F
22   2259   P2259     yes  T2:4    G3    N0       M0      M
23   2297   P2297     yes    Ta    G1    N0       M0      M
24   2307   P2307     yes  T2:4    G2    N0       M0      M
25   2335   P2335     yes    Ta    G1    N0       M0      M
26   2380   P2380     yes    T1    G3    N0       M0      M
27   2505   P2505     yes  T2:4    G3    N0       M0      M
28   2541   P2541     yes  T2:4    G3    N2       M0      M
29   2542   P2542      no  T2:4    G3    N2       M0      M
30   2571   P2571     yes  T2:4    G3    N1       M0      M
31    259    P259      no  T2:4    G3    N0       M0      M
32   2691   P2691     yes    Ta    G2    N0       M0      M
33   2816   P2816     yes    Ta    G1    N0       M0      M
34   2817   P2817     yes    T1 G2:G3    N0       M0      F
35   2821   P2821     yes    Ta    G1    N0       M0      M
36   3031   P3031     yes    T1 G2:G3    N0       M0      M
37   3046   P3046     yes    T1    G3    N0       M0      M
38   3161   P3161     yes    T1 G1:G2    N0       M0      F
39    338    P338     yes    Ta    G1    N0       M0      F
40   3395   P3395     yes  T2:4    G3    N1       M0      M
41   3554   P3554     yes    Ta    G2    N0       M0      M
42    407    P407     yes    Ta    G3    N0       M0      F
43    413    P413     yes  T2:4    G2    N0       M0      M
44    447    P447     yes  T2:4    G3    N2       Mx      M
45    448    P448     yes    T1    G3    N0       M0      M
46    506    P506      no    T1    G1    N0       M0      M
47    813    P813     yes    Ta    G1    N0       M0      M
48    824    P518      no    T1    G2    N0       M0      M
49    848    P848      no  T2:4    G3    Nx       M1      M
50    877    P877      no  T2:4    G2    N1       M0      F
51    908    P908     yes    Ta    G1    N0       M0      M
52    910    P910     yes  T2:4    G3    N0       Mx      M
53    924    P924     yes  T2:4    G3    N0       M0      M
54    943    P943      no  T2:4    G2    N0       M0      F
```

Code 21.8.

Make sure the order of sample names in the sample column is the same as the order of file names in the `pcos` object. You can view the order of sample names using the `sample.names` function (Code 21.9).

```
> sample.names(pcos)
 [1] "1033sproc.txt"   "10871sproc.txt"  "1210sproc.txt"   "12112sproc.txt"
 [5] "1212sproc.txt"   "1333sproc.txt"   "1343sproc.txt"   "1352sproc.txt"
 [9] "1379sproc.txt"   "1382sproc.txt"   "1410sproc.txt"   "1412sproc.txt"
[13] "1413sproc.txt"   "1448sproc.txt"   "1484sproc.txt"   "1498sproc.txt"
[17] "1512-6sproc.txt" "1528-7sproc.txt" "1533-1sproc.txt" "1541-4sproc.txt"
[21] "195sproc.txt"    "2259-1sproc.txt" "2297sproc.txt"   "2307sproc.txt"
[25] "2335sproc.txt"   "2380sproc.txt"   "2505sproc.txt"   "2541sproc.txt"
[29] "2542sproc.txt"   "2571sproc.txt"   "259sproc.txt"    "2691sproc.txt"
[33] "2816sproc.txt"   "2817sproc.txt"   "2821sproc.txt"   "3031sproc.txt"
[37] "3046sproc.txt"   "3161sproc.txt"   "338sproc.txt"    "3395sproc.txt"
[41] "3554sproc.txt"   "407sproc.txt"    "413sproc.txt"    "447sproc.txt"
[45] "448sproc.txt"    "506sproc.txt"    "813-1sproc.txt"  "824sproc.txt"
[49] "848sproc.txt"    "877sproc.txt"    "908sproc.txt"    "910sproc.txt"
[53] "924-1sproc.txt"  "943sproc.txt"
```

Code 21.9.

21.10 Finding Genomic Events: Chromosome Gains and Losses

We have now arrived at the point where we can determine genomic events within our samples, i.e., determining loss and gain events across each chromosome. Remember that the aCGH package uses a hidden Markov model approach to determine states. This is done by calling the `find.hmm.states` function, and the resulting data stored gets in a list in `pcos` called "hmm" (courtesy of the `hmm()` function). This code and output for the first sample are shown in Code 21.10. Unless you are familiar with the hidden Markov model method used, it is best to leave the `find.hmm.states` parameters at the default settings. This is a good time to sit back and take a coffee break, as R will need a few hours to process all the samples. If you do decide to sit and watch the output materialize, you will probably see warnings and error messages produced as R rolls on, but you can ignore these.

```
> hmm(pcos) <- find.hmm.states(pcos, aic = TRUE, delta = 1.5)
sample is  1    Chromosomes: 1  2  3  4  5  6  7  8  9  10  11  12  13  14  15
16  17  18  19  20  21  22  23
```

Code 21.10.

Once this process is complete, we need to apply the `mergeHmmStates()` function to the HMM results. This has the effect of merging contiguous segments whose mean values are extremely close to one another:

```
> hmm.merged(pcos) <- mergeHmmStates(pcos, model.use = 1, minDiff
= 0.25)
```

We must also estimate the variation in each sample, which assists in the process of calculating losses and gains from the computed states:

```
> sd.samples(pcos) <- computeSD.Samples(pcos)
```

That is all the preparatory work we need to do. Finally, with the states calculated and merging applied, we can use the find.genomic.events() function to do exactly what the function says! We combine this with the genomic.events() function to put the output into our pcos object (Code 21.11).

```
> genomic.events(pcos) <- find.genomic.events(pcos)
Finding outliers
Finding focal low-level aberrations
Finding transitions
Finding focal amplifications
Processing chromosome  1
Processing chromosome  2
Processing chromosome  3
Processing chromosome  4
Processing chromosome  5
Processing chromosome  6
Processing chromosome  7
Processing chromosome  8
Processing chromosome  9
Processing chromosome  10
Processing chromosome  11
Processing chromosome  12
Processing chromosome  13
Processing chromosome  14
Processing chromosome  15
Processing chromosome  16
Processing chromosome  17
Processing chromosome  18
Processing chromosome  19
Processing chromosome  20
Processing chromosome  21
Processing chromosome  22
Processing chromosome  23
```

Code 21.11.

21.11 Plotting and Summarizing Genomic States by Phenotype

In our analysis, we have a group of patients who differ according to several phenotype categories. In fact, if we look back at Code 21.8, we can see that we could separate our patients according to whether the tumor is a primary, tumor stage, tumor grade, nodal status, metastasis, or by gender. For the remainder of the analysis, we will focus on determining differences between our samples according to stage, but the methods we apply could be used for any phenotype.

Code 21.8 shows us that patients fall into one of three general categories for tumor staging. An incomplete list of stages is shown in Table 21.1 for readers unfamiliar with bladder cancer staging.

Table 21.1   Tumor Staging in Bladder Cancer.

Ta	Noninvasive papillary carcinoma
T1	Tumor invades connective tissue under the epithelium
T2	Tumor invades muscle
T3	Tumor invades perivesical (around the bladder) fatty tissue
T4	Tumor invades any of the following: prostate, uterus, vagina, pelvic wall, abdominal wall

Thus, our bladder cancer samples are grouped as noninvasive tumors (Ta), tumors that have not invaded the muscle or beyond (T1), and tumors that have invaded muscle and beyond (T1:4). We can count the number of tumors in each category (Code 21.12).

```
> table(pcos$phenotype$stage)

 T1 T2:4   Ta
  9   29   16
```

Code 21.12.

The most basic plotting tool available in the aCGH package is called using the plotGenome() function. This function allows you to select a sample and will then plot the log(2) ratios for each clone on every chromosome. We can plot examples of each stage. We can define the margins of the plot first:

```
> par(mar=c(2,1, 1, 0.5))
```

Now we can plot the log(2) ratios for each example (we can ignore X and Y chromosome values by declaring them as FALSE in the parameter list).

First, let us look at sample 51, which is classed as stage Ta (this tumor has the sample ID 908):

```
> plotGenome(pcos, samples = 1 X =  FALSE,Y = FALSE)
```

The plot is shown in Figure 21.1. An obvious region of genomic instability here is chromosome 9q, which appears to be lost.

The next example is for stage T1 tumors and is sample 46 in the list (sample ID 506):

```
> plotGenome(pcos, samples = 46 X =  FALSE,Y = FALSE)
```

The plot is displayed in Figure 21.2. We can see that for this tumor a number of chromosomes have losses, including 8, 9, and 11, and a few have gains, including 1, 13, and 15. If we focus on chromosome 9, we can see that this tumor appears to have lost this chromosome.

The example for stage T2:4 is sample 15 in the list (sample ID 1484):

```
> plotGenome(pcos, samples = 15 X =  FALSE,Y = FALSE)
```

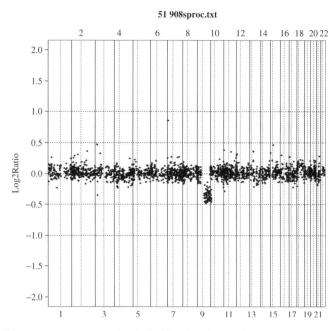

FIGURE 21.1   Using `plotGenome()` to plot the log(2) ratios of clones in sample 51.

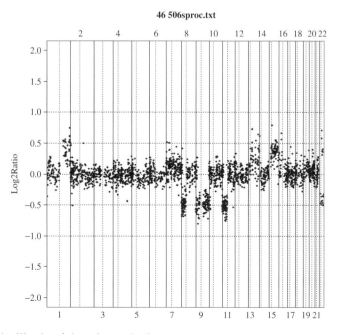

FIGURE 21.2   Log(2) ratios of clones in sample 46.

Again, as we can see in Figure 21.3, this tumor appears to have substantial genomic instability, with a number of chromosomes showing gains and losses, such as
the short arm of chromosome 4.

We can look at losses and gains on each chromosome for the previous examples
at a finer level using the `plotHmmStates()` function. This function, will plot the
log(2) ratios as well as the hmm states for regions of clones along the chromosome
(Code 21.13).

```
#Stage Ta - sample 51 (ID 908)
> plotHmmStates(pcos, sample.ind = 51, chr = 9)
#Stage T1 - sample 46 (ID 506)
> plotHmmStates(pcos, sample.ind = 46, chr = 8)
#Stage T2:T4 - sample 15 (ID 1484)
> plotHmmStates(pcos, sample.ind = 15, chr = 4)
```

Code 21.13.

The plots can be seen for each sample in Figures 21.4, 21.5, and 21.6. For our
stage Ta example, we can see that the long arm of chromosome 9q is called lost by the
hmm algorithm (bottom plot). The broken line represents the centromere position, and
the solid line on this plot shows where the loss (or gain) begins or ends. The stage T1
sample (Figure 21.5) shows two regions of loss on chromosome 8p. The stage T2:4
sample (Figure 21.6) shows regions lost on both chromosome arms.

To get summaries of numbers of clones lost or gained for each chromosome, we
can use the `summarize.clones function()`.

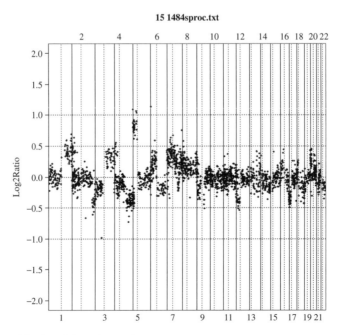

FIGURE 21.3   Log(2) ratios of clones in sample 15.

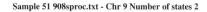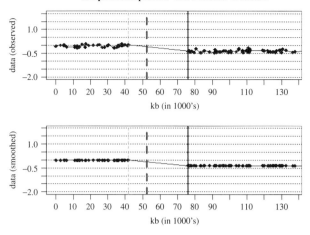

FIGURE 21.4 Chromosome 9 of sample 51.

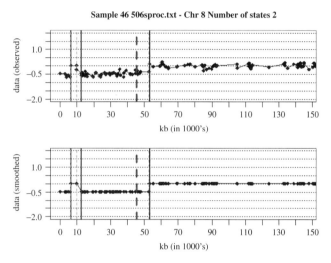

FIGURE 21.5 Chromosome 8 of sample 46.

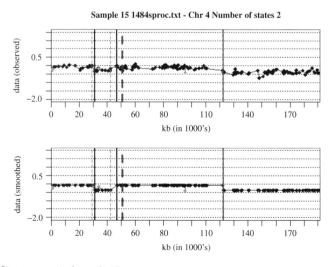

FIGURE 21.6 Chromosome 4 of sample 15.

Code 21.14 shows such a summary for the first 10 clones across the entire dataset.

```
> summarize.clones(pcos)[1:10,]
Stage All:
       Clone        Target    Chrom  kb  NumPresent NumGain
RP11-82D16  HumArray2H11_C9    1  1008       51        2
RP11-62M23 HumArray2H10_N30    1  3767       54        5
RP11-11105 HumArray2H10_B18    1  4738       52        6
RP11-51B4  HumArray2H10_Q30    1  6346       53        8
RP11-60J11 HumArray2H10_T30    1  7621       54        8
RP11-813J5 HumArray2H10_B19    1 10417       54        1
RP11-19901 HumArray2H10_W30    1 11074       51        1
RP11-188F7  HumArray2H9_C14    1 13178       54        0
RP11-178M15 HumArray2H9_F14    1 14564       53        0
RP11-219F4  HumArray2H9_I14    1 15607       54        4
  NumLost    PropPres PropGain PropLost
       12        0.94     0.04     0.24
        1        1.00     0.09     0.02
        1        0.96     0.12     0.02
        2        0.98     0.15     0.04
        1        1.00     0.15     0.02
        0        1.00     0.02     0.00
        1        0.94     0.02     0.02
        3        1.00     0.00     0.06
        2        0.98     0.00     0.04
        1        1.00     0.07     0.02
```

Code 21.14.

Furthermore, we can obtain a summary of losses and gains for each category of stage by simply parsing "pcos$phenotype$stage" to the "pheno" parameter of the function (Code 21.15). We could add this to our pcos object, but for ease of reference, we will store this data in a new object called pcos.stage.

```
> pcos.stage<-summarize.clones(pcos,pheno = pcos$phenotype$stage)
> pcos.stage[1:10,]
#Stage - All:
       Clone        Target  Chrom   kb NumPresent NumGain
RP11-82D16  HumArray2H11_C9    1  1008       51        2
RP11-62M23 HumArray2H10_N30    1  3767       54        5
RP11-11105 HumArray2H10_B18    1  4738       52        6
RP11-51B4  HumArray2H10_Q30    1  6346       53        8
RP11-60J11 HumArray2H10_T30    1  7621       54        8
RP11-813J5 HumArray2H10_B19    1 10417       54        1
RP11-19901 HumArray2H10_W30    1 11074       51               1
RP11-188F7  HumArray2H9_C14    1 13178       54        0
RP11-178M15 HumArray2H9_F14    1 14564       53        0
RP11-219F4  HumArray2H9_I14    1 15607       54        4

NumLost PropPresent PropGain PropLost
     12        0.94     0.04     0.24
      1        1.00     0.09     0.02
      1        0.96     0.12     0.02
      2        0.98     0.15     0.04
      1        1.00     0.15     0.02
      0        1.00     0.02     0.00
      1        0.94     0.02     0.02
      3        1.00     0.00     0.06
      2        0.98     0.00     0.04
      1        1.00     0.07     0.02
```

```
#Stage - T1
NumPresent NumGain NumLost PropPresent PropGain PropLost
         9       1       3           1     0.11     0.33
         9       2       1           1     0.22     0.11
         9       2       1           1     0.22     0.11
         9       3       1           1     0.33     0.11
         9       2       0           1     0.22     0.00
         9       1       0           1     0.11     0.00
         9       1       1           1     0.11     0.11
         9       0       1           1     0.00     0.11
         9       0       1           1     0.00     0.11
         9       1       0           1     0.11     0.00

#Stage - T2:4
NumPresent NumGain NumLost PropPresent PropGain PropLost
        26       0       5        0.90     0.00     0.19
        29       2       0        1.00     0.07     0.00
        27       3       0        0.93     0.11     0.00
        28       4       1        0.97     0.14     0.04
        29       5       1        1.00     0.17     0.03
        29       0       0        1.00     0.00     0.00
        26       0       0        0.90     0.00     0.00
        29       0       2        1.00     0.00     0.07
        28       0       1        0.97     0.00     0.04
        29       3       1        1.00     0.10     0.03
Stage - T2:4
NumPresent NumGain NumLost PropPresent PropGain PropLost
        16       1       4           1     0.06     0.25
        16       1       0           1     0.06     0.00
        16       1       0           1     0.06     0.00
        16       1       0           1     0.06     0.00
        16       1       0           1     0.06     0.00
        16       0       0           1     0.00     0.00
        16       0       0           1     0.00     0.00
        16       0       0           1     0.00     0.00
        16       0       0           1     0.00     0.00
        16       0       0           1     0.00     0.00
```

Code 21.15.

You would not want to check all the gains and losses for every chromosome in each sample simply by scanning the output by eye. Using the data held in the `pcos.stage` object, we can generate a graphic summary of the distribution of losses and gains for each chromosome, either for all samples combined or by phenotype using boxplots.

In our analysis, we are interested in surveying differences in genomic events between tumors according to staging, so we can generate different boxplots according to stage category. By looking at the table in Code 21.15, we can see that `summarize.clones()` outputs columns detailing the proportion of clones gained and lost for each chromosome in each stage. Proportion of clone gains columns are 27, 15, and 21 for stages Ta, T1, and T2:T4, respectively, and columns 28, 16, and 22 have proportion of clone losses for Ta, T1, and T2:T4, respectively.

To create boxplots for the losses and gains, all we have to do is create a vector by splitting the proportion columns according to chromosome in our `pcos.stage` object using the `split()` function within the `boxplot()` function (Code 21.16).

To generate boxplots for the distribution of gains in each chromosome for each stage separately, we first need to tell R to generate three plots in the same plot window:

```
> par(mfrow = c(3, 1))
```

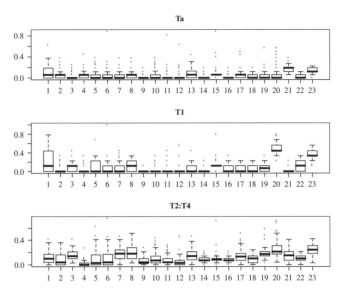

FIGURE 21.7   Distribution of chromosome gains for each stage.

Then we use the `boxplot()` function in combination with `split()` to draw the plots (Figure 21.7). By examining the first line of code in Code 21.16, we can see that the `split()` function splits the proportion of gains data in column 27 of the `pcos.stage` data frame according to the chromosome number designation held in column 3.

```
# stage: Ta
> boxplot(split(pcos.stage [,27], pcos.stage [,3]) ,main="Ta")
# stage: T1
> boxplot(split(pcos.stage [,15], pcos.stage [,3]) ,main="T1")
# stage: T2:T4
> boxplot(split(pcos.stage [,21], pcos.stage [,3]) ,main="T2:T4")
```

Code 21.16.

We can use exactly the same approach to generate boxplots (Figure 21.8) for chromosome loss events in each stage category (Code 21.17).

```
# stage: Ta
> boxplot(split(pcos.stage [,28], pcos.stage [,3]) ,main="Ta")
# stage: T1
> boxplot(split(pcos.stage [,16], pcos.stage [,3]) ,main="T1")
# stage: T2:T4
> boxplot(split(pcos.stage [,22], pcos.stage [,3]) ,main="T2:T4")
```

Code 21.17.

This plotting approach is actually a simple yet powerful way of visualizing genomic events for phenotypic categories of samples. For instance, you can quickly

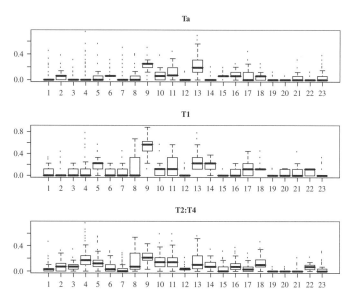

FIGURE 21.8 Distribution of chromosome losses for each stage.

compare the variation in losses for each chromosome between stage categories; for example, losses in chromosome 4 in stage Ta are evidently at a lower frequency than at later tumor stages. Assessment of chromosome 9 tells us that losses occur at a relatively higher level in all stages.

Visualization of genomic gains in this way is a useful first step to assessing differences between stages and gives the researcher a good overall feel for differences within data. The next step in the analysis involves determining any association between clone losses/gains and stage, and determining whether differences in clones between stages are significant.

21.12 Finding Association Between Clones and Phenotype

Before we look for association between losses or gains for each clone with different stage categories, we need to consider that nagging problem of multiple hypothesis testing. When testing many variables simultaneously for significant association or difference between conditions, there is an increased risk of generating false-positives. With more than 2000 clones to test in our example at the significance level of 1% ($P = 0.01$), there is a chance that more than 20 clones will wrongly be classed as significant gains/losses. This is called the family wise error rate (FWER). At the 5% significance level, that could be more than 100 clones.

Therefore, the `aCGH` package allows you to call the `maxT` function from the `multtest` package, which is a step-down method that computes adjusted P Values by permutation while controlling the FWER. Before we apply the test, we need to quickly set up a couple of objects that we can parse to the `maxT` function to prevent errors from occurring. This is because some R functions cannot handle data that contains "NA" elements, and although our stage data is complete, this step will be useful for other datasets where phenotypic data is not as complete. Create an object called `stage` to hold the stage data:

```
> stage<- phenotype(pcos)$stage
```

Then create a vector (of type logical) called `stage.na` that will hold true or false values for each sample, depending on whether stage data is available (it is for all samples in our data, but if an NA was present, it would return a FALSE for that sample):

```
> stage.na <- !is.na()
```

We then can create a new `aCGH` object called `pcos.na` that would only contain data for samples where stage designation was available:

```
> pcos.na <- pcos [ ,stage.na, keep = TRUE ]
```

Now we can use the `maxT` function, which is called `mt.maxt()`, to compute significant differences between stages for genomic events. This function takes as input a number of parameters. First we need to provide an `aCGH` object, which is `pcos.na` in our case. Second, we need to state the type of test we want to perform. The `mt.maxt()` function can perform parametric and nonparametric tests, including varieties of *t*-tests and Wilcoxon rank sum, but we have three groups so we will perform a one-way analysis of variance (ANOVA) by parsing "f" to the "test" parameter. Third, we need to state the number of permutations to perform, which will be 1000 in this case. Results are stored in an object called `assoc.stage`, and both the code and output are displayed in Code 21.18.

```
> assoc.stage <- mt.maxT(pcos.na$log2.ratios, stage[stage.na], test = "f",
B = 1000)
b=10     b=20     b=30     b=40     b=50     b=60     b=70     b=80     b=90     b=100
b=110    b=120    b=130    b=140    b=150    b=160    b=170    b=180    b=190    b=200
b=210    b=220    b=230    b=240    b=250    b=260    b=270    b=280    b=290    b=300
b=310    b=320    b=330    b=340    b=350    b=360    b=370    b=380    b=390    b=400
b=410    b=420    b=430    b=440    b=450    b=460    b=470    b=480    b=490    b=500
b=510    b=520    b=530    b=540    b=550    b=560    b=570    b=580    b=590    b=600
b=610    b=620    b=630    b=640    b=650    b=660    b=670    b=680    b=690    b=700
b=710    b=720    b=730    b=740    b=750    b=760    b=770    b=780    b=790    b=800
b=810    b=820    b=830    b=840    b=850    b=860    b=870    b=880    b=890    b=900
b=910    b=920    b=930    b=940    b=950    b=960    b=970    b=980    b=990    b=1000
```

Code 21.18.

The results returned by `mt.maxt()` to the `assoc.stage` object are displayed for the first five clones in Code 21.19.

```
> assoc.stage[1:5,]
            index   teststat   rawp   adjp
RP11-47P10   544   12.190526  0.001  0.066
RP11-264N7   451   10.778366  0.002  0.135
RP11-11M9    431   10.146352  0.001  0.184
RP11-94J9    533    9.600282  0.001  0.237
GS1-20208    858    9.186419  0.001  0.289
```

Code 21.19.

The key result is the adjusted P Value in the final column (adjp). If we decide to focus on clones with adjusted P Values less than 0.01, to view these clones we would need to order the `assoc.stage` data frame by the adjp column. We can order the clones using the `order()` function, where we tell the function to order by column 4 in ascending order (Code 21.20).

```
> assoc.stage<-assoc.stage[order(assoc.stage[,4]),]
> assoc.stage[1:10,]
             index  teststat   rawp  adjp
RP11-47P10     544 12.190526 0.001 0.066
RP11-264N7     451 10.778366 0.002 0.135
RP11-11M9      431 10.146352 0.001 0.184
RP11-94J9      533  9.600282 0.001 0.237
GS1-20208      858  9.186419 0.001 0.289
RP11-113F10    452  9.178542 0.001 0.291
RP11-118C24    435  9.146112 0.001 0.297
CTD-2238P18   2172  8.807218 0.002 0.341
RP11-238L9     432  8.511288 0.001 0.393
GS-2B19       2177  8.224284 0.002 0.453
```

Code 21.20.

As is often the case when correcting for multiple testing, you may be a bit disheartened to see that not one single clone has a significant adjusted P Value. Instead, we will have to turn to the raw P Values that are also provided in the `assoc.stage` object in column 3 (rawp). We can display how many clones have a raw P Value less than or equal to 0.01 as a table (Code 21.21).

```
> table(assoc.stage[,3]<=0.01)

FALSE   TRUE
 2229     89
```

Code 21.21.

This time we find that there are 89 clones deemed significant for either loss or gain. We will reorder the `assoc.stage` data frame by raw P Value:

```
> assoc.stage<-assoc.stage[order(assoc.stage[,3]),]
```

To view the list of clone names significant by grade, we can create a vector called `assoc.stage.sig` by selecting clone names from `assoc.stage` that have a P Value less than or equal to 0.01 and print them (Code 21.22). From our list, we can see that the most significant clone is RP11-47P10.

```
> assoc.stage.sig<-assoc.names[assoc.stage$rawp<=0.01]
pcos$clones.info[assoc.stage$rawp<=0.01,1]
> assoc.stage.sig
 [1] "RP11-47P10"  "RP11-11M9"   "RP11-94J9"   "GS1-20208"   "RP11-113F10"
"RP11-118C24" "RP11-238L9"
```

```
 [8] "RP11-264N7"   "CTD-2238P18" "GS-2B19"     "RP11-169A6"  "RP11-24K24"
"RP11-1E23"    "RP11-18F11"
[15] "RP11-85P1"    "RP11-5P14"   "CTD-2271H24" "CTB-16305"   "RP11-18A21"
"CTD-2086L13" "RMC20B4130"
[22] "RP11-5K16"    "RP11-101N17" "RP11-26D15"  "RP11-25305"  "RP11-131K16"
"RP11-243D3"   "GS1-183H7"
[29] "RP11-266E24" "RP11-28I11"  "RP11-213A19" "RP11-75022"  "RP1-81F12"
"RP11-170L13"  "RP11-130M1"
[36] "RP11-218I24" "RP11-182G5"  "GS1-54J22"   "RP11-27C19"  "RP11-17H7"
"RMC20P117"    "RP11-73G16"
[43] "RP11-15M15"  "RMC20B4087"  "RP11-93J6"   "CTD-2019H1"  "RP11-84M15"
"RP11-53C1"    "GS1-213H12"
[50] "RP11-137E8"  "RP11-239C17" "CTD-2031K20" "RP11-34N11"  "CTD-2222B22"
"RP11-256M13" "RP11-35E16"
[57] "RP11-75C17"  "RP11-213E22" "RP11-26M17"  "CTD-2006L14" "RP11-218A2"
"RP11-55D2"    "RMC20P090"
[64] "RP11-111K19" "CTD-2026B15" "CTB-134019"  "RP11-32D4"   "RP11-105F21"
"RP11-55P9"    "RP11-136D2"
[71] "CTB-159E10"  "RP11-2208"   "RP11-6E9"    "RP11-205N12" "RP11-208D7"
"RP11-49N15"   "RP11-100E15"
[78] "RP11-165014" "RP11-93I2"   "RP11-124D1"  "RP11-128J2"  "RP11-118C13"
"RP11-100L12" "RP11-92I18"
[85] "RP11-30F23"  "RP11-125M9"  "RP11-358D14" "RP11-82L8"   "RP11-252N11"
```

Code 21.22.

21.13 Associating Clones with Chromosomes

We now have a list of those clones that have a significant association for genomic events with tumor stage, and the final step of our analysis is to find out which chromosomes these clones relate to.

This is simple and can be determined with a single line of code. We place into a vector called `assoc.stage.chr` the chromosome assignment for each significant clone name held in the `assoc.stage.sig` vector. To do this, we use the `match()` function to match the clone names in `assoc.stage.sig` with clone names in column 1 of the `pcos$clones.info` data frame. If there is a match, the function retrieves the value from column 3 of `pcos$clones.info`, which is the chromosome number the clone relates to (Code 21.23).

```
> assoc.stage.chr<-pcos$clones.info[match(assoc.stage.sig,
pcos$clones.info[,1]),3]
> assoc.stage.chr
 [1]  4  4  4  7  4  4  4 20 20 20 20 20  4  4 20 19 17  9  4 20  4  4  4  4
 4  4  7  4 11  4  7 20  5
[35]  4  4  4  7  4  7 20  4 20 20  7 20  4  4  7  7  4 20  7 11  4  4 20  7  4
 9  4 20 20  4 20  5  9  4
[69]  9  4  4  4  4  4 17  7  4  9  4 20  4  9 17 10 20  4 20  4 17
```

Code 21.23.

We can already see a pattern emerging with regard to the frequency of some chromosomes having a higher proportion of genomic events associated with tumor stage. We can summarize the frequencies in a table (Code 21.24).

```
> table(assoc.stage.chr)
assoc.stage.chr
 4  5  7  9 10 11 17 19 20
42  2 11  6  1  2  4  1 20
```

Code 21.24.

The table reveals that chromosomes 4 and 20 have a high proportion of genomic events, but does not reveal whether these are losses or gains. However, we can visualize this information using the `plotfs()` function, which is a modified version of the `plotFreqStat()` function provided with the `aCGH` package. This function plots the positions of significant gains and losses along the genome on the *x*-axis and the frequencies of these genomic events along the *y*-axis (greater than 0 for gain and less than 0 for loss).

Before we create the plot, you may need to alter the margins of the plot window. The settings I have used are:

```
> par(mar=c(4, 4, 2, 0.5))
```

Now we can plot the significant genomic events:

```
> plotfs(pcos.na, assoc.stage, stage[stage.na], factor = 3,
X = FALSE, Y = FALSE)
```

Plots are drawn to show the frequencies of genomic events for samples grouped by tumor staging (Figure 21.9). A fourth plot is also drawn where the height of each bar along the *x*-axis represents the test statistic for each clone. This plot confirms our

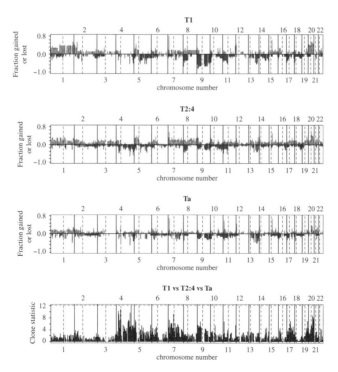

FIGURE 21.9 Frequencies of genomic events grouped by tumor staging.

earlier observation that both chromosomes 4 and 20 have a higher proportion of genomic events between tumor stage categories. The plots above this for each stage tell us that the genomic events in chromosome 4 are an increase in losses in stages T1 and T2 to T4, and the events in chromosome 20 are an increase in gains in these stages. As we saw in Code 21.23, significant losses and gains occur between stage categories at seven other chromosomes, including 5, 7, 9, 10, 11, 17, and 19.

21.14 Summary

In this chapter we have explored how we can process and analyze BAC array-CGH data in R using just the aCGH package. First, the analysis of the bladder cancer dataset should have provided the reader with an introduction to the processing, statistical analysis, and plotting capabilities of aCGH. Second, by working through the example, you will have learned the necessary skills and requirements for processing, analyzing, and visualizing array-CGH data—particularly the need to segment clones into states across chromosomes.

The interested reader may be interested in scouring the literature to compare the significant genomic events associated with tumor stage of bladder cancer uncovered in our study with those already published (a useful starting point might be Blaveri et al.[51]). Although we only focused on data output in SPROC format, it is not too much work to take data from other output types and create the necessary lists, such as log(2) ratios and phenotype data, in an aCGH object. There are also other useful functions in aCGH that we did not touch on. Two functions worth mentioning are threshold.func(), which allows you to declare losses or gains at clones as –1 or 1, respectively, and the clusterGenome() function, which provides a heatmap and dendrogram for samples clustered by genomic events at selected chromosomes. Other R packages the reader may want to investigate for array CGH data are DNAcopy and Glad.

21.15 Questions

1. What advantages does arrayCGH have over standard CGH technology?

2. What term do we use to describe a region of chromosome that has been lost or gained?

3. What operator can be used to access components in a list?

4. What R function allows us to automate the process of reading in many arrayCGH files in one go?

5. What are the components of an aCGH object?

6. What type of modeling procedure is used by the aCGH package to determine loss and gain events, and what R function achieves this?

7. What functions allow us to visualize genomic events that have occurred?

8. What is a phenotype in relation to genomic events?

9. What is multiple testing, and what is a drawback to applying this procedure in a statistical analysis?

22 | XML for Storing and Sharing Data

22.1 Background

In 2001, the Medical Research Council (MRC) in the United Kingdom launched an initiative to "promote new and extended use of" data generated by researchers that received their grants (http://www.mrc.ac.uk/PolicyGuidance/EthicsAndGovernance/DataSharing/index.htm). In 2006, it became compulsory for all applicants for MRC grants to include a statement explaining the researchers' strategy for preserving the data they generate for sharing and reuse. In the United States, this policy had already been adopted in 2003 by the National Institutes of Health (NIH) (http://grants.nih.gov/grants/policy/data_sharing/). Many journals now implement a data-sharing policy too for new manuscript submissions. This was quite a culture shock for biomedical researchers, who had never been overly keen on sharing data with their rivals.

The health-related funding bodies demand a maximum return for their investment in research, and that return is always improved benefits to the health of the population. It is their view that publicly funded research data is a "public good" and must be made available for "new research uses in a timely, responsible manner." Whereas it is realized that the researchers must benefit from their intellectual investment in the short term, scientific advancement is compromised if data is kept for their exclusive use. Many of the benefits to data sharing are obvious, including the ability to validate original research findings, integrate individual datasets, generate new hypotheses, apply different analysis approaches, and help avoid duplicating research. We have seen earlier in this book the benefits of integrating biological and clinical data with other data such as treatment and outcome. Researchers should also be encouraged that data sharing has a positive impact on citation rate.[52]

By "data" we are referring not just to biological or medical observational data, but also results produced from statistical and data mining analyses. Data sharing can be facilitated in a number of ways, including publication in the literature and the availability of secure data repositories for public access. To share data efficiently, one needs standards. Standards provide a mechanism to enable data integration as well as to overcome interoperability problems associated with different software and data formats. These often include rules and a vocabulary that allows one to describe data in a uniform way. Standards need to be flexible enough to cater for a variable research community and capable of adjusting as both the science and technologies change. However, we all have a different way of doing things, and the introduction and development of standards to describe and share data takes an extraordinary amount of effort. More than anything, the adoption of a standard can only occur if the blueprint is

337

acceptable to the research community. Perhaps more than any other domain, the bio-medical world has witnessed the difficulties involved in developing common standards for describing and sharing data. In his book *Biomedical Informatics*, Jules Berman provides an elegant and honest overview of developing biomedical standards, high-lighting the cons as well as the pros.[53]

Let us consider some examples. Imagine all the computer systems in place across all the departments in a single hospital. Each deals with complex pieces of healthcare data that often describe the same piece of information in different ways. Flexible stan-dards and guidelines (i.e., rules) that allow healthcare data to be exchanged in a uni-form way have been introduced by the Health Level 7 (HL7) organization, and these standards have been implemented worldwide (http://www.hl7.org/). An international organization called the Microarray and Gene Expression Data (MGED) Society over-sees the development of standards for microarray and functional genomics data (http://www.mged.org/). One of the standards defined by MGED is the minimum in-formation about a microarray experiment (MIAME), which is a set of guidelines to al-low unambiguous interpretation of data. The success of MIAME has led to MGED introducing a working group to use the model to implement a standard for in situ hy-bridization and immunohistochemistry experiments (MISFISHIE). Yet another organ-ization is the Clinical Data Interchange Standards Consortium (CDISC) devoted to the development of platform-independent standards for acquisition, exchange, submis-sion, and storage of clinical research data and metadata (http://www.cdisc.org/).

What is common about each set of standards is that they use a markup language to describe, share, and store information. The term "markup" is derived from the world of publishing whereby an author or editor adds marks in the margins of a manuscript that describe the style and formatting of pieces of text. Thus, markup means annota-tion, and a markup language is one that includes a defined set of annotation terms (vo-cabulary) for describing the structure, layout, and format of text or data. The contents and format of web pages are encoded using markup languages, and most readers will have heard of the Hypertext Markup Language (HTML). HTML is a predefined set of tags that define how an element on a web page is displayed—e.g., a tag can be used to specify that a piece of text appears in bold type. In 1998, the Extensible Markup Language (XML) was introduced that allowed a developer to define new tags (hence the "extensible") to describe any data element, and it was soon realized that XML would be a powerful mechanism for describing and sharing data. XML is itself an open standard that allows individuals to specify a custom markup language suited to their needs. Many standards organizations, including HL7, MGED, and CDISC, have defined XML-based languages for describing and exchanging data. One cannot over-state the current and future importance of XML as a data sharing mechanism within biomedical informatics. The ability to exchange data analysis results is just as impor-tant as sharing observational and experimental data. We have used R to generate pre-dictive models in earlier chapters, and each model provides a wealth of information. Interpretation of your results data is vital to a global understanding and acceptance of your model. By using an XML-based markup language, you can specify and describe each element of your model in a standardized way, which greatly enhances its inter-pretation. Recognizing this fact, the software-vendor-led Data Mining Group (DMG) has introduced the XML-based Predictive Model Markup Language (PMML) to describe statistical and data mining models (http://www.dmg.org/).

This chapter begins by giving a (very) brief introduction to XML. Cramming a comprehensive introduction to XML into a single chapter is impossible, so we will just cover necessary methods and approaches. Armed with enough knowledge, we will then focus on how we can use XML and R to store and share the results of data analyses in a structured and defined way. Although PMML is far from complete, we will look at how we can use this markup language to describe the decision tree model we generated in Section 15.9. As you work through the chapter, keep asking yourself what the benefits might be of preparing your results in this way.

22.2 Introduction to XML

In the last decade, XML has become as important a tool in biomedicine as any of the exciting new high-throughput technologies that provide so much data. XML is an unsung hero and provides us with a simple means of annotating the data we generate. We annotate our data using metadata—data that describes data. Metadata are just labels or tags that serve as descriptors. As humans, either consciously or subconsciously, we name and label everything we encounter in one way or another. We tag things in our minds, and this helps us to stay organized. A markup language based on XML is simply an organized structure of metadata tags that hold and describe your data.

Consider the data and statistical analysis detailed in Chapter 11 for systolic blood pressure measurements in a reduced alcohol trial.

Our data consists of scores from two patient groups, each with 20 subjects and statistical parameters as output from a *t*-test. We would like to write this data using a format that can be understood by anybody that we wish to share it with. So that others can easily understand each bit of datum, we need to be clear in how we phrase the metadata descriptors. It is not difficult to create a series of metadata to describe all the information related to the study (Table 22.1).

As I created this list, I tried to structure it so that the different types of information were grouped in an intuitive way. This led to me creating a hierarchical structure whereby the study data is sectioned into details, patients, and analysis. To write this information in XML format, we create XML tags that reflect the metadata. An XML tag is constructed by placing the metadata label inside angle brackets:

```
<p_value>
```

The tag can either be a start tag, such as the aforementioned one, or an end tag:

```
</p_value>
```

In its simplest form, the start and end tags will lie on both sides of some text or numeric value (i.e., the data that the tags describe) to form what is called an XML element:

```
<p_value>0.0002881</p_value>
```

Here, `<p_value>` is the opening tag, 0.0002881 is the data element, and `</p_value>` is the end tag, signified by the backslash. Elements can also contain other

Table 22.1   Data and Metadata for Blood Pressure and *T*-Test Parameters.

Metadata	Data
study	
details	
details: title	"Systolic blood pressure measurements for reduced alcohol"
details: researcher	"Paul Lewis"
details: analyst	"Nathan Hill"
patients	
patient: group	
patient: group: name	"pretrial", "posttrial"
patient: group: N	20, 20
patient: group: measurements	"147,155,140,165,151,160,158,144,143,162,154,171,148, 168,141,148,170,156,148,161", "134,145,141,155,151,148,148,128,137,160,140,160,155, 145,138,147,164,158,144,145"
analysis	
analysis: test	"t-test"
analysis: statistic	4.4289
analysis: df	19
analysis: p-value	0.0002881
analysis: lower 95%CI	3.876539
analysis: upper 95%CI	10.823461

elements, but the simple rule exists that an element starts and ends with a tag. Let us jump straight in and look at the XML elements we can create to describe the study with just a few blood pressure measurements added in each group (List 22.1).

List 22.1 XML code for blood pressure study.

```
<?xml version="1.0"?>
<study>
  <details>
    <title>Systolic blood pressure measurements for reduced alcohol</title>
    <researcher>Paul Lewis</researcher>
    <analyst>Nathan Hill</analyst>
  </details>
  <patients>
   <group>
     <name>pretrial</name>
     <N>20</N>
     <measurement>147</measurement>
     <measurement>155</measurement>
     <measurement>140</measurement>
```

```
      <measurement>165</measurement>
    </group>
    <group>
      <name>posttrial</name>
      <N>20</N>
      <measurement>134</measurement>
      <measurement>145</measurement>
      <measurement>141</measurement>
      <measurement>155</measurement>
    </group>
  </patients>
  <analysis>
    <test>t-test</test>
    <statistic>4.4289</statistic>
    <df>19</df>
    <p_measurement>0.0002881</p_measurement>
    <lower_95_CI>3.876539</lower_95_CI>
    <upper_95_CI>10.823461</upper_95_CI>
  </analysis>
</study>
```

Note the hierarchical tree-like structure of our XML code and how some elements are nested in other elements higher up the order. The first element tells us that the file is XML. The `study` element encompasses all other elements, and we begin with this element as we traverse the tree. We refer to this element as being the root element. The `study` element has three nested elements, which are `details`, `patients`, and `analysis`. The `details` element itself has three nested elements (`title`, `researcher`, and `analyst`). The `patients` element is a bit more complex because this contains a single element called `group`, which contains three further elements (`name`, `N`, and `measurement`). Take some time familiarizing yourself with how element start and end tags are used.

XML elements can also contain attributes in their start tag. An attribute is just an additional piece of information about the element. Consider the `patient` element shown previously. This contains a child called `group`, which itself contains a child called `name`. We could have actually combined the `group` and `name` elements by inserting the name information as an attribute into `group`:

```
<group name = 'pretrial'>
```

The World Wide Web Consortium (W3C) recommends a set of rules that, if followed, ensures that your XML is well formed (http://www.w3.org/). First, the code must include data elements between opening and closing metadata tags and the file must be plain text. The tags must consist of alphanumeric strings with no spaces. To separate words used in a tag, you could use an underscore character. XML can be written using a simple text editor. Software capable of extracting data from an XML document use what is called a parser. Popular web browsers also allow you to open well-formed XML files and will display the code in a well-spaced and nested format.

Hopefully, you can see by now how useful XML can be in describing data in a standardized way.

22.3 XML Schemas

If we were to develop a standard for reporting *t*-test results on biomedical data, we would need to make a specification that defines the rules on what each element can be and how it is used. Any researchers who wish to use the markup language to report their data can then refer to this specification as they construct their file. Indeed, a valid XML document is said to be well formed if the XML file is structured according to a specification laid out in either a Document Type Definition (DTD) or a schema. Schemas are more versatile than DTDs and will only be considered here. The schema contains a definition of the structure, content, and semantics of the XML document. The schema thus dictates the rules on how defined elements are used. Schemas are written in an XML format and can be saved in a different file with an ".xsd" extension. Schemas should be made available to anybody who wants to use your markup language and are often found as web pages, to which they can be referenced. A schema is itself composed of XML, and the content is used by parser software to learn about the data structure before it can get on with pulling out data. The schema for our blood pressure study XML file could look like that shown in List 22.2.

List 22.2 Schema for blood pressure study XML file.

```
<?xml version="1.0"?>
<xs:schema xmlns:xsi="http://www.w3.org/2001/XMLSchema">

<xs:element name="study">
 <xs:complexType>
  <xs:sequence>
   <xs:element name="details">
    <xs:complexType>
     <xs:sequence>
      <xs:element name="title" type="xs:string"/>
      <xs:element name="researcher" type="xs:string"/>
      <xs:element name="analyst" type="xs:string"/>
     </xs:sequence>
    </xs:complexType>
   </xs:element>
```

```
<xs:element name="patients">
  <xs:element name="group">
   <xs:complexType>
    <xs:sequence>
     <xs:element name="name" type="xs:string"/>
     <xs:element name="N" type="xs:positiveInteger"/>
     <xs:element name="measurement" type="xs:decimal"/>
    </xs:sequence>
   </xs:complexType>
  </xs:element>
 </xs:element>
 <xs:element name="analysis">
  <xs:complexType>
   <xs:sequence>
    <xs:element name="test" type="xs:string"/>
    <xs:element name="statistic" type="xs:string"/>
    <xs:element name="df" type="xs:positiveInteger"/>
    <xs:element name="p_value" type="xs:decimal"/>
    <xs:element name="lower_95_CI" type="xs:decimal"/>
    <xs:element name="upper_95_CI" type="xs:decimal"/>
   </xs:sequence>
  </xs:complexType>
 </xs:element>
 </xs:sequence>
 </xs:complexType>
</xs:element>
</xs:schema>
```

You will have immediately noticed that each element tag begins with "xs," which is common practice. The first element after the XML version declaration contains the `xs:schema` tag, which points to the w3.org specifications for writing schemas. The remaining elements describe the organization of elements and nested elements that we saw in our XML document. Each element allowed is specified using the `xs:element` tag, and the element name and type are given. If an element such as the root contains nested elements, you need to use the `xs:complexType` tag to enclose the nested elements, whose declarations are themselves enclosed using the `xs:sequence` tag.

We can now go back to our XML document and alter it slightly to include a pointer to our schema file at a given URL (List 22.3). This additional information is actually placed within the start tag of the root element (in this case, the `study` tag).

List 22.3 Amending the XML document to point to a schema.

```
<?xml version="1.0"?>
<!—created by Paul Lewis —>
<study
xmlns:xsi="http://www.w3.org/2001/XMLSchema"
xsi:schemaLocation="http://cancerinformatics.swansea.ac.uk/schema.xsd">
```

This has only been a brief introduction to XML, but enough information was provided to allow us to understand the XML variant called PMML to describe the data held in models that we build.

22.4 Using XML in R

The major resource for creating and processing XML in R is the XML package developed by the Omega Project for Statistical Computing (Omegahat).

22.4.1 The XML Package

The XML package provides the R user with an extremely comprehensive suite of tools for the manipulation of XML. The scope of XML is mirrored by the many functions provided by the package. Fortunately, the package author, Professor Duncan Temple Lang, has provided excellent help pages as well as additional documents and tutorials that can be found on the Omegahat website (http://www.omegahat.org/). The package can be obtained via the CRAN repository:

```
> install.packages("XML")
```

There are a number of ways you can represent an XML document in R, and all require a bit of time to learn. These functions include `xmlOutputBuffer()`, `xmlOutputDOM()`, and `xmlTree()`, which provide further functionality for adding new tags and retrieving contents. However, there is another function called `xmlHashTree()` that creates a "flat tree" environment. Although less functional, this approach is probably easier to learn and feels more transparent to a user becoming familiar with XML for the first time. `xmlHashTree()` stores nodes as objects of the class `xmlNode` created using the `xmlNode()` function. Similarly, text nodes and comment nodes can be created using `xmlTextNode()` and `xmlCommentNode()`. Objects created form part of an environment that allows us to access the contents of any node.

22.4.2 Creating an XML Document in R

We will cover here just the basics of what we need to know to create, manipulate, read, and store XML documents. In addition to `xmlHashTree()`, R provides us with another useful tool for creating and manipulating XML—the built-in R text editor available from the File menu. The R Editor can be opened and placed alongside the R Console, allowing you to open and view XML files.

Let us use R to create the XML file shown in Code 22.1.

```
> library(XML)
> tree = xmlHashTree()
> study = addNode(xmlNode("study"), character(), tree)
> details = addNode(xmlNode("details"), study, tree)
> title = addNode(xmlNode("title"), details, tree)
> addNode(xmlTextNode("Systolic blood pressure measurements for reduced
alcohol"), title, tree)
> researcher = addNode(xmlNode("researcher"), details, tree)
> addNode(xmlTextNode("Paul Lewis"), researcher, tree)
> analyst = addNode(xmlNode("analyst"), details, tree)
> addNode(xmlTextNode("Nathan Hill"), analyst, tree)
> patients = addNode(xmlNode("patients"), study, tree)

> group1 = addNode(xmlNode("group"), patients, tree)
> name1 = addNode(xmlNode("name"), group1, tree)
> addNode(xmlTextNode("pretrial"), name1, tree)
> N1 = addNode(xmlNode("N"), group1, tree)
> addNode(xmlTextNode("20"), N1, tree)
> measurement11 = addNode(xmlNode("measurement"),group1, tree)
> addNode(xmlTextNode("147"), measurement11, tree)
> measurement12 = addNode(xmlNode("measurement"), group1, tree)
> addNode(xmlTextNode("155"), measurement12, tree)
> measurement13 = addNode(xmlNode("measurement"), group1, tree)
> addNode(xmlTextNode("140"), measurement13, tree)
> measurement14 = addNode(xmlNode("measurement"), group1, tree)
> addNode(xmlTextNode("165"), measurement14, tree)

> group2 = addNode(xmlNode("group"), patients, tree)
> name2 = addNode(xmlNode("name"), group2, tree)
> addNode(xmlTextNode("posttrial"), name2, tree)
> N2 = addNode(xmlNode("N"), group2, tree)
> addNode(xmlTextNode("20"), N2, tree)
> measurement21 = addNode(xmlNode("measurement"),group2, tree)
> addNode(xmlTextNode("134"), measurement21, tree)
> measurement22 = addNode(xmlNode("measurement"), group2, tree)
> addNode(xmlTextNode("145"), measurement22, tree)
> measurement23 = addNode(xmlNode("measurement"), group2, tree)
> addNode(xmlTextNode("141"), measurement23, tree)
> measurement24 = addNode(xmlNode("measurement"), group2, tree)
> addNode(xmlTextNode("155"), measurement24, tree)

> analysis = addNode(xmlNode("analysis"), study, tree)
> test = addNode(xmlNode("test"), analysis, tree)
> addNode(xmlTextNode("t-test"), test, tree)
> statistic = addNode(xmlNode("statistic"), analysis, tree)
> addNode(xmlTextNode("4.4289"), statistic, tree)
> df = addNode(xmlNode("df"), analysis, tree)
> addNode(xmlTextNode("19"), df, tree)
> p_value = addNode(xmlNode("p_value"), analysis, tree)
> addNode(xmlTextNode("0.0002881"), p_value, tree)
> lower_95_CI = addNode(xmlNode("lower_95_CI"), analysis, tree)
> addNode(xmlTextNode("3.876539"), lower_95_CI, tree)
> upper_95_CI = addNode(xmlNode("upper_95_CI"), analysis, tree)
> addNode(xmlTextNode("10.823461"), upper_95_CI, tree)
```

Code 22.1.

After loading the XML library, we create a new environment called `tree` using `xmlHashTree()`. If you were to type "tree" at this point, an empty list would be returned. Using a function called `addNode()`, we then add our first node to the `tree` list and call it "study." This will be the root node. `addNode()` takes as its arguments the new node (created by parsing the name of the node to a function called `xmlNode()`), the parent node for which the new node will be a child, and the `tree` object. Because this first node we create will be a root node, there is no parent so we need to parse `character()`. Note, too, that we are creating an object (of class `xmlNode`) called `study`. By doing this, we can manipulate the node object later on that will have a direct effect on the information held within the `tree` object. Indeed, we can see this usage in the next line of code. We create a child node for study called `details` using `addNode()` by parsing the study node as the parent argument. We then proceed to work methodically through the tree, adding nodes and values in a hierarchical manner relative to how we want information stored. Nodes that hold values are added using the `xmlTextNode()` function. We do not create an `xmlNode` in this instance because there is no requirement for adding children at any point.

A number of nodes in the XML document will share the same tag (e.g., `group`). When we create the `xmlNode` objects for these cases, we need to name each object in a systematic way. For example, we need to create four `xmlTextNode` objects representing measurements in each of the two patient groups. There are two patient groups, so the `xmlNode` objects for these are named `group1` and `group2`. The node objects for each measurement can then be named according to what group they fall into (e.g., `measurement 12` refers to the second measurement in group 1 and `measurement 24` refers to the fourth measurement in group 2).

Once all of the nodes have been sequentially added, the `tree` object can be saved using the `saveXML()` function:

```
> saveXML(tree, file="blood.xml")
```

22.4.3 Reading, Viewing, and Editing an XML Document

If you are working in the R Console, you can quickly view an XML document by loading the text into the R Editor (remember to change the "Files of type" option to "All files (*.*)"). Once in the R Editor, you can quickly add or remove nodes and then re-save the document.

You can also load the XML directly into an object of class XML doc using `xmlTreeParse()`. There are other ways and functions for creating environments to hold the XML in R, but `xmlTreeParse()` is quite practical and easy to grasp. We can create an environment for the "blood.xml" document and then access the XML via one of the slots called `doc`, which itself contains a slot called `children`. This procedure and partial output are given in Code 22.2.

```
> tree<-xmlTreeParse("xmltest.xml")
> tree$doc$children
$study
<study>
 <details>
  <title>Systolic blood pressure measurements for reduced alcohol</title>
  <researcher>Paul Lewis</researcher>
  <analyst>Nathan Hill</analyst>
```

Code 22.2.

A quicker way of accessing the XML is to apply the `xmlRoot()` function to the XML doc object returned, which will create an `xmlNode` object containing the root node and its children:

```
> tree = xmlRoot(xmlTreeParse("xmltest.xml"))
```

You can then navigate the tree using the `[[` operator, as you would for list-like objects. For example, to retrieve the node describing the study researcher, you can index the appropriate element in the tree. This node is the second child node of the first child node of the root. Thus, the index for this node is `[[1]][[2]]` (Code 22.3).

```
> researcher<-tree[[1]][[2]]
> researcher
<researcher>Paul Lewis</researcher>
```

Code 22.3.

For larger XML files, however, this system becomes impractical, as navigating nodes using `[[` can become confusing; for smaller XML documents, it is both quick and convenient.

By knowing the index of a node, we can retrieve information on child nodes or values from text nodes using a number of functions, as well as available slots for `xmlNode`. A list of child nodes can be retrieved using the children slot of an `xmlNode` object (Code 22.4).

```
> tree[[1]]$children
$title
<title>Systolic blood pressure measurements for reduced alcohol</title>

$researcher
<researcher>Paul Lewis</researcher>

$analyst
<analyst>Nathan Hill</analyst>
```

Code 22.4.

To retrieve the tag name of a node, we can use the `name` slot (Code 22.5).

```
> doc[[1]][[2]]$name
[1] "researcher"
```

Code 22.5.

To return the data or value held in a node, we use the `xmlValue()` function (Code 22.6).

```
>  xmlValue(tree[[1]][[2]])
[1] "Paul Lewis"
```

Code 22.6.

We can also apply a function such as xmlValue across a list of nodes by parsing the xmlNode object and function to xmlSApply() (Code 22.7).

```
>  xmlSApply(tree[[1]], xmlValue)
                                                      title
"Systolic blood pressure measurements for reduced alcohol"
                                                  researcher
                                               "Paul Lewis"
                                                    analyst
                                               "Nathan Hill"
```

Code 22.7.

We can use the append.xmlNode() function to add new child nodes to a node (Code 22.8).

```
doc[[2]][[2]]<-append.xmlNode(doc[[2]][[2]], xmlNode("measurement ", 151))
>  doc[[2]][[2]]
<group>
 <name>posttrial</name>
 <N>20</N>
 <measurement>134</value>
 <measurement>145</value>
 <measurement>141</value>
 <measurement>155</value>
 <measurement>151</value>
</group>
```

Code 22.8.

We have added a child node with a "measurement" tag with the next value in our list of post-trial blood pressure measurements. It is straightforward to change the value of a node using the xmlNode() function:

```
>  tree[[2]][[2]][[4]]<-xmlNode("measurement ", 160)
```

22.5 Describing Data Mining Models with PMML

One of the major reasons that the DMG introduced PMML was because one often needs to interchange models between software tools. PMML allows for models to be described and shared in a standardized way, regardless of the tool used to create it. The members of the DMG include some of the big statistical software vendors. PMML is XML-based and is the specification for a number of data mining models. It is by no means complete with regard to coverage, but already has specifications for describing some of the more common machine learning and statistical approaches, such as decision trees, support vector machines, and logistic regression.

A number of commercial software packages have facilities for outputting a data mining model in PMML format, but at the time of this writing, Weka has not implemented this facility. As far as R is concerned, efforts have begun to create packages for converting models into PMML. The PMML package has been released by Togaware, and although only a few model types have been implemented so far, this package will undoubtedly grow (http://rattle.togaware.com). Unfortunately, we built some useful models in an earlier chapter using Weka (via RWeka), but as yet there is no tool available to convert our models to PMML. And here lies the beauty of XML. As certain models are already specified in PMML and R is capable of creating and reading XML, we have at our disposal a useful tool for generating PMML to describe our models.

We will use the Weka decision tree model (Chapter 15) as an example of how to create PMML, as trees are already well specified. Before we code, however, we need to familiarize ourselves with the necessary XML specification.

22.5.1 The PMML Specification for Decision Trees

At the time of this writing, the current version of PMML is 3.2. A PMML document can contain one or more models, and the root element of the document is always of type PMML. The PMML element should contain the child elements Header, Data-Dictionary, and one for the model (List 22.4).

List 22.4 Structure of a PMML document.

```
<?xml version="1.0"?>

<PMML version="3.2">

  <Header copyright = http://cancerinformatics.swansea.ac.uk description = "J48
decision tree model"/>

  <DataDictionary/>

  <...a model...>

          ClusteringModel

          GeneralRegressionModel

      MiningModel

          NaiveBayesModel

          NeuralNetwork

          RegressionModel

          RuleSetModel

          SequenceModel

          SupportVectorMachineModel

          TextModel

          TreeModel

  <...a model.../>

</PMML>
```

Many PMML schemas specify different components of a model, and each has a number of elements. Navigating all the schemas for PMML can be a nightmare, and it is easy to get confused about what elements should be included when describing your model. For our decision tree, we will only discuss elements that should definitely be included.

The `Header` child element of the PMML root is straightforward and often only contains a brief description of the model and copyright information. The `DataDictionary` child element is used to define the variables (fields) in the dataset used to generate the model. This dictionary provides attributes and elements to list and describe, among others, the variable name and data type (`optype` = categorical, ordinal, or continuous). A data dictionary may be shared between models and need not be specific to the model described. Other elements contained within the PMML root can be `MiningBuildTask`, which describes the configuration of the training run, and `TransformationDictionary`.

Currently, 11 model types are specified in PMML 3.2, including the `TreeModel` type. The schema for the `TreeModel` element is shown in List 22.5.

List 22.5 The PMML specification for decision trees.

```
<xs:element name="TreeModel">
  <xs:complexType>
   <xs:sequence>
    <xs:element ref="Extension" minOccurs="0" maxOccurs="unbounded"/>
    <xs:element ref="MiningSchema"/>
    <xs:element ref="Output" minOccurs="0" />
    <xs:element ref="ModelStats" minOccurs="0"/>
    <xs:element ref="Targets" minOccurs="0" />
    <xs:element ref="LocalTransformations" minOccurs="0" />
    <xs:element ref="Node"/>
    <xs:element ref="ModelVerification" minOccurs="0"/>
    <xs:element ref="Extension" minOccurs="0" maxOccurs="unbounded"/>
   </xs:sequence>
   <xs:attribute name="modelName" type="xs:string" />
   <xs:attribute name="functionName" type="MINING-FUNCTION"
use="required" />
   <xs:attribute name="algorithmName" type="xs:string" />
   <xs:attribute name="missingValueStrategy" type="MISSING-VALUE-
STRATEGY" default="none"/>
   <xs:attribute name="missingValuePenalty" type="PROB-NUMBER"
default="1.0"/>
   <xs:attribute name="noTrueChildStrategy" type="NO-TRUE-CHILD-
STRATEGY" default="returnNullPrediction" />
   <xs:attribute name="splitCharacteristic" default="multiSplit">
```

```
<xs:simpleType>
  <xs:restriction base="xs:string">
    <xs:enumeration value="binarySplit"/>
    <xs:enumeration value="multiSplit"/>
  </xs:restriction>
</xs:simpleType>
</xs:attribute>
</xs:complexType>
</xs:element>
```

Not all attributes and elements are commonly used for decision tree models, but those that are will have a brief description. The `modelName`, `functionName`, and `algorithmName` attributes are useful for describing the procedure used in model building and can sit alongside the `TreeModel` tag. Another attribute that we would use is `splitCharecteristic`, which dictates whether the nodes in the tree have no children, two children (binarysplit), or more than two children (multisplit).

The `MiningSchema` element has a child called `MiningField` that lists the fields (variables) used to build the model. Each field must have a unique name attribute, and an enumeration value (usagetype) can be supplied to state whether the field is a predictor variable (active) or an outcome variable (predicted).

The next important child element of `TreeModel` is called `Node`. It is `Node` that provides us with the actual structure of the tree in PMML. The first important `Node` attribute required is `score`, which describes the predicted class when a record "chooses" that node (e.g., "alive"). Another attribute is `recordcount`, which gives the overall number of records that chose the node. One of the child elements of node is `ScoreDistribution`, which can be used to give the actual number of records predicted for a particular class.

A decision tree is constructed using nodes that ask questions about a variable's value. These questions require a Boolean answer—TRUE or FALSE. More technically speaking, each tree node contains a function that, more often than not, asks whether the value is greater than, less than, or equal to another value. This function is termed a predicate in computing speak. More precisely, this type of function is a simple predicate, as there is only a comparison of two values. If more than two values are used in the function, this is a compound predicate. A simple predicate in PMML is described using the `SimplePredicate` element and this is a child of `Node`. Sometimes one sees the tag `Predicate` used. Attributes for a predicate element include `field`, which is the variable value to be compared; `operator` (List 22.6); and `value`, which is the value to compare against.

List 22.6 Operator values allowed for a predicate.

```
<xs:attribute name="operator" use="required">
    <xs:simpleType>
      <xs:restriction base="xs:string">
        <xs:enumeration value="equal"/>
```

```
        <xs:enumeration value="notEqual"/>
        <xs:enumeration value="lessThan"/>
        <xs:enumeration value="lessOrEqual"/>
        <xs:enumeration value="greaterThan"/>
        <xs:enumeration value="greaterOrEqual"/>
        <xs:enumeration value="isMissing"/>
        <xs:enumeration value="isNotMissing"/>
      </xs:restriction>
    </xs:simpleType>
  </xs:attribute>
```

We now have enough information on the PMML elements required to build an XML document for the J48 decision tree we created in Section 15.9.

22.5.2 Creating a PMML-Specified J48 Decision Tree Model in R

The decision tree we want to describe is shown in List 22.7.

List 22.7 The J48 decision tree to predict survival in breast cancer patients.

```
nodespos <= 0
|  size <= 20: 0 (2485.0/631.0)
|  size > 20
|  |  grade = 1: 0 (43.0/13.0)
|  |  grade = 2
|  |  |  size <= 25: 0 (162.0/69.0)
|  |  |  size > 25: 1 (218.0/97.0)
|  |  grade = 3: 1 (636.0/314.0)
|  |  grade = 4
|  |  |  size <= 38: 0 (63.0/27.0)
|  |  |  size > 38: 1 (20.0/5.0)
nodespos > 0
|  nodespos <= 5
|  |  size <= 23
|  |  |  grade = 1: 0 (70.0/18.0)
|  |  |  grade = 2
|  |  |  |  size <= 19: 0 (255.0/101.0)
|  |  |  |  size > 19: 1 (131.0/56.0)
|  |  |  grade = 3: 1 (430.0/181.0)
|  |  |  grade = 4: 0 (61.0/29.0)
|  |  size > 23: 1 (1073.0/298.0)
```

| nodespos > 5: 1 (1106.0/175.0)

Number of Leaves :   14

Size of the tree :   23

The R code to build the tree is shown in Code 22.9. The most arduous part of creating the PMML document is translating the hierarchical nature of the decision tree into XML nodes. Fortunately, decision trees and XML are hierarchical structures, and the PMML node structure follows that of the nodes in the decision tree. It is important to study the tree, each node, and the relationship between parents and children. It also helps to number the nodes in order.

```
#create the tree
> tree = xmlHashTree()

#create the "PMML" root element
> pmml = addNode(xmlNode("PMML", attrs=c(version="3.2")), character(), tree)

#create Header element
> addNode(xmlNode("Header", attrs=c(copyright =
"http://cancerinformatics.swansea.ac.uk", description = "J48 decision tree
model")), pmml, tree)

#create DataDictionary element
> datadictionary = addNode(xmlNode("DataDictionary"), pmml, tree)
> addNode(xmlNode("DataField", attrs=c(name="size", optype="continuous")),
datadictionary, tree)
> addNode(xmlNode("DataField", attrs=c(name="nodepos", optype="continuous")),
datadictionary, tree)
> datafield3 = addNode(xmlNode("DataField", attrs=c(name="grade",
optype="continuous")), datadictionary, tree)
> addNode(xmlNode("Value", attrs=c(value="1")), datafield3, tree)
> addNode(xmlNode("Value", attrs=c(value="2")), datafield3, tree)
> addNode(xmlNode("Value", attrs=c(value="3")), datafield3, tree)
> addNode(xmlNode("Value", attrs=c(value="4")), datafield3, tree)
datafield4 = addNode(xmlNode("DataField", attrs=c(name="alivestatus",
optype="continuous")), datadictionary, tree)
> addNode(xmlNode("Value", attrs=c(value="0")), datafield4, tree)
> addNode(xmlNode("Value", attrs=c(value="1")), datafield4, tree)

#create TreeModel element
> treemodel = addNode(xmlNode("TreeModel", attrs=c(modelname="J48",
functionName="classification", algorithm="weka.classification.tree.j48",
splitCharacteristic="binarySplit")), pmml, tree)

#create MiningSchema element
> miningschema = addNode(xmlNode("MiningSchema"), treemodel, tree)
> addNode(xmlNode("MiningField", attrs=c(name = "size", usageType="active")),
miningschema, tree)
> addNode(xmlNode("MiningField", attrs=c(name = "nodepos",
usageType="active")), miningschema, tree)
> addNode(xmlNode("MiningField", attrs=c(name = "grade", usageType="active")),
miningschema, tree)
> addNode(xmlNode("MiningField", attrs=c(name = "alivestatus",
usageType="predicted")), miningschema, tree)

#Create Node elements
> node1 = addNode(xmlNode("Node", attrs=c(score="alive", recordCount="6952")),
treemodel, tree)
```

```
> addNode(xmlNode("True"), node1, tree)
> addNode(xmlNode("ScoreDistribution", attrs=c(value="alive",
recordCount="3476")), node1, tree)
> addNode(xmlNode("ScoreDistribution", attrs=c(value="dead",
recordCount="3476")), node1, tree)

> node2 = addNode(xmlNode("Node", attrs=c(score="alive", recordCount="4783")),
node1, tree)
> addNode(xmlNode("SimplePredicate", attrs=c( field="nodespos",
operator="greaterThan", value="0")), node2, tree)
> addNode(xmlNode("ScoreDistribution", attrs=c(value="alive",
recordCount="3169")), node2, tree)
> addNode(xmlNode("ScoreDistribution", attrs=c(value="dead",
recordCount="1614")), node2, tree)

> node3 = addNode(xmlNode("Node", attrs=c(score="alive", recordCount="3117")),
node2, tree)
> addNode(xmlNode("True"), node3, tree)
> addNode(xmlNode("ScoreDistribution", attrs=c(value="alive",
recordCount="631")), node3, tree)
> addNode(xmlNode("ScoreDistribution", attrs=c(value="dead",
recordCount="2485")), node3, tree)

> node4 = addNode(xmlNode("Node", attrs=c(score="alive", recordCount="1667")),
node2, tree)
> addNode(xmlNode("SimplePredicate", attrs=c( field="size",
operator="lessOrEqual", value="20")), node4, tree)
> addNode(xmlNode("ScoreDistribution", attrs=c(value="alive",
recordCount="684")), node4, tree)
> addNode(xmlNode("ScoreDistribution", attrs=c(value="dead",
recordCount="983")), node4, tree)

> node5 = addNode(xmlNode("Node", attrs=c(score="alive", recordCount="56")),
node4, tree)
> addNode(xmlNode("True"), node5, tree)
> addNode(xmlNode("ScoreDistribution", attrs=c(value="alive",
recordCount="13")), node5, tree)
> addNode(xmlNode("ScoreDistribution", attrs=c(value="dead",
recordCount="43")), node5, tree)

> node6 = addNode(xmlNode("Node", attrs=c(score="alive", recordCount="1611")),
node4, tree)
> addNode(xmlNode("SimplePredicate", attrs=c( field="grade", operator="equal",
value="1")), node6, tree)
> addNode(xmlNode("ScoreDistribution", attrs=c(value="alive",
recordCount="641")), node6, tree)
> addNode(xmlNode("ScoreDistribution", attrs=c(value="dead",
recordCount="970")), node6, tree)

> node7 = addNode(xmlNode("Node", attrs=c(score="alive", recordCount="546")),
node6, tree)
> addNode(xmlNode("True"), node7, tree)
> addNode(xmlNode("ScoreDistribution", attrs=c(value="alive",
recordCount="259")), node7, tree)
> addNode(xmlNode("ScoreDistribution", attrs=c(value="dead",
recordCount="287")), node7, tree)

> node8 = addNode(xmlNode("Node", attrs=c(score="dead", recordCount="315")),
node7, tree)
> addNode(xmlNode("True"), node8, tree)
> addNode(xmlNode("ScoreDistribution", attrs=c(value="alive",
recordCount="97")), node8, tree)
> addNode(xmlNode("ScoreDistribution", attrs=c(value="dead",
recordCount="218")), node8, tree)

> node9 = addNode(xmlNode("Node", attrs=c(score="alive", recordCount="231")),
node7, tree)
> addNode(xmlNode("SimplePredicate", attrs=c( field="size",
operator="greaterThan", value="25")), node9, tree)
```

```
> addNode(xmlNode("ScoreDistribution", attrs=c(value="alive",
recordCount="162")), node9, tree)
> addNode(xmlNode("ScoreDistribution", attrs=c(value="dead",
recordCount="69")), node9, tree)

> node10 = addNode(xmlNode("Node", attrs=c(score="alive", recordCount="1065")),
node6, tree)
> addNode(xmlNode("SimplePredicate", attrs=c( field="grade", operator="equal",
value="2")), node10, tree)
> addNode(xmlNode("ScoreDistribution", attrs=c(value="alive",
recordCount="382")), node10, tree)
> addNode(xmlNode("ScoreDistribution", attrs=c(value="dead",
recordCount="683")), node10, tree)

> node11 = addNode(xmlNode("Node", attrs=c(score="dead", recordCount="1050")),
node10, tree)
> addNode(xmlNode("True"), node11, tree)
> addNode(xmlNode("ScoreDistribution", attrs=c(value="alive",
recordCount="314")), node11, tree)
> addNode(xmlNode("ScoreDistribution", attrs=c(value="dead",
recordCount="636")), node11, tree)

> node12 = addNode(xmlNode("Node", attrs=c(score="alive", recordCount="115")),
node10, tree)
> addNode(xmlNode("SimplePredicate", attrs=c( field="grade", operator="equal",
value="3")), node12, tree)
> addNode(xmlNode("ScoreDistribution", attrs=c(value="alive",
recordCount="68")), node12, tree)
> addNode(xmlNode("ScoreDistribution", attrs=c(value="dead",
recordCount="47")), node12, tree)

> node13 = addNode(xmlNode("Node", attrs=c(score="dead", recordCount="25")),
node12, tree)
> addNode(xmlNode("True"), node13, tree)
> addNode(xmlNode("ScoreDistribution", attrs=c(value="alive",
recordCount="5")), node13, tree)
> addNode(xmlNode("ScoreDistribution", attrs=c(value="dead",
recordCount="20")), node13, tree)

> node14 = addNode(xmlNode("Node", attrs=c(score="alive", recordCount="90")),
node12, tree)
> addNode(xmlNode("SimplePredicate", attrs=c( field="size",
operator="greaterThan", value="38")), node14, tree)
> addNode(xmlNode("ScoreDistribution", attrs=c(value="alive",
recordCount="63")), node14, tree)
> addNode(xmlNode("ScoreDistribution", attrs=c(value="dead",
recordCount="27")), node14, tree)

> node15 = addNode(xmlNode("Node", attrs=c(score="alive", recordCount="3984")),
node1, tree)
> addNode(xmlNode("True"), node15, tree)
> addNode(xmlNode("ScoreDistribution", attrs=c(value="alive",
recordCount="1096")), node15, tree)
> addNode(xmlNode("ScoreDistribution", attrs=c(value="dead",
recordCount="2888")), node15, tree)

> node16 = addNode(xmlNode("Node", attrs=c(score="dead", recordCount="1281")),
node15, tree)
> addNode(xmlNode("True"), node16, tree)
> addNode(xmlNode("ScoreDistribution", attrs=c(value="alive",
recordCount="175")), node16, tree)
> addNode(xmlNode("ScoreDistribution", attrs=c(value="dead",
recordCount="1106")), node16, tree)

> node17 = addNode(xmlNode("Node", attrs=c(score="alive", recordCount="2703")),
node15, tree)
> addNode(xmlNode("SimplePredicate", attrs=c( field="nodespos",
operator="greaterThan", value="5")), node17, tree)
```

```
> addNode(xmlNode("ScoreDistribution", attrs=c(value="alive",
recordCount="921")), node17, tree)
> addNode(xmlNode("ScoreDistribution", attrs=c(value="dead",
recordCount="1782")), node17, tree)

> node18 = addNode(xmlNode("Node", attrs=c(score="dead", recordCount="1371")),
node17, tree)
> addNode(xmlNode("True"), node18, tree)
> addNode(xmlNode("ScoreDistribution", attrs=c(value="alive",
recordCount="298")), node18, tree)
> addNode(xmlNode("ScoreDistribution", attrs=c(value="dead",
recordCount="1073")), node18, tree)

> node19 = addNode(xmlNode("Node", attrs=c(score="alive", recordCount="1332")),
node17, tree)
> addNode(xmlNode("SimplePredicate", attrs=c( field="size",
operator="greaterThan", value="23")), node19, tree)
> addNode(xmlNode("ScoreDistribution", attrs=c(value="alive",
recordCount="623")), node19, tree)
> addNode(xmlNode("ScoreDistribution", attrs=c(value="dead",
recordCount="709")), node19, tree)

> node20 = addNode(xmlNode("Node", attrs=c(score="alive", recordCount="88")),
node19, tree)
> addNode(xmlNode("True"), node20, tree)
> addNode(xmlNode("ScoreDistribution", attrs=c(value="alive",
recordCount="70")), node20, tree)
> addNode(xmlNode("ScoreDistribution", attrs=c(value="dead",
recordCount="18")), node20, tree)

> node21 = addNode(xmlNode("Node", attrs=c(score="alive", recordCount="1244")),
node19, tree)
> addNode(xmlNode("SimplePredicate", attrs=c( field="grade", operator="equal",
value="1")), node21, tree)
> addNode(xmlNode("ScoreDistribution", attrs=c(value="alive",
recordCount="553")), node21, tree)
> addNode(xmlNode("ScoreDistribution", attrs=c(value="dead",
recordCount="691")), node21, tree)

> node22 = addNode(xmlNode("Node", attrs=c(score="alive", recordCount="1611")),
node21, tree)
> addNode(xmlNode("True"), node22, tree)
> addNode(xmlNode("ScoreDistribution", attrs=c(value="alive",
recordCount="641")), node22, tree)
> addNode(xmlNode("ScoreDistribution", attrs=c(value="dead",
recordCount="970")), node22, tree)

> node23 = addNode(xmlNode("Node", attrs=c(score="dead", recordCount="187")),
node22, tree)
> addNode(xmlNode("True"), node23, tree)
> addNode(xmlNode("ScoreDistribution", attrs=c(value="alive",
recordCount="56")), node23, tree)
> addNode(xmlNode("ScoreDistribution", attrs=c(value="dead",
recordCount="131")), node23, tree)

> node24 = addNode(xmlNode("Node", attrs=c(score="alive", recordCount="356")),
node22, tree)
> addNode(xmlNode("SimplePredicate", attrs=c( field="size",
operator="greaterThan", value="19")), node24, tree)
> addNode(xmlNode("ScoreDistribution", attrs=c(value="alive",
recordCount="255")), node24, tree)
> addNode(xmlNode("ScoreDistribution", attrs=c(value="dead",
recordCount="101")), node24, tree)

> node25 = addNode(xmlNode("Node", attrs=c(score="alive", recordCount="701")),
node21, tree)
> addNode(xmlNode("SimplePredicate", attrs=c( field="grade", operator="equal",
value="2")), node25, tree)
```

```
> addNode(xmlNode("ScoreDistribution", attrs=c(value="alive",
recordCount="242")), node25, tree)
> addNode(xmlNode("ScoreDistribution", attrs=c(value="dead",
recordCount="459")), node25, tree)

> node26 = addNode(xmlNode("Node", attrs=c(score="dead", recordCount="611")),
node25, tree)
> addNode(xmlNode("True"), node26, tree)
> addNode(xmlNode("ScoreDistribution", attrs=c(value="alive",
recordCount="181")), node26, tree)
> addNode(xmlNode("ScoreDistribution", attrs=c(value="dead",
recordCount="430")), node26, tree)

> node27 = addNode(xmlNode("Node", attrs=c(score="alive", recordCount="90")),
node25, tree)
> addNode(xmlNode("SimplePredicate", attrs=c( field="grade", operator="equal",
value="3")), node27, tree)
> addNode(xmlNode("ScoreDistribution", attrs=c(value="alive",
recordCount="61")), node27, tree)
> addNode(xmlNode("ScoreDistribution", attrs=c(value="dead",
recordCount="29")), node27, tree)
```

Code 22.9.

After we create an XMLHashTree object called tree to hold the XML, we need to create the root element, called PMML. Using the attrs argument, we set the version attribute to "3.2" to state which version of PMML we are using. Then we create the Header element, parsing attributes for copyright and description and setting the parent to the PMML element. The DataDictionary element is created in the same way, also with the parent set to PMML. This element also contains four child elements called DataField that have attributes set for name and optype. The third and fourth DataField elements have Value children that tell us the different values for the grade and alivestatus variables.

The TreeModel element is created, setting attributes for modelname, functionname, algorithm, and splitCharacteristic. The MiningSchema element, a child of TreeModel, holds MiningField elements for each of the four variables. What follows is a lengthy piece of code to create XML elements for each of the decision tree nodes. Each Node element contains the score and recordcount attributes. In total, we create 27 Node elements to represent the nodes in the tree. The xmlNode created for each Node element is coded as node1, node2, node3, etc.

The Node elements each contain three child elements. Look closely at the four lines of code associated with each Node and child elements, and you will see that there are actually two types of code for the children. At this point, let us distinguish these as two sets of code, where the first set contains a child element called True and the second set has the SimplePredicate child. Both sets of code have two ScoreDistribution elements that hold attributes describing the number of cases predicted to be alive or dead at the point the tree branches.

The two sets of code arise because of the way a decision tree is explained in PMML. As we work through our decision tree, we ask if a variable value is greater than or equal to another value. The answer is either yes or no, which leads us to two child nodes (and the next questions), or a leaf node (predicting an outcome). For example, the first node asks if the number of nodes positive is greater than 0. If it is 0, we go on to ask a question about size. But look at the way this process is structured in PMML. The first `Node` element just states that we begin with 6,952 cases in the model, where 3,476 are classed as alive and the same number classed as dead. The second node (node2) set has the same structure, but has a `SimplePredicate` element defining the question asked by the parent node. The other child node set (node3) has a `True` element instead. What we are seeing is a parent element with a branch to one child node set (which contains the question via a `SimplePredicate` element) if the answer is no and a branch to a `True` child node set if the answer is yes. The entire pattern of `Node` elements is based on just these two types of node code sets. Arrive at a node, ask the question, and go to a node set with a `True` element if the answer is yes, or go to the alternative node set with a `SimplePredicate` element if the answer is no.

For any element in these node sets, the parent is defined by the penultimate argument in the `addNode()` function. We only need to create an `xmlNode` object if the node is a parent. Otherwise, we just perform the `addNode()` function without storing the result, as the child is automatically appended to the parent specified. The PMML output is shown in Code 22.10.

```
> tree
<PMML version="3.2">
 <Header copyright="http://cancerinformatics.swansea.ac.uk" description="J48
decision tree model"/>
 <DataDictionary>
  <DataField name="size" optype="continuous"/>
  <DataField name="nodepos" optype="continuous"/>
  <DataField name="grade" optype="continuous">
  <Value value="1"/>
  <Value value="2"/>
  <Value value="3"/>
  <Value value="4"/>
  </DataField>
  <DataField name="alivestatus" optype="continuous">
  <Value value="0"/>
  <Value value="1"/>
  </DataField>
 </DataDictionary>
 <TreeModel modelname="J48" functionName="classification"
algorithm="weka.classification.tree.j48" splitCharacteristic="binarySplit">
  <MiningSchema>
   <MiningField name="size" usageType="active"/>
   <MiningField name="nodepos" usageType="active"/>
   <MiningField name="grade" usageType="active"/>
   <MiningField name="alivestatus" usageType="predicted"/>
  </MiningSchema>
  <Node score="alive" recordCount="6952">
  <True/>
  <ScoreDistribution value="alive" recordCount="3476"/>
  <ScoreDistribution value="dead" recordCount="3476"/>
  <Node score="alive" recordCount="4783">
   <SimplePredicate field="nodespos" operator="greaterThan" value="0"/>
   <ScoreDistribution value="alive" recordCount="3169"/>
   <ScoreDistribution value="dead" recordCount="1614"/>
   <Node score="alive" recordCount="3117">
    <True/>
    <ScoreDistribution value="alive" recordCount="631"/>
    <ScoreDistribution value="dead" recordCount="2485"/>
   </Node>
```

```
<Node score="alive" recordCount="1667">
 <SimplePredicate field="size" operator="lessOrEqual" value="20"/>
 <ScoreDistribution value="alive" recordCount="684"/>
 <ScoreDistribution value="dead" recordCount="983"/>
 <Node score="alive" recordCount="56">
  <True/>
  <ScoreDistribution value="alive" recordCount="13"/>
  <ScoreDistribution value="dead" recordCount="43"/>
 </Node>
 <Node score="alive" recordCount="1611">
  <SimplePredicate field="grade" operator="equal" value="1"/>
  <ScoreDistribution value="alive" recordCount="641"/>
  <ScoreDistribution value="dead" recordCount="970"/>
  <Node score="alive" recordCount="546">
   <True/>
   <ScoreDistribution value="alive" recordCount="259"/>
   <ScoreDistribution value="dead" recordCount="287"/>
   <Node score="dead" recordCount="315">
    <True/>
    <ScoreDistribution value="alive" recordCount="97"/>
    <ScoreDistribution value="dead" recordCount="218"/>
   </Node>
   <Node score="alive" recordCount="231">
    <SimplePredicate field="size" operator="greaterThan" value="25"/>
    <ScoreDistribution value="alive" recordCount="162"/>
    <ScoreDistribution value="dead" recordCount="69"/>
   </Node>
  </Node>
  <Node score="alive" recordCount="1065">
   <SimplePredicate field="grade" operator="equal" value="2"/>
   <ScoreDistribution value="alive" recordCount="382"/>
   <ScoreDistribution value="dead" recordCount="683"/>
   <Node score="dead" recordCount="1050">
    <True/>
    <ScoreDistribution value="alive" recordCount="314"/>
    <ScoreDistribution value="dead" recordCount="636"/>
   </Node>
   <Node score="alive" recordCount="115">
    <SimplePredicate field="grade" operator="equal" value="3"/>
    <ScoreDistribution value="alive" recordCount="68"/>
    <ScoreDistribution value="dead" recordCount="47"/>
    <Node score="dead" recordCount="25">
     <True/>
     <ScoreDistribution value="alive" recordCount="5"/>
     <ScoreDistribution value="dead" recordCount="20"/>
    </Node>
    <Node score="alive" recordCount="90">
     <SimplePredicate field="size" operator="greaterThan" value="38"/>
     <ScoreDistribution value="alive" recordCount="63"/>
     <ScoreDistribution value="dead" recordCount="27"/>
    </Node>
   </Node>
  </Node>
 </Node>
</Node>
</Node>
<Node score="alive" recordCount="3984">
 <True/>
 <ScoreDistribution value="alive" recordCount="1096"/>
 <ScoreDistribution value="dead" recordCount="2888"/>
 <Node score="dead" recordCount="1281">
  <True/>
  <ScoreDistribution value="alive" recordCount="175"/>
  <ScoreDistribution value="dead" recordCount="1106"/>
 </Node>
 <Node score="alive" recordCount="2703">
  <SimplePredicate field="nodespos" operator="greaterThan" value="5"/>
  <ScoreDistribution value="alive" recordCount="921"/>
```

```
    <ScoreDistribution value="dead" recordCount="1782"/>
    <Node score="dead" recordCount="1371">
     <True/>
     <ScoreDistribution value="alive" recordCount="298"/>
     <ScoreDistribution value="dead" recordCount="1073"/>
    </Node>
    <Node score="alive" recordCount="1332">
     <SimplePredicate field="size" operator="greaterThan" value="23"/>
     <ScoreDistribution value="alive" recordCount="623"/>
     <ScoreDistribution value="dead" recordCount="709"/>
     <Node score="alive" recordCount="88">
      <True/>
      <ScoreDistribution value="alive" recordCount="70"/>
      <ScoreDistribution value="dead" recordCount="18"/>
     </Node>
     <Node score="alive" recordCount="1244">
      <SimplePredicate field="grade" operator="equal" value="1"/>
      <ScoreDistribution value="alive" recordCount="553"/>
      <ScoreDistribution value="dead" recordCount="691"/>
      <Node score="alive" recordCount="1611">
       <True/>
       <ScoreDistribution value="alive" recordCount="641"/>
       <ScoreDistribution value="dead" recordCount="970"/>
       <Node score="dead" recordCount="187">
        <True/>
        <ScoreDistribution value="alive" recordCount="56"/>
        <ScoreDistribution value="dead" recordCount="131"/>
       </Node>
       <Node score="alive" recordCount="356">
        <SimplePredicate field="size" operator="greaterThan" value="19"/>
        <ScoreDistribution value="alive" recordCount="255"/>
        <ScoreDistribution value="dead" recordCount="101"/>
       </Node>
      </Node>
      <Node score="alive" recordCount="701">
       <SimplePredicate field="grade" operator="equal" value="2"/>
       <ScoreDistribution value="alive" recordCount="242"/>
       <ScoreDistribution value="dead" recordCount="459"/>
       <Node score="dead" recordCount="611">
        <True/>
        <ScoreDistribution value="alive" recordCount="181"/>
        <ScoreDistribution value="dead" recordCount="430"/>
       </Node>
       <Node score="alive" recordCount="90">
        <SimplePredicate field="grade" operator="equal" value="3"/>
        <ScoreDistribution value="alive" recordCount="61"/>
        <ScoreDistribution value="dead" recordCount="29"/>
       </Node>
      </Node>
     </Node>
    </Node>
   </Node>
  </Node>
 </Node>
</TreeModel>
</PMML>
```

Code 22.10.

We do not have to stop at just describing the structure of the tree. We can use an element called Extension as a child of TreeModel to describe the confusion matrix and other aspects of model validation. There are also PMML elements for univariate statistics and graph descriptions.

22.6 Summary

In this chapter we have introduced XML and how one can create and process XML using R. There are a number of reasons for processing XML in R, particularly if there is a need to store data, statistics, and data mining models in a standardized way. PMML is an XML-based markup language that has been developed for describing data mining models. Describing models in such a standardized way permits sharing of models between software packages and researchers. Because XML uses metadata tags to describe information, it is ideally suited for biomedical research so that other researchers can understand and interpret the results of a study.

22.7 Questions

1. How does XML facilitate both storage and sharing of biomedical data?

2. Why do we require data standards in research, and how does XML benefit the notion of standardized data?

3. What do we mean by the term "metadata"?

4. Why do you think PMML is important to biomedicine?

5. What is the difference between an XML element and an XML node?

6. Do you think that XML would be useful for describing any of the other datasets presented in this book?

References

1. R Development Core Team (2008). *R: A language and environment for statistical computing*. R Foundation for Statistical Computing, Vienna, Austria. ISBN 3-900051-07-0, URL: http://www.R-project.org

2. Ripley BD (2001). Using databases with R. R News, 1(January),18–20, URL: http://cran.r-project.org/doc/Rnews/

3. Department of Health and Human Services (2000). 45 CFR (Code of Federal Regulations), Parts 160 to 164. Standards for Privacy of Individually Identifiable Health Information (Final Rule). *Federal Register*, 65(250), 82461–82510, URL: http://aspe.hhs.gov/adminsimp/final/PvcPre01.htm

4. Department of Health and Human Services (1991). 45 CFR (Code of Federal Regulations), 46. Protection of Human Subjects (Common Rule). *Federal Register*, 56(250), 28003-28032, URL: http://www.hhs.gov/ohrp/humansubjects/guidance/45cfr46.htm

5. Vollset SE, Tverdal A, and Gjessing HK (2006). Smoking and deaths between 40 and 70 years of age in women and men. *Ann Intern Med*, 144, 381–389.

6. Hair JF, Anderson RE, Tatham RL, and Black WC (1995). Multivariate data analysis. Prentice Hall, New Jersey, pp. 364–483.

7. Pielou EC (1975). *Ecological Diversity*. Wiley & Sons.

8. Jackson DA (1993). Stopping rules in principal components analysis: a comparison of heuristical and statistical approaches. *Ecology,* 74, 2204–2214.

9. Kaufman L and Rousseeuw PJ (1990). *Finding Groups in Data: An Introduction to Cluster Analysis*. Wiley, New York.

10. Blamey RW, Davies CJ, Elston, CW, Johnson J, Haybittle JL, and Maynard, PV (1979). Prognostic factors in breast cancer—the formation of a Prognostic Index. *Clin Oncol*, 5, 227–236.

11. Haybittle JL, Blamey RW, Elston CW, Johnson J, Doyle PJ, and Campbell FC (1982). A Prognostic Index in primary breast cancer. *Br J Cancer*, 45, 361–366.

12. Blamey RW, Ellis IO, Pinder SE, Lee AH, Macmillan RD, Morgan DA, Robertson JF, Mitchell MJ, Ball GR, Haybittle JL, and Elston CW (2007). Survival of invasive breast cancer according to the Nottingham Prognostic Index in cases diagnosed in 1990–1999. *Eur J Cancer*, 43, 1548–1555.

13. van 't Veer LJ, Dai H, van de Vijver MJ, He YD, Hart AA, Mao M, Peterse HL, van der Kooy K, Marton MJ, Witteveen AT, et al. (2002). Gene expression profiling predicts clinical outcome of breast cancer. *Nature*, 415, 530–536.

14. Spiess AN, Feig C, Schulze W, Chalmel F, Cappallo-Obermann H, Primig M, and Kirchhoff C (2007). Cross-platform gene expression signature of human

spermatogenic failure reveals inflammatory-like response. *Hum Reprod.*, 22, 2936–2946.

15. Chen HY, Yu SL, Chen CH, Chang GC, Chen CY, Yuan A, Cheng CL, Wang CH, Terng HJ, Kao SF, Chan WK, Li HN, Liu CC, Singh S, Chen WJ, Chen JJ, and Yang PC (2007) A five-gene signature and clinical outcome in non-small-cell lung cancer. *N Engl J Med,* 356, 11–20.

16. Delen D, Walker G, and Kadam A (2005). Predicting breast cancer survivability: a comparison of three data mining methods. *Artif Intell Med,* 34, 113–127.

17. Venkataraman G, Heinze G, Holmes EW, Ananthanarayanan V, Bostwick DG, Paner GP, Bradford-De La Garza CM, Brown HG, Flanigan RC, and Wojcik EM (2007). Identification of patients with low-risk for aneuploidy: comparative discriminatory models using linear and machine-learning classifiers in prostate cancer. *Prostate*, 67, 1524–1536.

18. Polat K, Yosunkaya S, and Güneş S (2008). Comparison of different classifier algorithms on the automated detection of obstructive sleep apnea syndrome. *J Med Syst*, 32, 243–250.

19. Edén P, Ritz C, Rose C, Fernö M, and Peterson C (2004). "Good old" clinical markers have similar power in breast cancer prognosis as microarray gene expression profilers. *Eur J Cancer*, 40, 1837–1841.

20. Witten IH and Frank E (2005). *Data Mining: Practical Machine Learning Tools and Techniques,* 2nd ed, Morgan Kaufmann, San Francisco.

21. Kubat M, Holte RC, and Matwin S (1998). Machine learning for the detection of oil spills in satellite radar images. *Machine Learning*, 30, 2–3.

22. Vapnik V (1998). *Statistical Learning Theory*. Wiley, New York.

23. Breiman L (2001). Random Forests. *Machine Learning*, 45, 5–32.

24. Foresight (2006). Infectious diseases: preparing for the future. Executive Summary. Office of Science and Innovation, London.

25. Paquet C, Coulombier D, Kaiser R, and Ciotti M (2006). Epidemic intelligence: a new framework for strengthening disease surveillance in Europe. *Eurosurveillance*, 11, Issue 12.

26. Skowronski DM, Astell C, Brunham RC, Low DE, Petric M, Roper RL, Talbot PJ, Tam T, and Babiuk L (2005). Severe acute respiratory syndrome (SARS): a year in review. *Annu Rev Med*, 56, 357–381.

27. Elliott A. (2005). Medical imaging. *Nuclear Instruments and Methods in Physics Research A*, 546, 1–13.

28. Editorial (2005). Medical imaging informatics research and development trends—an editorial. *Computerized Medical Imaging and Graphics*, 29, 91–93.

29. Graham RNJ, Perriss RW, and Scarsbrook AF (2005). DICOM demystified: a review of digital file formats and their use in radiological practice. *Clinical Radiology*, 60, 1133–1140.

30. Dale A, Liu A, Fischl B, Buckner R, Belliveau J, Lewine J, and Halgren E (1995). Dynamic statistical parametric mapping combining fMRI and MEG for high-resolution imaging of cortical activity. *Neuron*, 26, 55–67.

31. Tabelow K, Polzehl J, Voss HU, and Spokoiny V (2006). Analyzing fMRI experiments with structural adaptive smoothing procedures. *Neuroimage*, 15, 55–62.

32. Worsley KJ, Liao CH, Aston J, Petre V, Duncan GH, Morales F, and Evans AC (2002). A general statistical analysis for fMRI data. *NeuroImage*, 15, 1–15.

33. Calhoun VD and Adali T (2006). Unmixing fMRI with independent component analysis. *Engineering in Medicine and Biology Magazine*, 25, 79–90.

34. McKeown MJ, Hansen LK, and Sejnowsk TJ (2003). Independent component analysis of functional MRI: what is signal and what is noise? *Curr Opin Neurobiol*, 13, 620–629.

35. Brassen S, Weber-Fahr W, Sommer T, Lehmbeck JT, and Braus DF (2006). Hippocampal–prefrontal encoding activation predicts whether words can be successfully recalled or only recognized. *Behav Brain Res*, 171, 271–278.

36. Gentleman RC, Carey VJ, Bates DM, et al. (2004). Bioconductor: open software development for computational biology and bioinformatics. *Genome Biology*, 5, R80.

37. Pepper SD, Saunders EK, Edwards LE, Wilson CL, and Miller CJ (2007). The utility of MAS5 expression summary and detection call algorithms. *BMC Bioinformatics*, 8, 273.

38. Seo J and Hoffman EP (2006). Probe set algorithms: is there a rational best bet? *BMC Bioinformatics*, 7, 395.

39. Barrett T and Edgar R (2006). Gene expression omnibus (GEO): microarray data storage, submission, retrieval, and analysis. *Methods Enzymol*, 411, 352–369.

40. Gautier L, Cope L, Bolstad BM, and Irizarry RA (2004). Affy-analysis of Affymetrix GeneChip data at the probe level. *Bioinformatics*, 20, 307–315.

41. Wilson CL and Miller CJ (2005). Simpleaffy: a Bioconductor package for Affymetrix quality control and data analysis. *Bioinformatics*, 21, 3683–3685.

42. The Gene Ontology Consortium (2000). Gene ontology: tool for the unification of biology. *Nature Genetics*, 25, 25–29.

43. Lengauer C, Kinzler KW, and Vogelstein B. (1998). Genetic instabilities in human cancers. *Nature*, 396, 643–649.

44. Knudson AG (1971). Mutation and cancer: statistical study of retinoblastoma. *Proc Natl Acad Sci USA*, 68, 820–823.

45. Kallioniemi A, Kallioniemi OP, Sudar D, Rutovitz D, Gray JW, Waldman F, and Pinkel D (1992). Comparative genomic hybridization for molecular cytogenetic analysis of solid tumors. *Science*, 258, 818–821.

46. Weiss MM, Hermsen MA, Meijer GA, van Grieken NC, Baak JP, Kuipers EJ, and van Diest PJ (1999). Comparative genomic hybridisation. *Mol Pathol*, 52, 243–251.

47. Mantripragada KK, Buckley PG, Diaz de Stahl T, and Dumanski JP (2004). Genomic microarrays in the spotlight. *Trends in Genetics*, 20, 87–93.

48. Wicker N, Carles A, Mills IG, Wolf M, Veerakumarasivam A, Edgren H, Boileau F, Wasylyk B, Schalken JA, Neal DE, Kallioniemi O, and Poch O (2007). A new look towards BAC-based array-CGH through a comprehensive comparison with oligo-based array CGH. *BMC Genomics*, 8, 84.

49. Jain AN, Tokuyasu TA, Snijders AM, Segraves R, Albertson DG, and Pinkel D (2002). Fully automatic quantification of microarray image data. *Genome Research*, 12, 325–332.

50. Stransky N, Vallot C, Reyal F, Bernard-Pierrot I, de Medina SG, Segraves R, de Rycke Y, Elvin P, Cassidy A, Spraggon C, Graham A, Southgate J, Asselain B, Allory Y, Abbou CC, Albertson DG, Thiery JP, Chopin DK, Pinkel D, and Radvanyi F (2006). Regional copy number-independent deregulation of transcription in cancer. *Nat Genet*, 38, 1386–1396.

51. Blaveri E, Brewer JL, Roydasgupta R, Fridlyand J, DeVries S, Koppie T, Pejavar S, Mehta K, Carroll P, Simko JP, and Waldman FM (2005). Bladder cancer stage and outcome by array-based comparative genomic hybridization. *Clin Cancer Res*, 11, 7012–7022.

52. Piwowar HA, Day RS, and Fridsma DB (2007). Sharing detailed research data is associated with increased citation rate. *PLoS ONE* 2,e308.

53. Berman JJ (2007). *Biomedical Informatics,* Jones and Bartlett, Sudbury, MA.

Appendix

A.1 Statistical Hypothesis Testing

In biomedical research, we are nearly always concerned with finding some new factor that can benefit the health of the population in some way. The population in this sense could be everybody in the country or just patients with a subtype of a disease. We cannot use an entire population for our research, so we take samples where we generate or collect data from a finite number of eligible people. We always hope that our sample represents the entire population for whatever variable we may be interested in. Parameters that we measure in our study can be of many different types, such as the mean or the median.

When we make measurements on a change in a population (e.g., how patients respond after a new drug therapy), whether it be using one, two, or more samples, we need to be sure that the change we see is truly due to the actual variable(s) we assume are causing the effect. Alternatively, the change may simply be due to chance.

In our research, we usually have a "hypothesis" that some independent variable of interest appears to have an effect. We then set out to measure some parameter derived from the variable (e.g., a mean) to see if it differs within a sample of people or between different samples (e.g., blood pressure for patients given a placebo vs patients treated with a new drug). To actually test the difference with a degree of confidence, we need to apply statistical methods to our sample data. To do this we specify a null hypothesis (H_0) about the parameters we are measuring. The null hypothesis is usually stated as there being no difference between the value of the parameter in one sample relative to that in one or more other samples. Indeed, the null hypothesis is often opposite to what your own hypothesis might be, referred to as the alternative hypothesis (H_1). Take the example in Chapter 11 where we were interested in whether there was a difference in mean tumor size between old and young breast cancer patients. We would state our null and alternative hypotheses as:

H_0: There is **NO** difference in mean tumor size between breast cancer patients age 70 or over and age 45 or under.

H_1: There **IS** a difference in mean tumor size between breast cancer patients age 70 or over and age 45 or under.

A statistical test will generate a test statistic, which is then used to test the viability of the null hypothesis. We did not state in our alternative hypothesis whether we believed that tumor size was greater or smaller in one group relative to the other. In this situation, we apply a two-tailed test to allow for a difference to be detected in either direction. If we hypothesized that tumor size was greater in one group relative to the other, we would have to apply a one-tailed test. Once the statistical test is performed and a test statistic calculated, we calculate a probability value (P Value), which is the probability of generating our data if the null hypothesis were true. Different test statistics have probability distributions, which are used to calculate a P Value given the value of the test statistic. Usually, if the P Value is less than levels of 0.05 or 0.01, we reject the null hypothesis and say that there is a significant difference between the parameters we are measuring across groups.

What follows is a brief introduction to inferential statistical tests. The purpose is to provide some general background to the statistical tests covered in Chapter 11. The order of the tests are set depending on whether the data is numerical or categorical and whether you have one, two, or more samples in your dataset. Before we look at each test, we will begin with a brief introduction to probability distributions, particularly the normal distribution, as well as summarize commonly used parameters used to provide measurements for our data.

It is best if you refer to the application of each test in Chapter 11 as you work through the math in each case. Although some of the test descriptions that follow contain additional example data, others (such as the t-test) do not, as it is simple to take the data from the corresponding section in Chapter 11 and plug it into the equations and tables. A single appendix is no substitute for a statistical textbook, and the reader is strongly advised to have one on hand when analyzing data.

A.2 A Brief Introduction to Probability Distributions

The probability distribution for any discrete variable tells us the probability with which any value can occur. You may often see a probability distribution referred to as a probability function. Probability distributions are often visualized using either a table or graph (or sometimes via an equation). Perhaps the most important of all probability distributions is the normal distribution.

Consider the following example. A doctor collects data for the number of cigarettes smoked over the last 24 hours by 50 smokers who have attended his surgery. He lists the number of cigarettes smoked by each patient as an ordered vector:

[0,1,1,2,3,3,4,4,4,5,5,5,6,6,7,7,8,8,8,8,9,9,9,9,10,10,10,10,10,11,11,11,12,12,12,12, 14,14,14,14,15,15,15,15,16,17,18,18,19,20]

He creates categories for numbers of cigarettes smoked and presents the data as a table, as shown in Table A.1.

Thus, you can think of a probability distribution for a discrete variable as being a list of probabilities for each of the possible values that the variable can be. Table A.1 also shows the cumulative probability distribution, which is calculated by adding the preceding probabilities as we move down the table.

The cumulative probability distribution can be quite useful. For instance, we may be interested in knowing the probability that a patient smokes at least 11 or 12 cigarettes per day. This also brings us to the concept of percentile ranks. A patient who smoked 18 cigarettes within the last day smoked more than or the same amount as the

Table A.1   Probability Distribution of Number of Cigarettes Smoked by Smokers at Clinic.

Cigarettes smoked	Frequency of occurrence	Probability of number of cigarettes smoked	Cumulative probability
0	1	1/50 (0.02)	1/50 (0.02)
1–2	3	3/50 (0.06)	4/50 (0.08)
3–4	5	5/50 (0.10)	9/50 (0.18)
5–6	5	5/50 (0.10)	14/50 (0.28)
7–8	6	6/50 (0.12)	20/50 (0.40)
9–10	9	9/50 (0.18)	29/50 (0.58)
11–12	7	7/50 (0.14)	36/50 (0.72)
13–14	4	4/50 (0.08)	40/50 (0.80)
15–16	5	5/50 (0.10)	45/50 (0.90)
17–18	3	3/50 (0.06)	48/50 (0.96)
20	2	2/50 (0.04)	50/50 (1.00)

number of cigarettes smoked by 96% of the remaining patients. This person is said to fall into the 96th percentile.

The doctor could also represent the probability distribution graphically, as shown as a bar plot in Figure A.1.

Probability distributions commonly used in statistics are bionomial, Poisson, and, as we shall see in Section A.4, normal.

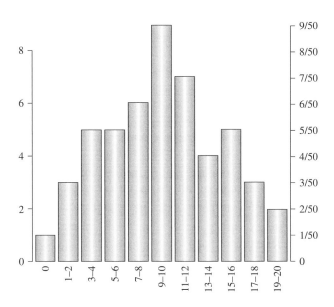

FIGURE A.1   Bar plot representation of a probability distribution.

A.3 Mean, Median, and Variance

Simple statistics, such as the mean and median, allow you to describe and summarize your data. Typically, the mean and median are used as measures of central tendency of the data distribution (i.e., the middle of the distribution). In the following sections, when we talk about the population, we mean the entire population of people (e.g., all smokers in the United Kingdom) that we can take a sample (e.g., all smokers who visit a single doctor's surgery in the United Kingdom) of values from.

A.3.1 Mean

The arithmetic mean (or average) of a sample is a simple statistic to calculate. You just need to add up all the values in the sample and divide by the number of values:

$$\mu = \Sigma X/N$$

where μ is referred to as the population mean, X is the values in the sample, and N is the number of values. More generally, if the scores come from a sample representing the population, the symbol M replaces the actual population mean μ.

The mean is useful if the data distribution is roughly asymmetric, but must be treated with caution if the distribution is skewed toward high or low values.

The mean number of cigarettes smoked by patients in the example in Section A.2 is 9.72.

A.3.2 Median

A more useful measurement in skewed distributions is the median. If the number of values for a sample is odd, the median is the middle value of the ordered numbers. For example, the median of values (2, 4, 5, 8, 10) is 5. The median of a sample with an even number of values is calculated by taking the mean of the middle two numbers. For example, the median of values (2, 4, 5, 8, 10, 12) is 6.5 because 6.5 lies halfway between 5 and 8.

A.3.3 Variance

Sometimes just knowing the central tendency of a distribution is not enough. We often need to understand how our data is spread out across the distribution. Take the example of cigarettes smoked in Section A.2. We already know that the mean number of cigarettes smoked is 9.72, but we may want to know whether the majority of smokers actually smoke a number of cigarettes close to the mean or whether there is quite a spread in number. In other words, how much variability (variance) is there in the number of cigarettes smoked by these patients.

The variance is calculated as the average squared deviation from the mean of all values. The general formula is written as:

$$\sigma^2 = \frac{\Sigma(X - \mu)^2}{N}$$

The general formula for variance calculated for a sample from the population with the sample mean is:

$$S^2 = \frac{\Sigma(X - M)^2}{N}$$

The parameter S^2 is said to be biased in the sense that it is often an overestimate or underestimate of the true population variance. The formula for an unbiased estimate of variance is:

$$S^2 = \frac{\Sigma(X - M)^2}{N - 1}$$

For the smoking example, we would calculate the variance (S^2) as:

$$S^2 = \frac{0 + (1 - 9.72)^2 + (1 - 9.72)^2 + (2 - 9.72)^2 + (3 - 9.72)^2 + ... (20 - 9.72)^2}{50 - 1} = 25.51$$

The square root of the variance is called the standard deviation (σ), which is a useful parameter when dealing with data that follows a normal distribution. Like the variance, the standard deviation provides an estimate (S) of the spread of values in a sample. The standard deviation for our example is 5.05 (i.e., $\sqrt{25.51}$).

A.4 The Normal Distribution and Standard Deviation

Normal distributions model continuous variables. Textbooks will tell you that a continuous variable with values between $-\infty$ and ∞ follows a normal distribution if the values conform to the probability density function:

$$f(x) = \frac{1}{\sigma\sqrt{2\pi}} \times \exp\{-0.5[(x - \mu)/\sigma]^2\}$$

where x represents the values in the sample, μ is the mean, and σ is the standard deviation.

If we plot the probability function for a range of values, we see that it has a "bell shape" (Figure A.2). There is actually a whole family of normal distributions whose curves all show variations of the characteristic bell shape. If we look again at the probability density function, we see that it is defined by the mean (μ) and standard deviation (σ). The most simple normal distribution has a mean of 0 and a standard deviation of 1.

The normal distribution is important because it describes many real-life biological variables.

A.4.1 Standard Deviation

The standard deviation of a normal distribution is important because you can use it to calculate the percentile rank of a particular score. Approximately 68% of scores fall

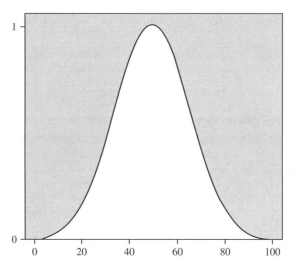

FIGURE A.2   Bell shape normal curve for the normal distribution.

within one standard deviation of the mean (Figure A.3). We also see that about 96% of scores fall within two standard deviations and about 99.7% of scores within three standard deviations. To calculate the percentile rank of a score, we first need to work out a value called the standard score:

for a population: $Z = \dfrac{x - \mu}{S}$ or for a sample: $Z = \dfrac{x - M}{S}$

Then we can use a conversion table (found in most statistical textbooks and on websites) to find the percentile of our score. If we calculate a standard score for 18 cigarettes smoked:

$Z = (18 - 9.72)/5.05 = 1.64$

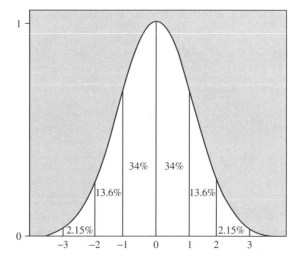

FIGURE A.3   Normal curve showing standard deviations.

We can look this up in a table and see that the percentile is roughly 95%, which approximates to the score of 96% we saw in Section A.2 when we simply looked this value up in Table A.1.

A.4.2 Standard Error of the Mean

The standard deviation also allows us to calculate the standard error of the mean. The standard error of any statistic is an estimate of the error of that particular sample statistic. To calculate the standard error (SE) of the mean, we just divide the sample standard deviation by the square root of the total number of values in the sample:

$$SE_M = \frac{S}{\sqrt{N}}$$

For the smoking example, the standard error is 0.71 ($5.05/\sqrt{50}$).

A.4.3 Ninety-Five Percent Confidence Interval for the Mean

A confidence interval (CI) is the range of scores that should contain a statistic of interest with a given degree of certainty. If we have a 95% confidence interval for our sample mean, we are 95% certain that the mean lies within the upper and lower limits of this range.

We can calculate the upper and lower limits of the range from the mean and standard error. Previously, we mentioned that plus or minus 2 standard deviations from the mean covers roughly 96% of the data, but it is known that plus or minus 1.96 standard deviations covers 95%. To obtain the upper and lower 95% confidence limits for our mean, we just need to multiply the standard error of the mean by 1.96 and then add or subtract this value from the mean:

$$95\% \ CI = M \pm (1.96 \times SE)$$

For our smoking sample, the lower 95% CI is 8.33 ($9.72 - 1.96 \times 0.71$) and the upper 95% CI is 11.11 ($9.72 + 1.96 \times 0.71$).

A.5 Numerical Data: Statistical Tests for Differences Between Groups

Table A.2 details the statistical tests outlined in this appendix, with a brief description of when to use each test and the required R function.

A.5.1 One-Sample Tests

One-Sample *T*-Test

The *t*-tests are a family of parametric statistical methods that compare the means of two groups in relation to the variance. They can be applied if the population from which the samples are derived follows a normal distribution. The type of *t*-test you apply to your data will depend on whether you have one or two samples, equal or unequal variance in two samples, or whether two samples have data that are paired or

Table A.2   Summary of Statistical Tests, When They Should Be Applied, and Corresponding R Functions.

Test	When to use	R function
T-test: one sample	Compare mean from a sample against population mean where the data is numerical and normally distributed.	`t.test(x, mu)`
Wilcoxon signed rank test: one sample	Compare median from a sample against population median where data is numerical but not normally distributed.	`wilcox.test(x, mu)`
T-test: two sample, unpaired, equal variance	Compare means from two groups where data is numerical and normally distributed. Groups are unpaired and the variance is assumed to be equal.	`t.test(x,y,var.equal = TRUE)`
T-test: two sample, unpaired, unequal variance	Compare means from two groups where data is numerical and normally distributed. Groups are not paired and the variance is unknown or known to be unequal.	`t.test(x, y)`
T-test: two sample, paired	Compare means from two groups where data is numerical and normally distributed. Corresponding values in each group are paired.	`t.test(x,y,paired=TRUE)`
Mann-Whitney U-test	Compare groups where data is numerical but not normally distributed. Nonparametric equivalent to unpaired, unequal variance t-test and uses ranking of values.	`wilcox.test(x, y)`
F-test	Compare variances between groups where data is numerical and normally distributed.	`var.test(x,y)`
ANOVA	Compare means of two or more samples where data is numerical and normally distributed.	`anova(lm(formula))`
Kruskal-Wallis rank sum test	Compare two or more samples where data is numerical but not normally distributed. A nonparametric alternative to ANOVA.	`kruskal.test(list())`
Chi-square test	Test for association between two factors. Data is categorical and presented as a contingency table.	`chisq.test(x)`
Fisher's exact test	Test for association between two factors with small sample size. Data is categorical and presented as a contingency table.	`fisher.test(x)`
Pearson's correlation	Measure the association between variables for a sample where data is numerical and normally distributed.	`cor.test(x,y)`
Spearman's rank correlation	Measure the association between variables for a sample where data is not normally distributed. A nonparametric alternative to Pearson's correlation.	`cor.test (x,y, method="spearman")`
Linear regression	Test for whether a linear relationship exists between two variables and generate a linear model for predicting values of one variable given the value of another. Data should be normally distributed.	`lm(formula)`
Cohen's kappa test	Measure the level of agreement between two raters. Data is categorical and presented in a contingency table.	`kappa2(x)`
Odds ratio	Find the ratio of odds of a condition in one group compared to another.	`oddsratio(x)`
Relative risk	Compare the risk of an outcome in one group relative to another.	`riskratio(x)`

unpaired. A *t*-statistic is calculated, which is then tested for significance according to the number of degrees of freedom and the significance level (α).

The simplest of *t*-tests to calculate is the one-sample *t*-test. The one-sample *t*-test is generally used to test for difference between a sample mean and the known (or hypothesized) mean of the population:

$$t = \frac{\bar{x} - \mu}{S/\sqrt{n}}$$

S is the standard deviation of the sample, and $\bar{x} - \mu$ calculates the difference between the mean of the sample ($\bar{x}$) and the population mean (μ). The degrees of freedom for t in this case are $n - 1$.

Wilcoxon Signed Rank Test

The Wilcoxon signed rank test can be used to compare the median of a sample with the median of the population. It does not assume a normal distribution of the data and is thus considered a nonparametric test. The data should be ordinal, interval, or ratio, but not nominal. The data should also be continuous and symmetric around the median.

Before we calculate the test statistic z, we perform the following tasks:

- Remove any values in our sample that are equal to the population median (with which we wish to compare our sample median λ).
- Count up the values in our sample greater than λ.
- Count up the values less than λ.
- Make r equal to the lowest value of the two counts (n).

We can then calculate z as:

$$z = \frac{|r - (n/2)| - 0.5}{\sqrt{n/2}}$$

Notice here that we are taking the absolute value of the difference between r and half the selected count. If we have 10 or fewer values in our sample, r actually becomes the test statistic.

A.5.2 Two-Sample Tests

T-Test for Two Unpaired Samples with Equal Variance

You can use this test if you have two samples with normally distributed data where it is known that the variance in each sample is the same. There are two variations of this test, depending on whether the sample sizes are the same or different. If the sample sizes are equal, the *t*-statistic can be calculated as follows:

$$t = \frac{\bar{x}_1 - \bar{x}_2}{S_{x_1 x_2} \times \sqrt{(2/n)}} \quad \text{where: } S_{x_1 x_2} = \frac{\sqrt{(S^2_{x_1} + S^2_{x_2})}}{2}$$

Here we are calculating the ratio of the difference in means of the two groups ($\bar{x}_1 - \bar{x}_2$) to the standard error of the difference between the means ($S_{x_1 x_2}$ is the pooled standard deviation for both groups, and n is the total number of values across both groups).

If the sample sizes are different, t is calculated as:

$$t = \frac{\bar{x}_1 - \bar{x}_2}{S_{x_1 x_2} \times \sqrt{(1/n_1 + 1/n_2)}} \quad \text{where: } S_{x_1 x_2} = \frac{\sqrt{(n_1 - 1) \times S^2_{X_1} + (n_2 - 1) \times S^2_{X_2}}}{n_1 + n_2 - 2}$$

$S_{x_1 x_2}$ is an unbiased estimator of the variance based upon the product of the variance (S^2_x) and degrees of freedom ($n_x - 1$) for each group, as well as the total degrees of freedom ($n_1 + n_2 - 2$).

T-Test for Two Unpaired Samples with Unequal Variance

If your two samples of normal data are unpaired and it is assumed that the variance is not equal, you apply a test often referred to as Welch's t-test. You might have noticed in the formulas for the t-tests assuming equal variance that the variances calculated in the denominators were estimates of pooled variances ($S_{x_1 x_2}$). If the variances in both samples differ, we need to use an equation to calculate t that addresses this issue in the denominator:

$$t = \frac{\bar{x}_1 - \bar{x}_2}{\sqrt{(S^2_1/n_1) + (S^2_2/n_2)}}$$

S^2_1 and S^2_2 represent the variance in groups 1 and 2, respectively. The degrees of freedom required to calculate a probability that the t-statistic is significant has to be calculated using a complex formula. The degrees of freedom will be slightly smaller than those for the preceding t-tests.

T-Test for Two Paired Samples

To generate a t-statistic for paired samples, we need to calculate the mean difference between the paired values in the first and second samples and divide this by the standard error of the differences:

$$t = \frac{\bar{d}}{S_d \times \sqrt{n}}$$

Mann-Whitney U-Test

In general, the Mann-Whitney U-test can be considered the nonparametric equivalent of the two-sample t-test where both samples are independent.

The test statistic calculated for Mann-Whitney is called U and is relatively easy to calculate after some preliminary steps:

- Pool all the values from both samples and create a ranked list of these values in the order of lowest to highest.
- Add the ranks for all values that were taken from sample 1 (R_1).
- Add the ranks for all values that were taken from sample 2 (R_2).

Now you can calculate U for both groups using:

$$U_1 = R_1 - \frac{n_1(n_1 + 1)}{2}$$

$$U_2 = R_2 - \frac{n_2(n_2 + 1)}{2}$$

The smaller of the two U values is then taken as the test statistic.

The F-Test

The F-test, also called the variance ratio test, is a simple procedure for comparing the variances of two samples with data following a normal distribution:

$$F = \frac{S_1^2}{S_2^2}$$

A.5.3 Tests for More Than Two Samples

Analysis of Variance (ANOVA)

If we need to evaluate whether a parameter such as the mean varies in more than two samples (groups) that follow a normal distribution, we need to use ANOVA to generate our test statistic. The name is derived from the fact that the method separates the total variance across all samples into two components. The first component (between-group variation) accounts for variation due to differences between cases of different groups. The second component (within-group variation) accounts for the variation between cases within a group. Our goal is to derive a statistic called F that is a ratio of the between-group variance to the within-group variance.

Calculating the F-statistic relies on a series of steps. The first step is to determine the sums of squares within groups, between groups, and across all groups:

$$SS_{TOT} = \Sigma X_T^2 - \frac{(\Sigma X_T)^2}{n_T}$$

$$SS_{BG} = \frac{(\Sigma X_1)^2}{n_1} + \frac{(\Sigma X_2)^2}{n_2} + \cdots + \frac{(\Sigma X_K)^2}{n_K} - \frac{(\Sigma X_T)^2}{n_T}$$

$$SS_{WG} = \Sigma X_T^2 - \frac{(\Sigma X_1)^2}{n_1} + \frac{(\Sigma X_2)^2}{n_2} + \cdots + \frac{(\Sigma X_K)^2}{n_K}$$

SS_{TOT} is the total sum of squares, SS_{BG} is the between-group sum of squares, and SS_{WG} is the within-group sum of squares. n_1, n_2, and n_K are the number of values in each of the K groups, whereas n_T is the total number of values across all groups. Next we calculate the within-group, between-group, and total degrees of freedom using the same notation:

$$df_{BG} = K - 1$$

$$df_{WG} = n_T - K$$

$$df_{TOT} = df_{BG} + df_{WG}$$

In the third step we use the sum of squares and degrees of freedom to calculate the mean square values for within groups and between groups:

$$MS_{BG} = \frac{SS_{BG}}{df_{BG}}$$

$$MS_{WG} = \frac{SS_{WG}}{df_{WG}}$$

Now we have the necessary information to calculate the F-statistic:

$$F = \frac{MS_{BG}}{MS_{WG}}$$

A table for the F-distribution is then used to calculate a P Value using df_{BG} and df_{WG}.

Kruskal-Wallis Rank Sum Test

If your multigroup data does not follow a normal distribution, you can use the Kruskal-Wallis test as an alternative to ANOVA. To find the test statistic, H, we must first:

- Rank the values from all groups together.
- Calculate the sum of the ranks for each group ($R_i, R_i, \ldots, R_i$).

$$H = \frac{12}{n(n + 1)} \times \Sigma R^2_i/n_i - 3(n + 1)$$

where n is the total number of values in all groups and n_i is the number of values in each group i.

A.6 Categorical Data: Testing for Association Between Factors

Biomedical categorical data is qualitative, and often the data structure is represented as different samples (or groups) of two or more categories. Data is usually presented in a contingency table as counts or proportions according to group and category. In this type of data, we are actually dealing with two factors: the groups and the categories. We want to know whether there is an association between the two factors, i.e., between groups and counts or proportions in each category. An example is that given in Section 11.8 where counts of breast cancer patients are displayed in a contingency table according to age group and ER status (shown again in Table A.3). Age and ER status are our factors, with age having two categories (young and old) and ER having three categories (1, 2, and 3).

Pearson's chi-square (χ^2) test is the most widely known of a series of χ^2 "goodness of fit tests" that can be used to test for association between two such factors. The null hypothesis in this scenario is that there is no association between factors.

Table A.3   A 2 × 3 Contingency Table.

	ER status		
	1	2	3
Group 1	57	39	4
Group 2	84	16	0

Table A.4	Contingency Table Showing the Column and Row Totals.

	ER status			
	1	**2**	**3**	**Total**
Group 1	57	39	4	100
Group 2	84	16	0	100
Total	141	55	4	200

A.6.1 Chi-Square Test

The first step of a χ^2 test is to calculate what the expected value should be for each count in each cell of the contingency table under the assumption that the null hypothesis is true. To do this, we first need to add up the counts across each of the rows and each of the columns in our contingency table (Table A.4).

The expected value for each table cell is the product of the corresponding row total and column total divided by the grand total ($=$ sum of row totals or columns).

For example, the expected value of the first cell in our contingency table (with an observed count of 57) is:

$$\frac{100 \times 141}{200} = 70.5$$

We then repeat this for each cell, as shown in Table A.5.

Now we can work out the χ^2 statistic using the following formula:

$$\chi^2 = \frac{\Sigma(E - O)^2}{E}$$

where E is the expected and O is the observed count in each cell. As you can see from the formula, you sum the square of the difference between the expected and observed in each cell divided by the expected for that cell.

Once you have the χ^2 test statistic, you can look this value up in a χ^2 table to find a P Value of association, with degrees of freedom calculated as:

df $=$ (number of rows $-$ 1) $\times$ (number of columns $-$ 1)

The χ^2 test is generally applicable only if the expected number in every cell of the contingency table is greater than 10. As we can see from our table, there are two cells

Table A.5	Contingency Table Showing the Observed and Expected Counts.

	ER status			
	1	**2**	**3**	**Total**
Group 1	57 (70.5)	39 (27.5)	4 (2)	100
Group 2	84 (70.5)	16 (27.5)	0 (2)	100
Total	141	55	4	200

with expected values less than 10. In this situation, we should turn to Fisher's exact test designed for use on contingency tables built with small sample sizes. A description of the calculation steps involved in Fisher's exact test for contingency tables greater in size than 2×2 is beyond the scope of this appendix.

A.7 Statistical Tests for Measuring the Association Between Two Variables

To measure the degree of association between two variables of numerical data, we use correlation analysis. For general correlation analysis, the dataset will consist of pairs of scores for two variables measured on the same cases in a sample from the population. If both variables are normally distributed and we can assume a linear relationship between these variables, we calculate the Pearson product moment correlation coefficient (r) as our measure of association. If not, we use a nonparametric equivalent measure, such as Spearman's rank correlation coefficient. A linear relationship between these pairs of scores is usually first assessed using a scatterplot (Figure A.4).

A.7.1 Pearson's Correlation Coefficient

A linear relationship between values of two variables is best described by drawing a straight line through the points on a scatterplot (Figure A.4). The line is drawn so as to minimize the distance between all points and this "line of best fit." Pearson's correlation coefficient is actually an overall measure of how close the points are to the line (i.e., a measure of the overall fit). We calculate Pearson's correlation coefficient using the straightforward equation:

$$r = \frac{\Sigma(x - \bar{x})(y - \bar{y})}{\sqrt{[\Sigma(x - \bar{x})^2 \, \Sigma(y - \bar{y})^2]}}$$

where x represents values from the first variable, $\bar{x}$ is this variable's mean, and y represents values from the second variable having mean $\bar{y}$. The value of r will range from -1 to 1 through 0. An r value of 1 shows perfect positive correlation between the vari-

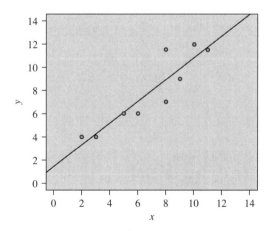

FIGURE A.4   Scatterplot of x and y with line of best fit.

ables (i.e., as the values of variable one increase, so do the values of variable two). A value close to 0 shows no correlation. A value close to -1 shows excellent negative correlation (i.e., as the values of variable one increase, the values of variable two decrease—or vice versa).

To test the significance of the correlation, we take the value of r_p (if our sample size is less than or equal to 150) or use r to calculate a t-statistic if the sample size is greater than 150:

$$t = \sqrt{(n - 2)/(1 - r^2)}$$

A P Value is then calculated using either a Pearson's correlation coefficient table for r or a t-distribution table for the t-statistic with $n - 2$ degrees of freedom.

If the sample size is small, we may prefer to calculate r^2, which is the proportion of variance of one variable explained by the linear relationship with the second variable. Thus, if an r value for a small sample size is 0.5 and significant, it is worth remembering that the relationship only explains 25% ($r^2 = 0.5 \times 0.5$) of variation for one of the variables.

A.7.2 Spearman's Rank Correlation Coefficient

If the data for two variables does not follow a normal distribution, if there is reason to believe that the relationship is not linear, or if the sample size is small, we can calculate Spearman's rank correlation coefficient as a measure of association.

Again, this test is straightforward and can be calculated in just a few steps:

- Rearrange the order of values in each variable from smallest value to highest and assign ranks, where tied values are assigned the mean value of the ranks that would have been given if there was no tie.

- Calculate Pearson's correlation coefficient between the ranks of each variable.

Be aware that you are not measuring a linear association between variables and that you cannot calculate r^2 in this situation.

A.8 Linear Regression

If correlation tells us whether we have a linear relationship between two normally distributed variables, regression can provide us with a model for that relationship. If the values of one variable (y) are dependent on the values of the other (x), we can use the model to predict the dependent variable given the independent variable.

In the previous section we discussed the concept of measuring the linear relationship between two variables using a line of best fit through the data plotted as a scatterplot. If there is a good linear fit, it is this line we use to predict a value for y given the value of x. The equation for a straight line is:

$$y = a + bx$$

where x represents values of the independent variable; y is the predicted value of the dependent variable; a is the intercept, or value of y when x is 0; and b is the gradient or slope of the line showing the rate at which y values increase with respect to x. The b parameter is referred to as the regression coefficient and is summarized diagrammatically in Figure A.5.

To generate the equation for a predictive model, we need to determine the values of a and b from a dataset. The a and b parameters are found easily using ordinary squares:

$$b = \frac{\Sigma(x - \bar{x})(y - \bar{y})}{\Sigma(x - \bar{x})^2}$$

$$a = \bar{y} - b\bar{x}$$

Now that we have the model parameters (just to recap), we are interested in two things:

- Testing the linear relationship between the x and y variables
- Predicting y values from x values using the model (equation)

To test the null hypothesis that there is no linear relationship between x and y, we test whether the slope (b) is different from 0. This can be done either by ANOVA to generate an F-statistic or by calculating a test statistic that follows the t-distribution:

$$t = \frac{b}{\text{SE}(b)}$$

SE(b) is calculated as:

$$\text{SE}(b) = \frac{S_{res}}{\sqrt{\Sigma(x - \bar{x})^2}} \text{ where } S_{res} = \sqrt{(\Sigma(y - \bar{y})^2)/(n - 2)}$$

A P Value for the t-test statistic can be calculated using the standard table for the t-distribution with $n - 2$ degrees of freedom.

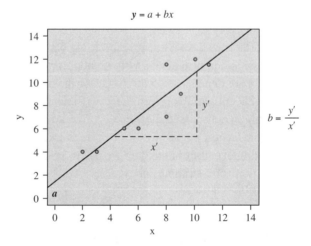

FIGURE A.5   Scatterplot with correlation coefficients a and b.

Predicting values of y from x using the linear equation is easy. You just plug in the x value of interest (your equation should already contain the values for a and b you worked out using least squares) and multiply it by b and a.

A.9 Measuring Agreement: Cohen's Kappa Test

Cohen's kappa test produces a coefficient that is a statistical measure of agreement between two people (raters). We can use it to test the reliability of raters. The test requires a series of categorical scores from each rater. The example in Chapter 11 used data (scores = 1, 2, or 3) for three scorers of an antibody for tumor proliferation protein. Two scorers were experienced pathologists and the other was an inexperienced student. Our task was to use the kappa test to measure agreement between each of the three scorers, ultimately testing the reliability of the student for scoring. Because the kappa test is not an inferential statistical test, we are not testing a null hypothesis.

The first step in the test is to generate a contingency table representing the agreement in scores between the two scorers. Let us look at the scores for the two pathologists given in vectors in Code 11.32:

```
pathologist 1:  1,1,2,1,1,3,2,1,1,1,2,1,1,2,2,2,1,1,1,3
pathologist 2:  1,1,2,1,1,1,2,1,1,1,2,1,1,2,2,2,1,1,1,3
```

We need to create a 3×3 contingency table because we have two scorers and three categories, where each column represents a category for pathologist 1 scores and each row a category for pathologist 2 scores. We then count up the number of times there is agreement or no agreement (dependent on the score type) between the scorers (Table A.6).

You can see that agreement between raters is represented in the diagonal line of cells. The next step is to calculate the expected counts for each cell in exactly the same way as we did for the chi-square test in Section A.6. The expected counts for our example are shown in Table A.7.

We then have a series of observed and expected counts for each cell. All we need to do now is sum the observed counts (O_d) along the diagonal and the same for the expected (E_d):

$$O_d = 12 + 6 + 1 = 19$$

$$E_d = 7.8 + 1.8 + 0.1 = 9.7$$

Table A.6   Contingency Table for Agreement in Scores.

		pathologist 1		
		1	2	3
	1	12	0	1
pathologist 2	2	0	6	0
	3	0	0	1

Table A.7   Contingency Table with the Expected Counts Added.

		pathologist 1			
		1	**2**	**3**	**Total**
	1	**12(7.8)**	0 (3.9)	1 (1.3)	13
pathologist 2	**2**	0 (3.6)	**6 (1.8)**	0 (0.6)	6
	3	0 (0.6)	0 (0.3)	**1 (0.1)**	1
	Total 12	6	2	20	

We then put these values into the following equation to calculate the κ statistic:

$$\kappa = (O_d/T - E_d/T)/(1 - E_d/T)$$

where T is the total of rows (or columns). If you put the values in from the previous contingency table, you should generate a κ statistic of 0.903. The κ statistic provides a score between 0 and 1, where generally it is considered that any score above 0.70 indicates good agreement between the two raters.

A.10 Odds Ratio and Relative Risk

The background, methodologies, and equations to calculate odds ratios and relative risk are actually detailed in Section 11.11 of Chapter 11, and there is no need to repeat these here. Note that the data is presented as contingency tables for which the statistics are calculated. Also note from the R output that P Values are calculated using the chi-square and Fisher's exact test (explained previously).

Index